Advances in Industrial Control

Advances in Industrial Control is a series of monographs and contributed titles focusing on the applications of advanced and novel control methods within applied settings. This series has worldwide distribution to engineers, researchers and libraries.

The series promotes the exchange of information between academia and industry, to which end the books all demonstrate some theoretical aspect of an advanced or new control method and show how it can be applied either in a pilot plant or in some real industrial situation. The books are distinguished by the combination of the type of theory used and the type of application exemplified. Note that "industrial" here has a very broad interpretation; it applies not merely to the processes employed in industrial plants but to systems such as avionics and automotive brakes and drivetrain. This series complements the theoretical and more mathematical approach of Communications and Control Engineering.

Indexed by SCOPUS and Engineering Index.

Proposals for this series, composed of a proposal form downloaded from this page, a draft Contents, at least two sample chapters and an author cv (with a synopsis of the whole project, if possible) can be submitted to either of the:

Series Editors

Professor **Michael J. Grimble**
Department of Electronic and Electrical Engineering, Royal College Building, 204 George Street, Glasgow G1 1XW, United Kingdom
e-mail: m.j.grimble@strath.ac.uk

Professor **Antonella Ferrara**
Department of Electrical, Computer and Biomedical Engineering, University of Pavia, Via Ferrata 1, 27100 Pavia, Italy
e-mail: antonella.ferrara@unipv.it

or the

In-house Editor

Mr. **Oliver Jackson**
Springer London, 4 Crinan Street, London, N1 9XW, United Kingdom
e-mail: oliver.jackson@springer.com

Proposals are peer-reviewed.

Publishing Ethics

Researchers should conduct their research from research proposal to publication in line with best practices and codes of conduct of relevant professional bodies and/or national and international regulatory bodies. For more details on individual ethics matters please see: https://www.springer.com/gp/authors-editors/journal-author/journal-author-helpdesk/publishing-ethics/14214

More information about this series at http://www.springer.com/series/1412

Carlos Bordons · Félix Garcia-Torres ·
Miguel A. Ridao

Model Predictive Control of Microgrids

Springer

Carlos Bordons
Systems Engineering and Automatic Control
Universidad de Sevilla
Seville, Spain

Félix Garcia-Torres
Microgrids Laboratory
Centro Nacional del Hidrógeno
Puertollano, Ciudad Real, Spain

Miguel A. Ridao
Systems Engineering and Automatic Control
Universidad de Sevilla
Seville, Spain

ISSN 1430-9491 ISSN 2193-1577 (electronic)
Advances in Industrial Control
ISBN 978-3-030-24569-6 ISBN 978-3-030-24570-2 (eBook)
https://doi.org/10.1007/978-3-030-24570-2

MATLAB® and Simulink® are registered trademarks of The MathWorks, Inc., 1 Apple Hill Drive, Natick, MA 01760-2098, USA, http://www.mathworks.com.

Mathematics Subject Classification (2010): 93C83, 93C35, 93C10, 49L20

This Springer imprint is published by the registered company Springer Nature Switzerland AG
The registered company address is: Gewerbestrasse 11, 6330 Cham, Switzerland

Series Editor's Foreword

Control systems engineering is viewed very differently by researchers and those that must implement designs. The former group develops general algorithms with a strong underlying mathematical basis, whilst the latter has more local concerns over the limits of equipment, quality of control and plant downtime. The series *Advances in Industrial Control* attempts to bridge this divide and hopes to encourage the adoption of more advanced control techniques when they are needed.

The rapid development of new control theory and technology has an impact on all areas of control engineering and applications. There are new theories, actuators, and sensor systems, computing methods, design philosophies and new application areas. This specialized monograph series encourages the development of more targeted control theory that is driven by the needs and challenges of applications. A focus on applications is essential if the different aspects of the control design problem are to be explored with the same dedication that the control synthesis problems have received. The series provides an opportunity for researchers to present an extended exposition of new work on industrial control, raising awareness of the substantial benefits that can accrue, and the challenges that can arise.

The authors are well known for their work on Model Predictive Control (MPC) and its applications, including the seminal text that Professor Bordons co-authored with Professor Eduardo Camacho (*Model Predictive Control* second edition, ISBN: 978-1-85233-694-3, Springer, 2004). The present text is concerned with the design of the MPC of microgrids. There are many real challenges in this area because of the changing nature of power production, storage and the distributed location of power generation systems. The user requirements are also changing because of the changing customer needs in areas such as electric vehicles.

There is a significant interest in developments in MPC and in applications. This book provides a clear exposition of the background and the control problems in microgrids. It also justifies the use of predictive control methods relative to other options. For those that are getting to know MPC there is a simple and clear introductory chapter. The models for the system and devices are also discussed in some detail since MPC does of course depend upon reasonable model knowledge.

The key problem of energy management is considered and the role of MPC is explained. The problem is clearly very complex but once in the MPC framework the equations are in a standard and familiar form. It is an advantage of MPC that engineers can then focus upon the wider problems of total system control rather than the peculiarities of individual subsystems. There are also benefits to using an optimization-based approach and these are described. The chapter on demand-side management and the integration of electric vehicles is very timely considering the huge changes in the automotive industry.

Perhaps one of the more valuable contributions a control scientist can make to the power engineering community is the way uncertainties of various kinds can be addressed in a formal mathematical framework. The chapter on uncertainties in microgrids should therefore be particularly valuable to those with more of a power engineering background since it reveals the type of solutions that may be applied in a range of uncertain modelling and control situations. The traditional control engineer should also benefit from the insights that the power engineering problem presents with an interesting discussion on power quality in microgrids and the methods for quality control.

This is therefore a text which deals with an increasingly important application area that has an influence on the environment and has the potential to affect people's lives drastically. The cross-fertilization of ideas between the power and control communities is also to be applauded. It is therefore a welcome addition to the series on *Advances in Industrial Control*.

April 2019

Mike J. Grimble
Glasgow, UK

Foreword

Our society is experimenting a transition from an energy system based on fossil fuels to a new energy system based on renewable energies and electric transportation systems. There is a need for new control algorithms to cope with the intermittent, stochastic, and distributed nature of the generation and with the new consumption patterns. Microgrids are receiving a lot of interest from the research community because they are going to play a major role in this transition. The control of microgrids brings significant challenges which need to be addressed with advanced control techniques. The book offers an actual and wide vision of the main problems encountered when controlling microgrids and how Model Predictive Control (MPC) can supply appropriate solutions. Although there are many techniques that can be used to control microgrids, MPC is one of the most promising technologies to be applied in this context because it can offer solutions at all levels, from the long horizon scheduling level, to the control of power converters.

MPC has been successfully applied in the industry but in this context, it can add solutions to problems derived from nature of the generation and demand and also to the need to operate with equipment from different nature such as geographically distributed energy resources. This book presents MPC techniques going from the more basic to more complex forms and using the appropriate technique for each of the microgrids control problems.

This book is a timely contribution of a very topical matter, integrating relevant control techniques in an emerging field. It can be of great interest for researchers and engineers working in the energy sector. The book is written with a practical viewpoint but not lacking the necessary rigor, including study cases both simulated and with real experimentation in a pilot plant. It is worth mentioning that the book is accompanied with a modular simulator which will allow the reader to follow the included examples in the book and also to model their own microgrid to test the control methods described in the book. Furthermore, since the book is easy to read and has many examples and a simulator, it is very well suited as the main text in an undergraduate, master, or Ph.D. course on the control of microgrids. Definitely, this book is a must-read for anyone interested in the design and implementation of advanced controllers for microgrids.

Eduardo F. Camacho
IEEE and IFAC Fellow
University of Seville
Seville, Spain

Preface

Control of microgrids is evolving considerably over the past few years. Microgrids, which are small-scale power systems with a cluster of loads, distributed generators, and storage units operating together, can be considered as the most innovative area in the electric power field today, so new control issues are appearing. This book aims at providing solutions to the operation of renewable energy microgrids by the use of Model Predictive Control (MPC). The range of problems to be addressed by the microgrid control system is very wide. Although there are many techniques that can be used for the control of microgrids, MPC provides a general framework to solve most of the issues using some common ideas in an integrated way. MPC solves an optimization problem incorporating a feedback mechanism, which allows the system to face uncertainty and disturbances. It can handle physical constraints and can incorporate the future behavior of the system, which is of crucial importance for microgrids.

The book gives a complete overview of the main control topics in microgrids, covering all the control levels, with special emphasis on energy management systems. Along the book, several control problems in microgrids will be dealt with, providing appropriate solutions using MPC. The book introduces the fundamentals of MPC, focusing on the techniques that are of interest for microgrids. A basic Energy Management System (EMS) is developed with a simple MPC, which is extended along the book to include economic optimization, electrical market, demand-side management, and integration of electric vehicles. Uncertainties management by stochastic MPC as well as distributed methods for the interconnection of microgrids is also addressed, and a special formulation of MPC for power quality control is developed.

The complete methodology and approach that the authors have followed in their own research are presented in this book. In this way, the readers are guided through the pathway from conception to implementation of the appropriate solution to microgrid control problems. Several examples, simulations, and experiments are included. MPC techniques are developed for case studies that include several renewable sources as well as hybrid storage. Some experimental results for a pilot-scale microgrid are presented, as well as simulations of scheduling in the electricity market and integration of electric vehicles into the microgrid. The results indicate that the development of the appropriate controllers will facilitate competitive participation of renewable energy in the new model of the electrical system.

Most of the controllers are demonstrated on a complete nonlinear model of a microgrid. The authors provide a modular simulator to be run in MATLAB/Simulink so that readers can create their own microgrids using the blocks supplied. This way, readers are encouraged to replicate the examples provided in the book, and they can also develop and validate control algorithms on existing or projected microgrids. The simulator offers a unique and reliable benchmark tool that has been developed and validated under a laboratory environment, providing a reliable simulation tool for testing the controllers.

The text is an excellent aid for undergraduate and postgraduate students on advanced control in the areas of Electrical/Electronic Engineering, Power Systems, and Energy. It will also be of interest to researchers and practitioners who want to keep updated in this fast-changing field. It can be of great value for practitioners responsible for the operation of microgrids or distribution grids, engineers of distribution and transmission system operators or working in the field of electric vehicles infrastructure.

Seville, Spain — Carlos Bordons
Puertollano, Spain — Félix Garcia-Torres
Seville, Spain — Miguel A. Ridao
April 2019

Acknowledgements

The authors want to express their gratitude to many people who have made this book possible. The book gathers many outcomes of the research conducted by the authors and their research groups over the past years. Many people have been involved in this research: In the first place, the authors want to thank Luis Valverde, who worked hard on the setup of the experimental plant Hylab during his Ph.D. work, developing the first MPC for this microgrid. The thesis of Paulo R. C. Mendes provided the results of electric vehicle integration, and the work done by Pablo Velarde for his thesis allowed the development of a stochastic MPC. The works done by Carlos Montero on power converters and Alejandro del Real on distributed control are also appreciated. Many researchers have had an active role and have supplied continuous support for the development of algorithms, simulations, and experiments. We want to thank Guillermo Teno, Juan J. Márquez, Pedro Fernández, and Rubén Galera for their enthusiastic work. The authors are especially grateful to colleagues who have revised the manuscript and have provided useful comments and suggestions: Sergio Vázquez, José L. Martinez-Ramos, Juan M. Escaño, José M. Maestre, and Kumars Rouzbehi. The compromise of Felipe Rosa has been crucial for the laboratory implementation, with the contributions of Javier Pino. The authors are also grateful to their home institutions (Universidad de Sevilla and Centro Nacional del Hidrógeno) and Junta de Comunidades de Castilla-La Mancha, Ministry of Economy, Industry and Competitiveness and European Commission for their support. Part of the work has been done in the framework of projects CONFIGURA and AGERAR, whose funding has contributed to make this book possible. Finally, the authors want to appreciate the leadership and motivation of Eduardo F. Camacho, who encouraged them to do research in the world of Automatic Control.

The authors particularly wish to thank their families for their patient and understanding.

Contents

Abbreviations

AC	Alternate Current
ARIMA	Autoregressive Integrated Moving Average
ARMA	Autoregressive Moving Average
BESS	Battery Energy Storage System
BMS	Battery Management System
BOP	Balance of Plant
CAES	Compressed-Air Energy Storage
CARIMA	Controlled Autorregressive Integrated Moving Average
CC-MPC	Chance-Constraints Model Predictive Control
CCS-MPC	Continuous Control Set Model Predictive Control
CERTS	Consortium for Electric Reliability Technology Solutions
CHP	Combined Heat and Power
CSC	Current Source Converter
DC	Direct Current
DER	Distributed Energy Resources
DG	Distributed Generation
DMPC	Distributed Model Predictive Control
DMS	Distribution Management System
DP	Dynamic Programming
DR	Demand Response
DS	Distributed Storage
DSM	Demand-Side Management
DSO	Distribution System Operator
EMS	Energy Management System
ESS	Energy Storage System
EV	Electric Vehicle
FBESS	Flow Battery Energy Storage System
FC	Fuel Cell
FCEV	Fuel Cell Electric Vehicle
FCS-MPC	Finite Control Set Model Predictive Control

FESS	Flywheel Energy Storage System
HBC	Hysteresis Band Control
HESS	Hydrogen Energy Storage System
HF	High Frequency
HV	High Voltage
HVAC	Heating, Ventilating, and Air-Conditioning
ICE	Internal Combustion Engine
ICT	Information and Communications Technologies
IGBT	Insulated Gate Bipolar Transistor
IPS	Intelligent Power Switch
LAN	Local Area Network
LMP	Locational Marginal Pricing
LOH	Level of Hydrogen
LP	Linear Programming
LQR	Lineal Quadratic Regulator
LTI	Linear Time-Invariant
LV	Low Voltage
MA	Microgrid Aggregator
MEA	Membrane Electrode Assembly
MEMS	Microgrid Energy Management System
MIBEL	Iberian Electricity Market
MIMO	Multiple-Input Multiple-Output
MINLP	Mixed-Integer Nonlinear Programming
MIQP	Mixed-Integer Quadratic Programming
MIQQ	Mixed-Integer Quadratic Programming with Quadratic Constraints
MLD	Mixed Logic Dynamical
MO	Market Operator
MPC	Model Predictive Control
MPPT	Maximum Power Point Tracking
MS-MPC	Multiple Scenarios Model Predictive Control
MV	Medium Voltage
O&M	Operation and Maintenance
OMIE	Iberian Electricity Market Operator
OPF	Optimal Power Flow
PCC	Point of Common Coupling
PEM	Polymer Electrolyte Membrane
PHS	Pumped-Hydro Storage
PI	Proportional Integral
PID	Proportional Integral Derivative
PLC	Programmable Logic Controller
PV	Photovoltaic
PWA	Piecewise Affine
PWM	Pulse Width Modulation
QP	Quadratic Programming
RES	Renewable Energy Source

RFB	Redox Flow Battery
RMS	Root Mean Square
RTO	Real-Time Optimization
SCADA	Supervisory Control And Data Acquisition
SEI	Solid Electrolyte Interphase
SISO	Single-Input Single-Output
SMES	Superconducting Magnetic Energy Storage
SMPC	Stochastic Model Predictive Control
SO	System Operator
SOC	State of Charge
SOH	State of Health
TB-MPC	Tree-Based Model Predictive Control
THD	Total Harmonic Distortion
TSO	Transmission System Operator
UESS	Ultracapacitor Energy Storage System
UPS	Uninterrupted Power System
VSI	Voltage Source Inverter
WAN	Wide Area Network

Chapter 1
Microgrid Control Issues

Abstract The evolution from the existing energy system based on fossil fuels to a new scheme with high penetration of renewable energy and electric transport systems introduces new challenges in architecture, control, and management of the electrical grid. This situation demands new schemes for the future electricity grids, where distributed generation, demand response, and energy storage systems may be easily integrated. The novel paradigm of microgrid that intends to provide a solution to these issues is presented in this chapter. The new control challenges that appear in microgrids are introduced, proposing Model Predictive Control (MPC) as a powerful tool to face them. This chapter presents an overview of the main topics on automatic operation and control of microgrids that will be tackled along the book, showing the most appropriate MPC technique to deal with them.

1.1 Microgrid as a New Paradigm for the Electrical System

The goal of reducing greenhouse gas emissions is shifting its focus toward more environmentally-friendly and sustainable sources of energy. Renewable energy is already playing an important role within a society that is not only more energy-dependent but also more aware of environmental problems. The adoption of Renewable Energy Sources (RESs) poses several challenges that arise from their inherent intermittency and the requirement to satisfy uncertain user demand. While the traditional means of energy generation, controllable from the source, allow the adjustment of production to demand, the implantation of new technologies based on renewable resources with uncertain and fluctuant profiles makes it necessary to provide new solutions to problems which had not arisen before. High penetration of RESs produces energy imbalances in the grid with the related problems in power quality and reliability.

One way to get over these problems is by including Energy Storage Systems (ESSs), such as batteries, ultracapacitors, hydrogen, flywheels, etc. This buffering capability can help avoid the consideration of renewable as undispatchable sources, due to its inherent forecast difficulties and variability. Redesigning the grid into smaller, more manageable units comes out as a solution to the outlined problems.

C. Bordons et al., *Model Predictive Control of Microgrids*,
Advances in Industrial Control, https://doi.org/10.1007/978-3-030-24570-2_1

In these units, stored energy compensates both the discontinuous nature of renewable generation and the randomness of the consumer's behavior. Distribution power grids are becoming highly active networks with more Distributed Energy Resources (DERs). Power flow is no longer flowing one way from the substation transformers to the end users, but instead, it is flowing two ways. This scheme reduces transport losses if the generation is produced next to the consumption.

A structural solution to the new challenges in the electrical grid is the splitting of the electrical network into smaller units based on distributed resources. Thus the concept of microgrid, introduced by Lasseter [48], becomes a powerful tool to improve the electrical system quality and reliability. According to [56], a microgrid can be seen as a set of loads, generators, and storage units whose management can be done independently or connected to the external network in a coordinated way to contribute to the whole system stability. It refers to a small-scale power system with a cluster of loads and distributed generators operating together with energy management, control and protection devices, and associated software [71]. Different accepted definitions of a microgrid exist. The Consortium for Electric Reliability Technology Solutions (CERTS) defines a microgrid as an aggregation of loads and microsources operating as a single system providing both power and heat [47]. The term microgrid is usually used to describe a customer-owned facility containing generation, as well as consumption, where there is a chance to manage the power exchange between the microgrid and the rest of the grid, with the capability of switching between islanded and grid-connected mode [10, 44].

During emergency situations (failures, large disturbances, blackouts, etc.), generators and loads can be disconnected from the distribution grid, keeping service without harming system integrity. Although originally associated to the electrical grid, the concept of microgrid has been extended to any set of loads and generators operating as a whole controllable system that can supply electricity as well as thermal energy or fuels (such as natural gas, hydrogen or biogas) to a certain area [78]. Nowadays, the operation of DERs together with controllable loads (home consumption or electric vehicles) and different storage technologies, such as batteries, ultracapacitors or hydrogen, constitute the core of the microgrid concept. The microgrid allows a successful interconnection among units, including, in an integrated way, distributed generation, local loads, and ESSs. A microgrid can operate connected to the main distribution grid, through the Point of Common Coupling (PCC), or in islanded mode, and it can even be connected to other microgrids, giving rise to more sophisticated systems. Microgrids can guarantee the quality of supply for local loads such as hospitals, offices, shopping malls, neighborhoods, university campuses, or industrial areas.

The paradigm of microgrids, with its own control, facilitates the scalable integration of local generation and loads in the current electrical networks, allowing a better penetration of DG and RESs [9]. The technical challenges associated with the design, operation, and control of microgrids are enormous [42]. Microgrids can operate both in islanded mode and connected to the main grid through the PCC and thus an appropriate control of the microgrid for a stable and economically efficient operation in both situations is necessary. The control system must regulate frequency

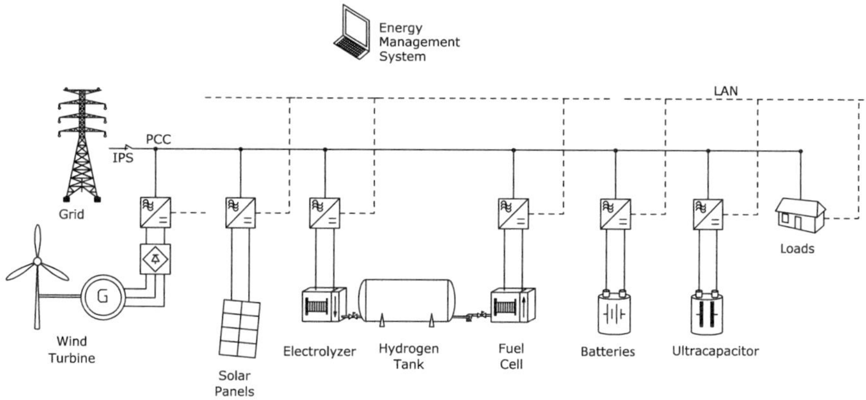

Fig. 1.1 Example of a microgrid

and voltage in any operation mode, must dispatch the load among the different DGs and ESSs, manage the power flow with the main grid and optimize operation costs with supply reliability. In the connected mode, frequency and voltage are imposed by the main grid, which has synchronous generators and great spinning reserves. When serious disturbances or failures occur, the microgrid will switch to islanded mode, so that it must supply power to the critical loads and manage frequency and voltage. An important point is the management of transients during switching (especially synchronization during the transition from islanded to grid-connected mode), as shown in [8].

Microgrids can be considered as one of the basic structures of the new electrical grids. This way, the electrical grid perceives the microgrid as an individual element that reacts to appropriate control signals and operates as a unique controllable system. Figure 1.1 shows an example of a microgrid with diverse types of generation and storage. The use of microgrids can permit the massive implementation of distributed generation (mainly RES) since the operation can be done in a decentralized way, reducing the need of a centralized coordination that would be extremely complex. Apart from this generic benefit for the electrical system, the local reliability of the microgrid is improved, the quality of service can be enhanced and the chance to access the electrical market as another agent is open for the microgrid. Hence, the design and implementation of appropriate advanced control strategies is the key factor for successful integration of microgrids into the electrical system.

1.1.1 Microgrids and Storage

The use of ESSs enables the opportunity to choose the microgrid adequate operation strategy both in islanded and grid-connected modes, being possible to manage the appropriate way to exchange energy among microgrid components and with the

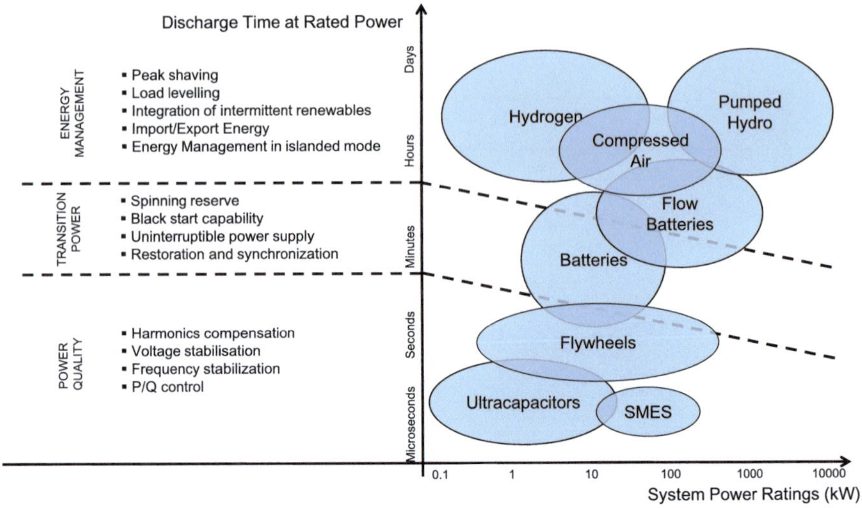

Fig. 1.2 Typical power rating, discharge time, and fields of application of several ESSs

external network. There are several technologies to store energy, and each ESS has its own advantages and disadvantages considering economic cost, energy and power rating, autonomy, time response, lifetime and degradation issues. The use of Hybrid Energy Storage Systems (HESSs), integrating several storage technologies, appears as a solution to mitigate the disadvantages of these technologies. The control strategy must have the ability to decide which ESS should be used at each moment in order to minimize overall costs.

In Fig. 1.2, the relationship between power rating and discharge time for different storage technologies can be observed. According to this relationship, different fields of application of the ESSs can be seen. To be noted is the fact that systems with low discharge time are used for grid stabilization in aspects of power quality, although they typically present low specific energy. One example includes ultracapacitor technology which shows fast response time with low energy capacity. At the other end of the spectrum, ESSs with the ability to store significant amounts of energy, e.g., pumped hydro or hydrogen systems, can be found. They have enough capacity to supply energy for periods of several days or more, compensate for power fluctuation, such as wind or solar generation, and are able to smooth peak demand of loads. The intermediate spectrum operates at intervals of minutes ensuring uninterruptible power supply, black start,[1] and spinning reserve [25].

The hybridization among energy storage technologies has aroused great interest in recent years [28, 38]. Specifically, the combination of hydrogen storage together with electrochemical batteries and ultracapacitors looks like an appropriate alliance for renewable generation. The use of hydrogen for storing electrical energy from

[1] Black start is the process of restoring microgrid to operation without relying on the external power transmission grid to recover from a total or partial shutdown.

renewable sources is based on producing hydrogen by electrolysis, store it in several forms (pressurized, metallic hydrides, etc.) and later using it to generate electricity through fuel cells. This combines the long-term storage capacity of hydrogen with the fast dynamic response of ultracapacitors.

Energy storage technologies will compensate the imbalances introduced by RESs fluctuations in the grid, providing the appropriate quality of the power supplied to the end loads. A grid architecture based on microgrids can be used to decide if each microgrid must work connected to the grid, or change to an islanded mode if there is a failure in main grid or when requested by the operator to reduce consumption [10, 44]. Storage is not only a technical solution for network management but also a way of efficiently utilizing renewable resources by avoiding generation shedding in times of overproduction and load shedding when generation is deficient. The design and development of an advanced control system are crucial for the convenient operation of hybrid ESSs. The control strategy can take advantage of the characteristics of each ESS, considering degradation issues and operation constraints, therefore it appears as a technological solution to increase the efficiency, autonomy, and lifetime.

1.1.2 Control Goals and Challenges

The primary goal of the microgrid control system is to ensure stable delivery of electrical power to its local load consumers using DERs and ESSs in an efficient and reliable way, both in normal conditions and during contingencies, regardless of the connection to the external grid. In addition to this, it must encompass cost optimization and prevent equipment damage. The economic and environmental benefits of microgrids and, consequently, its acceptance and degree of penetration in the electrical system are strongly linked to the capabilities and operational performance of the control system.

Microgrids pose new challenges that must be faced up in the design of their control systems, due to certain particular features. The most relevant are [56]:

- Energy flows: Unlike conventional grids, the integration of DG in low voltage can cause bidirectional power flows and give rise to complications in protection systems or undesired flow behavior.
- Stability: Local oscillations may appear due to the interaction of the control systems and the issues associated to the transitions between islanded and connected modes.
- Low inertia. The dynamic features of DG resources, especially those electronically adapted, are different of those based on big generation turbines. If adequate control mechanisms are not implemented, the low system inertia and lack of spinning reserves may give rise to considerable frequency and voltage deviations when working in islanded mode.
- Uncertainty. The uncertainty associated to demand and, especially, generation is significant in microgrids, since the use of RESs ties generation to environmental

conditions. Therefore, a reliable and economic operation must include generation, prices, and demand forecasts.
- Network model: The normally accepted assumptions of three equilibrated phases, inductive transmission lines and constant loads may no longer be valid, entailing the need of adapting the models to the new situation. A microgrid is inherently subject to load unbalancing by its DG units.

In order to cope with these challenges, the control system must guarantee a reliable operation of the microgrid. The main functions that can be requested to the control system are [8, 56]:

- Control of currents and voltages in the DG units, tracking references and appropriately damping oscillations.
- Frequency and voltage regulation both in islanded and grid-connected mode.
- Power balance, adapting to the changes both in load and generation while keeping frequency and voltage inside acceptable limits.
- Demand-Side Management (DSM) mechanisms that allow certain variation in the demand of a portion of the loads in order to adapt to microgrid needs [2, 79].
- Smooth transitions between operation modes, using the most adequate strategy in each case and quickly detecting the situations that cause the change. Resynchronization with the main grid.
- Economic dispatch, sharing power among the DGs and ESSs, reducing operational costs while keeping reliability. The optimization of the operational cost will include the maximization of the economic benefit in case of grid connection. This may include the provision of ancillary services.
- Management of power flows between the microgrid and the main grid and possibly with other microgrids.

Most of the challenges will be addressed in this book, providing solutions based on control strategies in the framework of Model Predictive Control.

1.1.3 Why Model Predictive Control?

The range of problems to be addressed by the microgrid control system is very wide. In small-scale microgrids, the main challenge is to achieve a safe operation, balancing generation and demand, while in others an optimal operation considering economic criteria is needed. In the first case, in general, microgrid energy management has been carried out by heuristic algorithms [39]. Several works [35, 43, 66, 82] use the Hysteresis Band Control (HBC) method for energy management, because of its conceptual simplicity and ease of implementation.

On the other hand, MPC solves an optimization problem at each sampling time in order to determine minimal running cost while meeting the demand and considering technical and physical limits. The following features make MPC a good candidate for microgrid control:

- It incorporates a feedback mechanism, which allows the system to face uncertainty and disturbances.
- It can handle physical constraints, such as storage capacity or generator slew-rate power limits.
- It can incorporate generation and demand forecasts.
- It is based on the future behavior of the system, which is of crucial importance for systems that depend on demand and renewable energy generation.

The main difference between MPC and HBC or other heuristic methods is that MPC guarantees optimality while HB does not. Regarding the performance advantages of MPC over improved HB method, there are several papers comparing these approaches and others [12, 32, 74]. An impressive reduction in operation costs can be observed when comparing both techniques [73].

Although in power systems it is common to solve unit commitment and economic dispatch of grids by using an Optimal Power Flow (OPF) formulation [16], this is static and open-loop optimization, and has some limitations. The open-loop nature of this optimization does not allow to compensate for inaccuracies originated from modeling uncertainties and disturbances (measurement noise or unexpected reactions of some components). On the other hand, the intrinsic closed-loop nature of MPC relies on the system response in order to apply fast corrective actions. Using measurements to update the optimization problem for the next time step introduces the key feature of feedback.

Nowadays there are many applications that use Model Predictive Control strategies. Generic MPC-based structures are seen in [5, 60]. Optimal control of distributed energy resources using MPC with battery storage system is developed in [54]. In the case of hybrid storage systems, MPC appears to be a good solution as shown in [21, 38, 73]. A hydrogen-based domestic microgrid being controlled by an MPC-based structure is presented in [75], and other works also refer to optimal generation for renewable microgrids considering hybrid storage systems [32, 63]. MPC has also been applied to energy management of microgrids connected to electric vehicle charging stations [29, 33, 52]. Examples of economic optimization are exposed in [13, 59, 61].

As can be seen, MPC is a technique that is being used in microgrids and has a great potential to solve many open issues in this field. Although there are other existing techniques that can be used for the control of microgrids, MPC provides a general framework to address most of the issues using some common ideas in an integrated way. The use of MPC for microgrids will be expanded along the remaining chapters of the book.

1.1.4 Hierarchical Control of Microgrids

As has been discussed above, the microgrid control system must address several aspects, involving diverse control perspectives and timescales. Fast electrical control of the phase, frequency, and voltage of individual resources must be done in

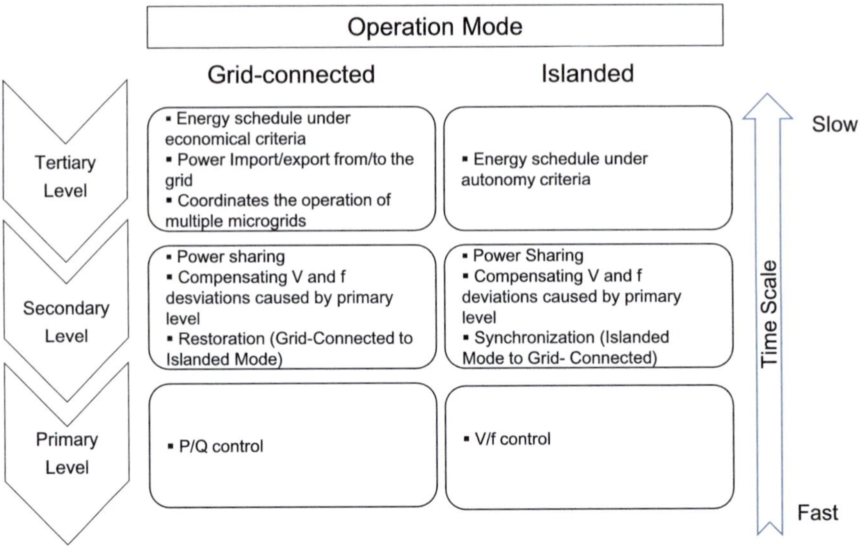

Fig. 1.3 Hierarchical control levels of a microgrid

timescales lower than second, while unit commitment, economic dispatch, demand-side optimization and energy exchanges with the utility grid are performed with longer timescales (minutes, hours, days or even months). Therefore, an extended approach is to develop a hierarchical control structure. From the control viewpoint, many authors [8, 56–58, 62, 70, 77] agree on the existence of three control levels. These levels are shown in Fig. 1.3 and are explained in depth in [19, 46, 76].

The primary level operates on a fast timescale and maintains the voltage and frequency stability during changes on the generation or demand, and after switching to islanded mode. This control is implemented locally in the DERs. The secondary level is responsible for guaranteeing that the voltage and frequency deviations are returned to zero after a load or generation change. It is responsible for eliminating any steady-state error introduced by the primary control and is also used for synchronization with the grid during the transition from islanded mode to grid connection. The tertiary control is used to control the power flow between the microgrid and the main grid (or other microgrids) and for the optimal operation on large timescales (planning and scheduling). This level may include several optimization strategies, according to the timescales. The term Energy Management System (EMS) refers to a system which addresses some of these issues (mainly scheduling and power sharing) and it therefore comprises secondary and tertiary levels.

This hierarchy is commonly accepted by the research community (see [3, 41]), although there are other approaches, like distributed schemes that use multi-agent systems [24, 42]. Figure 1.4 depicts the information flow among the control levels, which will be detailed in the following. This will be the scheme adopted in this book.

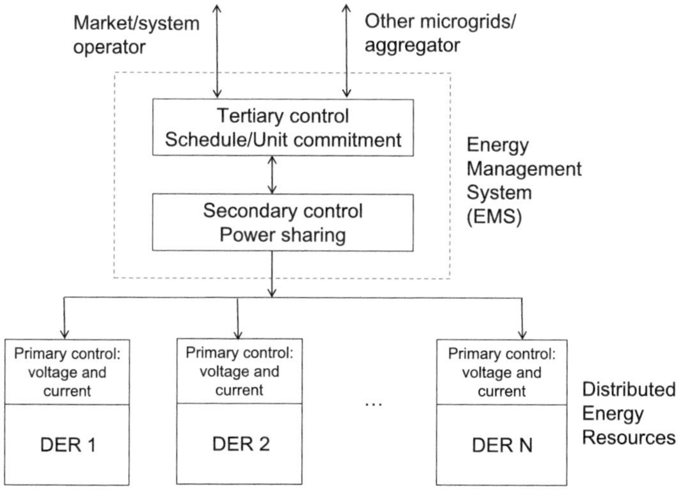

Fig. 1.4 Information flow among the control levels

Although this hierarchy is commonly used, the definition of the layers given by some authors is slightly different. The scheme proposed in [19] is also composed of a three-level control structure over the microgrid. The primary droop control ensures reliable operation in the microgrid. The second and third levels are in charge of power quality aspects and economic optimization through classical Proportional Integral (PI)-control methods.

The major issues and challenges in microgrid control are discussed in [56], where a review of the state of the art in control strategies and trends is presented; a general overview of the main control principles (such as droop control, model predictive control or multi-agent systems) is also included. Microgrid control strategies are classified into three levels: primary, secondary, and tertiary, where the first ones are associated with the operation of the microgrid itself, and tertiary level refers to the coordinated operation of the microgrid and the main grid.

Primary Control

The connections among the different components of the microgrid are done using power electronics converters, which are typically Voltage Source Inverters (VSI) in the case of AC microgrids [41, 49] and DC/DC power converters in DC microgrids. These power electronics devices are connected in parallel through the microgrid AC or DC bus. The objective of the primary level is to compensate for the instantaneous mismatch between scheduled and demanded power. Based on this requirement, it generates the voltage reference signals for the DERs. Droop control is the most common method used to solve the primary control level issues, which is explained in [41, 45, 50, 51].

In case of parallel operation of power converters, the droop control method consists of subtracting a term that depends on the output average active and reactive power to the frequency and voltage of each module in order to emulate virtual inertias. These control loops, called $P - f$ and $Q - U$ droops, are depicted by Eqs. (1.1) and (1.2) [18, 40]:

$$f - f^{ref} = G_P(s) \cdot (P - P^{ref}) \tag{1.1}$$

$$U - U^{ref} = G_Q(s) \cdot (Q - Q^{ref}) \tag{1.2}$$

where f and U are the frequency and amplitude of the output voltage, f^{ref} and U^{ref} are the references, P and Q denote the active and reactive power, and $G_P(s)$ and $G_Q(s)$ their corresponding transfer functions, which are typically substituted by their static gains. Although $P - f$ and $Q - U$ are the common couplings nowadays, since low-voltage grids are characterized by a high ratio of resistance to reactance, active power could be more appropriate for controlling voltage. This is done, for example, with dynamic voltage restorers in order to compensate voltage sags [42].

However, although this technique achieves high reliability and flexibility, it has several drawbacks that limit its application, particularly when nonlinear loads appear, since the control units should take into account harmonic currents, while balancing active and reactive power at the same time. Another important drawback of the droop method is its load-dependent frequency and amplitude deviations. In order to solve this problem, an upper controller is implemented in the microgrid central controller that can restore the frequency and amplitude in the microgrid [41].

Secondary Control

The objective of secondary control is to make sure that the power supplied by different sources is determined according to the one scheduled by the tertiary level. That is, the load must be shared among RESs. The secondary control ensures that the frequency and voltage deviations are regulated toward zero after any variation of generation or demand. The frequency and voltage amplitude levels in the microgrid are measured and compared with the references f^{ref} and U^{ref} and the tracking errors are sent to the microgrid components in order to restore frequency and voltage [41]. In this manner, corrective signals are sent to the controllers which act accordingly to ensure proportional power sharing among various DG units. This control can also be used for microgrid synchronization with the main grid [76].

Tertiary Control

This control layer can be considered as the economic level of the microgrid. No technology will be definitively integrated into society if it does not have economic benefits in comparison with its predecessor. Tertiary control decides the schedule of active and reactive power exchange with the external grid and among the different units of the microgrid. Based on inputs such as forecast, operational costs or prices, the tertiary controller prepares the generation and storage dispatch schedule, which is communicated to the secondary controller. Advanced control algorithms can be included in this level to provide optimal solutions taking into account economic,

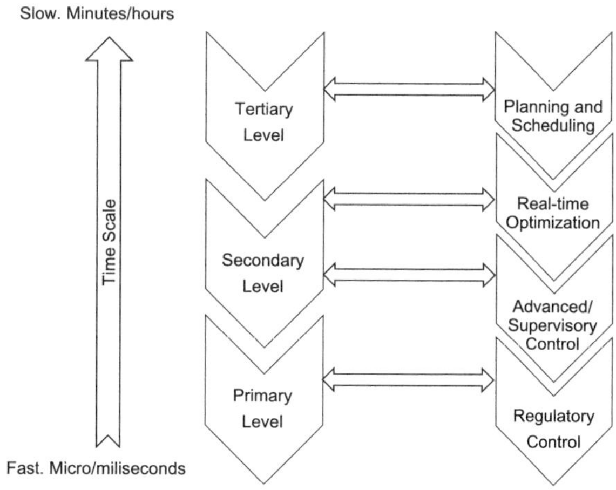

Fig. 1.5 Comparison of hierarchical control levels in microgrids and the process industry

degradation and environmental criteria. The optimization problem of energy in a renewable energy microgrid with different kinds of energy storage systems, which exchanges energy with the main grid, can be solved with MPC techniques [23]. The use of MPC techniques will allow the maximization of the economic benefit of the microgrid, minimizing degradation causes of each storage system and fulfilling the different system constraints at the same time [73]. There is an important number of articles dedicated to the management of interconnected microgrids such as [27, 55, 80]. This can be included in the tertiary level and this issue will be addressed in Chap. 8.

This hierarchical structure is similar to that used in power systems control, that is performed in a hierarchical scheme for both frequency and voltage and implemented in three steps: primary (at the generators level), secondary (at the control area level), and tertiary (at the system level), as depicted in [37]. There are clear interactions among these levels, each one having different objectives, time response, and geographical implications.

Notice that this hierarchy also resembles the one typically used in the process industry world, where usually a four-level pyramid is assumed, as shown in Fig. 1.5. In that case, primary level is for regulatory control (basic control loops), the secondary level is for advanced/supervisory control (set points calculation for the primary loops), the tertiary level is for Real-Time Optimization (RTO), and the top level is for scheduling and planning. Comparing those classifications, primary level in microgrids corresponds to regulatory control, and secondary level in microgrids includes advanced control and RTO. The top level in the industry is a kind of offline long-term planning similar to the tertiary layer in microgrids.

1.2 Overview of Control Topics in Microgrids

The automation of the traditional electrical system is still based on the design and operation as it was decades ago, but future electricity grids will have to deal with new environmental and economic challenges as well as social transformations. Security, environmental protection, power quality, economic benefit, and energy efficiency have to be addressed from a new point of view in order to adapt to changing needs in new electrical markets. The introduction of Information and Communications Technologies (ICT) will upgrade the power system automation (including transmission, distribution, and individual feeders) using the latest technology. These aspects will completely change the conception of the electrical power system, becoming more active where even the customers can monitor their own voltage and power and manage their energy consumption, for example, based on the electricity prices [10, 44]. But the change does not end here, the decreasing costs of renewable energies and the continuous introduction of energy storage technologies, linked to a virtual platform based on ICT, will allow end users not only to produce their own energy but also to decide if they prefer to sell to the grid, to buy from it, or to store it. With an increasing demand for improving productivity and efficiency, the operation requirements of energy systems are getting more restricted. This leads to the necessity of investing in automatic operation and control of microgrids aimed to achieve an enhanced efficiency of the overall system. There are many open topics in microgrid control, the most important are presented in the following and will be dealt with along the book, showing the appropriate MPC technique to address them.

1.2.1 Management of Hybrid Energy Storage Systems

An ESS which can respond properly in the short, medium, and long term still does not exist. The combination of different types of ESSs in one hybrid system can provide solution to storage problems in microgrids. The integration of several technologies in a unique ESS requires an efficient management with an optimal algorithm for sharing power, minimizing the overall cost and managing the different timescales.

The technological maturity of energy storage technologies with a short lifetime is still a barrier for their technological implantation. In this aspect, the correct use of the selected ESS where degradation causes are taken into account can improve their economic profitability as a result of the increment in their useful life. When several ESSs are integrated into a hybrid system, the problem of sharing power among them must be solved. The importance of power sharing in hybrid ESSs, taking advantage of the transient response and autonomy of each technology and respecting the degradation causes have been discussed in several studies [31, 69]. The existing literature exposes how the use of hybrid ESSs can satisfy a major range of applications, improving the lifetime of the single components. Each technology has its own degradation causes that establish the operating limits or the associated degradation costs of exposing the ESS to working conditions. Transient response is a key characteristic feature of ESSs, sometimes more critical than efficiency, due to the importance of supplying rapidly changing electric loads. Fast transient response is essential for autonomy in startup and fast power response, but long-term autonomy in order to have the lower level of dependence with the main grid must also be considered.

An adequate use of the hybrid ESS requires the development of a controller which takes into account all the constraints, limitations, degradation issues, and the economic cost of each ESS. The high number of constraints and variables to be optimized hinders the control problem making advanced control algorithms necessary.

An example of hybridization using ultracapacitors, battery, and hydrogen storage is described in [69] which is used in an electric vehicle and a microgrid taking into account dynamic and degradation aspects of each ESS. An ultracapacitor module, working as a fast dynamic and high power-density device, supplies energy to regulate bus voltage. A battery module, as a high energy density device, operates to supply energy to the ultracapacitor bank to keep it charged, and a fuel cell, as the slowest dynamic source in this system, operates to supply energy to the battery bank to keep it charged. Several applications use heuristic methods with classical controllers [81], managing renewable energy microgrids with hybrid ESSs considering the dynamics of each component. Other works manage the hybridization between different ESSs, using frequency filters which generate decent results [68].

The high number of constraints and variables to be optimized increases the complexity of the associated control problem, making it difficult to reach an optimal solution using traditional heuristic methods. The use of multi-objective cost function in MPC makes the controller able to quantify the operation cost of the ESSs according to their number of life cycles or hours also considering their degradation mechanisms. Applications of MPC for load sharing of hybrid ESSs composed of a fuel cell and an ultracapacitor, including some degradation issues are presented in [38, 72]. Similar developments have been done in the hybridization of a fuel cell and a battery in vehicle applications [4, 11]. In order to control the connection and disconnection of units (which strongly influences the lifetime), logical (binary) variables such as the startup/shutdown of the fuel cell and electrolyzer or charge/discharge states in the batteries and ultracapacitor are introduced [32, 60]. This will be further addressed in Chap. 5.

1.2.2 Economic Optimization

The electricity market is a complex process where the production system provides a total amount of electricity, at each instant, corresponding to a varying load consumption. The market rules determine the energy price in the day-ahead market, matching offers from generators to bids from consumers to develop a classic supply and demand equilibrium price, usually on an hourly interval. With the liberalization of electricity markets, renewable energy producers have the opportunity to dispatch their production through electricity pools. The main characteristic of these markets is that bids have to be done in advance and any imbalance (defined as the deviation between the actual production and the energy bid) is penalized in real-time markets. The difficulties for an accurate forecast of renewable energy systems joined to the penalty deviations used in the real-time markets make it difficult for clean energy to play an important role in the electricity market. Their inherited characteristic of intermittent generation is another limitation.

The need for reducing not only the peak of energy consumption on the demand side but the final price that end consumers have to pay is currently a concern of our society. The traditional scheme of an electric market based on the perspective of centralized generation system increases the energy price since it is generated to its final consumption due to the different conversion steps in the electrical transmission and distribution system. The increasing presence of distributed generation at the distribution side close to the end consumer opens the possibility of an active behavior in demand side where the consumer is becoming a prosumer[2] with the possibility to sell/purchase energy to/from the main grid. Microgrids will allow the end consumer not only to decide when to sell or purchase but also to store the energy or to supply the loads through ESSs [31, 32].

The optimization of the final energy price can be considered as one of the main goals in microgrids. This makes of great importance to have an accurate energy forecast algorithm from the point of view of generation and consumption, requiring an appropriate energy price prediction system. Microgrids have two main operation modes (grid-connected and islanded) with different operation goals. The optimization criteria in islanded mode should be the autonomy along the scheduled horizon, while the goal in grid-connected mode is a complex problem. Since real-operational scenario in renewable energy microgrids typically differs from the forecast computed by the economic dispatch, it is difficult to achieve the contracted schedule agreed with the grid market/system operator in the day-ahead and intraday markets. The economic optimization may include the possibility to provide ancillary services to the utility grid, receiving additional benefits in return [17]. The timescale, sampling time, as well as optimization criteria are different at each step of the electrical market and the optimization problem differs if the day-ahead market is considered from the real-time markets, but also different considerations have to be achieved in real-operational scenario of the microgrid. Similar problems with different timescales have to be considered in case of islanded mode.

The complexity of the associated control problem of integration of microgrids in the electrical market calls for advanced control algorithms [60]. The richness of the field of MPC appears as one powerful tool. Different weighting factors can be assigned to the diverse components of global optimization criteria formulated under a cost function which is minimized at each sample instant. The MPC methodology is especially very powerful for the design of hierarchical multilayer control systems made by a number of control algorithms operating at different timescales. MPC offers a wide range of choice in model structures, prediction horizons, and optimization

[2]The word is formed from the words "producer" and "consumer".

criteria ranging from high-speed computation in operational scenarios with short sampling times to complex optimization criteria with long schedule horizon and sampling periods.

The mathematical formulation used by MPC allows the inclusion in the same control problem of both logical and continuous variables, in order to consider different operation modes or energy scenarios (excess or deficit in generation can be considered in the same cost function just associating binary variables). All these MPC aspects are discussed, explained and developed in Chap. 5, where different case studies are also exposed.

1.2.3 *Power Quality*

Power quality will be an important factor in the transition toward the smart grid. According to the different regulation policies, generation should meet the growing demand cleanly, reliably, sustainably and at low cost. The future electrical grids will be mainly composed of renewable generation with the inherited characteristic of intermittency. From the point of view of consumers, traditional resistive loads are being replaced by the use of sensitive loads given by electronic devices.

High penetration of renewable energy produces imbalances in the grid with the related problems in power quality supply. Additionally, the increasing level of electronic loads in the grid requires an extraordinary effort in order to achieve acceptable levels of power quality. Both aspects open new challenges in the control problem of the future smart grid from the power quality point of view.

Energy storage technologies could compensate the imbalances introduced by RES fluctuations in the grid and, at the same time, provide the appropriate power quality to the end users. A new grid conception based on microgrids can even decide if the microgrid must work connected to the grid, or switch to islanded mode in case the power supply of the main grid is not satisfactory.

Microgrids appear to be a key solution to provide the required flexibility to the power system in a fully based renewable energy system. In this scenario, the enhanced power quality operation of microgrids should be included developing advanced power electronics for interfacing the ESSs, which minimize the effects of intermittency of renewable energy systems and compensate the presence of harmonics or unbalanced loads. Fast transition between grid-connected and islanded mode should be included in order to mitigate the effects of faults in the main grid. This makes it necessary to have an advanced control architecture to enhance power quality and reliability for the consumer.

Most of the existing literature for primary and secondary control in microgrids is based on classical PI (Proportional Integral)-PWM (Pulse Width Modulation) controllers. These type of controllers does not achieve decent results in the transient response which is highly dependent on the tuning of the parameters of the controllers. The use of advanced control techniques such as MPC can overcome some of these problems. The MPC technique applied to power converters is introduced in [65], being characterized by its fast dynamic response. MPC needs a model of the system to predict the output in the control horizon utilized by the controller.

Various features of MPC principles emerge as a promising alternative to enhance the power quality issues in microgrids. MPC can handle power converters with multiple switches and operation modes such as islanded, grid-connected or transition between both modes. MPC has the potential to replace involved control architectures, such as cascaded loops, by a unique controller, integrating multi-objective purposes using the same cost function, as will be shown in Chap. 9.

1.2.4 Interconnection of Microgrids

A massive introduction of microgrids and ESSs will bring a new paradigm in the way of managing the electrical system, and consequently, new network architecture, advanced management, and control strategies must be developed. The interconnection of different microgrids in a network introduces flexibility, but also complexity, to the system for both market and technical operation, and new agents come into play and must be coordinated with traditional agents, as Market Operator (MO) or System Operator (SO), which needs also to take new roles and functions in a microgrid-based network.

The penetration of DER and microgrids introduce an important complexity in the tasks to be carried out by the SO. Topologically the grid becomes active and the distributed generation involves bidirectional power flow, but also the connection and disconnection of microgrids will have an impact on quality and reliability. All these issues introduce challenges in Distribution Management System (DMS) functionalities, making management and control of the network more complex. This scenario is favorable for the emergence of new agents. On one side, the Microgrid Management System (MMS) must handle individual microgrids, including power flows among devices but also energy interchange with other microgrids or the main grid. On the other side, Microgrid Aggregator (MA) is an intermediate agent with a mission to coordinate and participate in the management and control of a set of microgrids and aggregate the information of the set of microgrids, acting as mediator with SO or MO.

The control and management of the set of microgrids can be managed in several ways by the different agents as a function of the implemented network architecture. The most simple situation is the decentralized management of each microgrid, but this solution does not take into account the complete network behavior with non-guaranteed results regarding economical and technical aspects. On the opposite hand, a centralized control by the MMS, MA or SO, solves the problems of a decentralized approach but the complexity of the management of multiple microgrids can be computationally intractable. Furthermore, in some cases, parts of the network belong to different organizations, which makes centralized control impossible to be implemented [22].

The distribution of the control effort among local control agents can be a very convenient choice. In a distributed management strategy, decisions are taken by local agents, but a communication among them is established, allowing the interchange of relevant data regarding the behavior of other microgrids to improve the global performance of the network. This communication can be implemented hierarchically through the SO or MA.

This book tackles the use of centralized and distributed approaches to interconnected microgrids. Distributed MPC (DMPC) is a growing research field [1, 15, 67] with application to several problems. Also, feasible solutions to MPC centralized problems will also be presented in the chapter devoted to interconnected microgrids, with a solution based on a distributed optimization of the global problem [14]. The implementation of different MPC techniques for this problem is addressed in Chap. 8.

1.2.5 Uncertainties in Microgrids

The intermittency of renewable energy and the random behavior of consumers adds a stochastic component to the control problem. All these variables are not fully controllable in practical applications but the knowledge of their time evolution is useful to improve microgrids management and control, especially if MPC approaches are applied. Typically, a prediction can be obtained from solar irradiance, wind forecasts or using historical data of atmospheric conditions, electrical prices, and load consumption. Later this information can be processed by a collection of techniques as statistical analysis, neural networks, genetic algorithms, machine learning, etc. Nevertheless, uncertainty in these values is unavoidable because of the practical difficulty in obtaining an accurate prediction due to the nature of renewable generation and the inherent variability of customer loads. These features of generation and demand can be overcome by balancing the power output using storage devices, demand-side management, or flexible dispatchable generation resources. But the difficulty in predicting generation and demand adds significant unavoidable uncertainty. To solve this challenge, a new approach in decision making in microgrid systems is introduced: deterministic decision making can be replaced by a stochastic solution, explicitly taking into account the uncertainty in the system.

The receding horizon technique of MPC provides some robustness to the control of systems with uncertainties, but when they are significant, more elaborate techniques are needed. With this objective in mind, a kind of family of MPC algorithms, named Robust MPC [7], has received significant interest last years. In Robust MPC, uncertainties are considered deterministic and bounded, and early solutions are based on min-max optimization problems. This solution is very conservative because it is based on the worst-case situation, and constraints must be satisfied for all possible uncertainty values.

MPC (or Robust MPC) is essentially a deterministic approach and can be inadequate for many systems where uncertainties are the main issue. Stochastic MPC [53] is based on a characterization of the uncertainties (i.e., probabilistic distribution of random variables), and including it in the optimization problem formulation. Unlike classical and Robust MPC, constraints are defined stochastically and typically some violations are allowed with preset probability criteria. Then, the obtained solutions are typically less conservative, and hence, the performance is better in terms of cost, as will be tackled in Chap. 7.

1.2.6 Microgrids and Electric Vehicles

The electrification of the transport system will have a great impact on the operation of the electrical system. The appearance of V2G (Vehicle-to-grid) systems, which consist of the use of electric car batteries during periods when they are not used as energy storage units for an electrical network, will change the way microgrids can interact with the main grid. Taking into account the current size of the fleet and the expectation of a gradual increase in the number of electric vehicles, it is expected that the energy storage capacity that can be provided in a near future will be sufficient to balance the supply and demand on a microgrid, and hence, improve the performance and stability of the network. The use of electric vehicles connected to the smart grid is a trend for the future and the development of algorithms for managing the use of vehicle batteries is a strategic research area.

It is estimated that an electric vehicle is in motion only 4% of time, so it could be available as an electrical energy storage unit during the remaining time [34]. Moreover, in normal use, the car batteries are recharged overnight (which is the period of low electricity demand) and are parked in the workplace during periods of high electrical demand, so this power could be used to meet the grid peak demand. This storage capacity is especially useful with renewable energy sources, as their fluctuating nature makes it harder to adjust production and demand.

Additionally, the V2G systems enable to establish new business models in which new actors appear, such as Load Managers that would be responsible for recharging infrastructures, providing service to vehicles, buying and selling electrical energy and building relationships with the network managers. In the last years, many control algorithms for charging electric vehicles in intelligent networks have appeared in literature. In [20, 64], the problem is solved by real-time optimization algorithms, whereas in [36] an MPC-based algorithm is presented. Also, solutions based on hierarchical distributed algorithms have been presented [6, 26, 30].

The integration of V2G systems can be a key factor in network stability guarantee against load fluctuations. In this framework, this book aims to contribute to this research area through the study and development of control algorithms applied to renewable energy microgrids. This issue will be treated in depth in Chap. 6.

1.3 Outline of the Chapters

This book aims to provide solutions based on MPC techniques to problems and challenges regarding control of microgrids. To achieve this objective, the rest of the book is organized as follows.

Chapter 2 introduces the fundamentals of MPC. The main MPC techniques will be described and formulated here, although other derivations of MPC will be developed when needed along the book. Chapter 3 deals with the elements that constitute the microgrid: generators, storage units and loads. Dynamical models of the components as well as of the whole microgrid are developed in order to have a complete nonlinear model that can be used to demonstrate the controllers. A modular simulator is provided so that readers can create their own microgrid using the blocks supplied. This way, the examples provided in the book can be replicated and it can also be used to develop and validate control algorithms on existing or projected microgrids.

Energy Management is treated in Chap. 4. This will be the first approach to this problem and the existing methods used nowadays are analyzed. A basic MPC strategy that only uses continuous variables is presented and implemented in this chapter. Also, a laboratory-scale microgrid is presented, which will be used along the book as a benchmark to illustrate most of the control techniques presented. Some methods are demonstrated on the simulator and experimentally validated on the laboratory benchmark.

Chapter 5 shows improvements with respect to the basic controller that can be achieved using the latest developments in hybrid MPC. The algorithm for the economic optimization of the microgrid using MPC is implemented here. Day-ahead and intraday markets, regulation service and real scenario load-sharing problems are addressed. In order to model both continuous/discrete dynamics and switching between different operating conditions, the microgrid is modeled within the framework of Mixed Logic Dynamical (MLD) systems. Taking into account the presence of integer variables, the MPC problem is solved as MIQP (Mixed-Integer Quadratic Programming).

The consideration of dynamic demand, such as manipulable loads and electric vehicles, is a topical subject which is addressed in Chap. 6. The possibility of load curtailment and time-shifting can help to improve the operation of the microgrid, provided the controller is able to manage this. Problems associated to the changing topology of the microgrid due to the connection and disconnection of a number of vehicles must be solved with the reconfiguration of the controller.

Chapter 7 deals with uncertainties that appear in microgrids related to the inherent randomness of renewable supply and load demand. Stochastic MPC is presented as an approach to address these issues. Different stochastic methodologies are developed and compared, providing a robust solution to the problem.

Distributed MPC techniques, valid for the optimization of large-scale systems, will be presented in Chap. 8. The integration of microgrids forming networks where neighbors have the possibility of interaction will greatly contribute to the smart grid. They can be considered as systems of systems, for whom dedicated distributed techniques are needed.

This book could not end without addressing power quality management using MPC, which is a very active topic in the research community. An optimal solution for power quality management is proposed in Chap. 9, where new control algorithms for power converters associated to DERs are developed for both cases: grid-connected and islanded mode. Since power converters have a finite number of switching states, MPC based on a finite set of control actions is presented and used to enhance the power quality of the energy supplied by the microgrid.

This way, the book covers an ample range of solutions based on MPC for the most important open topics in microgrid control.

References

1. Alvarado I, Limon D, Muñoz de la Peña D, Maestre JM, Ridao MA, Scheu H, Marquardt W, Negenborn RR, De Schutter B, Valencia F, Espinosa, J (2011) A comparative analysis of distributed MPC techniques applied to the HD-MPC four-tank benchmark. J Process Control 21(5):800–815. Special Issue on Hierarchical and Distributed Model Predictive Control
2. Alvial-Palavicino C, Garrido-Echeverria N, Jimenez-Estevez G, Reyes L, Palma-Behnke R (2011) A methodology for community engagement in the introduction of renewable based smart microgrid. Energy Sustain Develop 15(3):323–324
3. Anand S, Fernandes BG, Guerrero JM (2013) Distributed control to ensure proportional load sharing and improve voltage regulation in low-voltage DC microgrids. IEEE Trans Power Electron 28(4):1900–1913
4. Arce A, Del Real AJ, Bordons C (2009) MPC for battery/fuel cell hybrid vehicles including fuel cell dynamics and battery performance improvement. J Process Control 19(8):1289–1304
5. Arnold M, Negenborn RR, Andersson G, De Schutter B (2009) Model-based predictive control applied to multi-carrier energy systems. In: 2009 Power Energy Society General Meeting, PES '09. IEEE, pp 1–8
6. Bashash S, Fathy H (2011) Robust demand-side plug-in electric vehicle load control for renewable energy management. In: American Control Conference (ACC), pp 929–934
7. Bemporad A, Morari M (1999) Control of systems integrating logic, dynamics, and constraints. Automatica 35(3):407–427
8. Bidram A, Davoudi A (2012) Hierarchical structure of microgrids control systems. IEEE Trans Smart Grid 3:19631976
9. Bidram A, Lewis FL, Davoudi A (2014) Distributed control systems for small-scale power networks. IEEE Control Syst Mag 34(6):56–77
10. Bollen MHJ, Zhong J, Zavoda F, Meyer J, McEachern A, Lopez F (2010) Power quality aspects of smart grids. In: Proceedings of international conference on renewable energies and power quality
11. Bordons C, Ridao MA, Pérez A, Arce A, Marcos D (2010) Model predictive control for power management in hybrid fuel cell vehicles. In: 2010 IEEE Vehicle power and propulsion conference (VPPC). IEEE, pp 1–6
12. Bouzid AM, Guerrero JM, Cheriti A, Bouhamida M, Sicard P, Benghanem M (2015) A survey on control of electric power distributed generation systems for microgrid applications. Renew Sustain Energy Rev 44:751–766
13. Bozchalui MC, Hashmi SA, Hassen H, Cañizares CA, Bhattacharya K (2012) Optimal operation of residential energy hubs in smart grids. IEEE Trans Smart Grid 3(4):1755–1766
14. Camponogara E, de Lima ML (2012) Distributed optimization for MPC of linear networks with uncertain dynamics. IEEE Trans Autom Control 57(3):804–809
15. Camponogara E, Jia D, Krogh BH, Talukdar S (2002) Distributed model predictive control. IEEE Control Syst 22(1):44–52

16. Capitanescu F, Martinez Ramos JL, Panciatici P, Kirschen D, Marano Marcolini A, Platbrood L, Wehenkel L (2011) State-of-the-art, challenges, and future trends in security constrained optimal power flow. Electric Power Syst Res 81(8):1731–1741
17. Contreras SF, Cortés CA, Myrzik JMA (2018) Multi-objective probabilistic power resources planning for microgrids with ancillary services capacity. In: 2018 Power systems computation conference (PSCC), pp 1–8
18. De Brabandere K, Bolsens B, Van den Keybus J, Woyte A, Driesen J, Belmans R (2007) A voltage and frequency droop control method for parallel inverters. IEEE Trans Power Electron 22(4):1107–1115
19. De Brabandere K, Vanthournout K, Driesen J, Deconinck G, Belmans R (2007) Control of microgrids. In: 2007 IEEE Power engineering society general meeting. IEEE, pp 1–7
20. Deilami S, Masoum AS, Moses PS, Masoum M (2011) Real-time coordination of plug-in electric vehicle charging in smart grids to minimize power losses and improve voltage profile. IEEE Trans Smart Grid 2(3):456–467
21. del Real AJ, Arce A, Bordons C (2007) Hybrid model predictive control of a two-generator power plant integrating photovoltaic panels and fuel cell. In: IEEE conference on decision and control, pp 5447–5452
22. del Real AJ, Galus MD, Bordons C, Andersson G (2009) Optimal power dispatch of energy networks including external power exchange. In: 2009 European control conference (ECC), 3616–3621
23. Delfino F, Rossi M, Ferro G, Minciardi R, Robba M (2015) MPC-based tertiary and secondary optimal control in islanded microgrids. In: 2015 IEEE international symposium on systems engineering (ISSE), pp 23–28
24. Dimeas AL, Hatziargyriou ND (2005) Operation of a multiagent system for microgrid control. IEEE Trans Power Syst 20(3):1447–1455
25. Dötsch C (2009) Energy storage. In: Technology guide. Springer, pp 362–367
26. Fan Z (2012) A distributed demand response algorithm and its application to PHEV charging in smart grids. IEEE Trans Smart Grid 3(3):1280–1290
27. Fathi M, Bevrani H (2013) Statistical cooperative power dispatching in interconnected microgrids. IEEE Trans Sustain Energy 4(3):586–593
28. Ferrari-Trecate G, Gallestey E, Letizia P, Spedicato M, Morari M, Antonine M (2004) Modeling and control of co-generation power plants: a hybrid system approach. IEEE Trans Control Syst Technol 12(5):694–705
29. Galus MD, Andersson G (2009) Power system considerations of plug-in hybrid electric vehicles based on a multi energy carrier model. In: 2009 Power energy society general meeting, PES '09. IEEE, July 2009, pp 1–8
30. Galus MD, Andersson G, Art S (2012) A hierarchical, distributed pev charging control in low voltage distribution grids to ensure network security. In: 2012 IEEE Power and energy society general meeting, pp 1–8
31. Garcia-Torres F, Bordons C (2015) Optimal economical schedule of hydrogen-based microgrids with hybrid storage using model predictive control. IEEE Trans Ind Electron 62(8):5195–5207
32. Garcia-Torres F, Valverde L, Bordons C (2016) Optimal load sharing of hydrogen-based microgrids with hybrid storage using model predictive control. IEEE Trans Ind Electron 63(8):4919–4928
33. Garcia-Torres F, Vilaplana DG, Bordons C, Roncero-Sanchez P, Ridao MA (2019) Optimal management of microgrids with external agents including battery/fuel cell electric vehicles. IEEE Trans Smart Grid 10(4):4299–4308
34. Gautschi M, Scheuss O, Schluchter C (2009) Simulation of an agent based vehicle to grid (v2g) implementation. Electric Power Syst Res 120:177–183
35. Giaouris D, Papadopoulos AI, Ziogou C, Ipsakis D, Voutetakis S, Papadopoulou S, Seferlis P, Stergiopoulos F, Elmasides C (2013) Performance investigation of a hybrid renewable power generation and storage system using systemic power management models. Energy 61:621–635
36. Di Giorgio A, Liberati F, Canale S (2014) Electric vehicle charging control in smartgrids: a model predictive control approach. Control Eng Pract 22:147–162

37. Gomez-Exposito A, Conejo AJ, Cañizares C (2018) Electric energy systems: analysis and operation. CRC Press
38. Greenwell W, Vahidi A (2010) Predictive control of voltage and current in a fuel cell-ultracapacitor hybrid. IEEE Trans Ind Electron 57(6):1954–1963
39. Gu W, Wu Z, Yuan X (2010) Microgrid economic optimal operation of the combined heat and power system with renewable energy. IEEE Power Energy Soc General Meet, 1–6
40. Guerrero JM, Garcia de Vicuña L, Matas J, Castilla M, Miret J (2005) Output impedance design of parallel-connected ups inverters with wireless load-sharing control. IEEE Trans Ind Electron 52(4):1126–1135
41. Guerrero JM, Loh PC, Chandorkar M, Lee T (2013) Advanced control architectures for intelligent microgrids, part i: decentralized and hierarchical control. IEEE Trans Ind Electron 60(4):1254–1262
42. Hatziargyriou N (2014) Microgrids: architectures and control. Wiley-IEEE Press
43. Ipsakis D, Voutetakis S, Seferlis P, Stergiopoulos F, Papadopoulou S, Elmasides C (2008) The effect of the hysteresis band on power management strategies in a stand-alone power system. Energy 33(10):1537–1550
44. Janjic A, Stajic Z, Radovic I (2011) Power quality requirements for the smart grid design. Int J Circ Syst Signal Process
45. Katiraei F, Iravani MR (2006) Power management strategies for a microgrid with multiple distributed generation units. IEEE Trans Power Syst 21(4):1821–1831
46. Katiraei F, Iravani R, Hatziargyriou N, Dimeas A (2008) Microgrids management. IEEE Power Energy Mag 6(3):54–65
47. Lasseter R, Akhil A, Marnay C, Stephens J, Dagle J, Guttromson R, Meliopoulous A, Yinger R, Eto J (2002) The certs microgrid concept. White paper for Transmission Reliability Program, Office of Power Technologies, US Department of Energy
48. Lasseter RH (2002) Microgrids. IEEE Power Eng Soc Winter Meet 1:305–308
49. Lee K, Eidson JC, Weibel H, Mohl D (2005) IEEE 1588-standard for a precision clock synchronization protocol for networked measurement and control systems. In: Conference on IEEE, vol 1588
50. Marwali MN, Jung J, Keyhani A (2004) Control of distributed generation systems-part ii: Load sharing control. IEEE Trans Power Electron 19(6):1551–1561
51. Mehrizi-Sani A, Iravani R (2010) Potential-function based control of a microgrid in Islanded and grid-connected modes. IEEE Trans Power Syst 25(4):1883–1891
52. Mendes PRC, Valverde L, Bordons C, Normey-Rico JE (2016) Energy management of an experimental microgrid coupled to a v2g system. J Power Sour 327:702–713
53. Mesbah A (2016) Stochastic model predictive control: an overview and perspectives for future research. IEEE Control Syst Mag 36(6):30–44
54. Negenborn RR, Houwing M, De Schutter B, Hellendoorn J (2009) Model predictive control for residential energy resources using a mixed-logical dynamic model. In: International conference on networking, sensing and control, pp 702–707
55. Nunna HK, Doolla S (2013) Multiagent-based distributed-energy-resource management for intelligent microgrids. IEEE Trans Ind Electron 60(4):1678–1687
56. Olivares DE, Mehrizi-Sani A, Etemadi AH, Cañizares CA, Iravani R, Kazerani M, Hajimiragha AH, Gomis-Bellmunt O, Saeedifard A, Palma-Behnke R, Jimenez-Estevez GA, Hatziargyriou ND (2014) Trends in microgrid control. IEEE Trans Smart Grid 5(4):1905–1919
57. Palizban O, Kauhaniemi K (2015) Hierarchical control structure in microgrids with distributed generation: Island and grid-connected mode. Renew Sustain Energy Rev 44:797–813
58. Palizban O, Kauhaniemi K, Guerrero JM (2014) Microgrids in active network management part i: Hierarchical control, energy storage, virtual power plants, and market participation. Renew Sustain Energy Rev 36:428–439
59. Parisio A, Glielmo L (2011) A mixed integer linear formulation for microgrid economic scheduling. In: 2011 IEEE International conference on smart grid communications (SmartGridComm). IEEE, pp 505–510

60. Parisio A, Rikos E, Glielmo L (2014) A model predictive control approach to microgrid operation optimization. IEEE Trans Control Syst Technol 22(5):1813–1827
61. Parisio A, Rikos E, Tzamalis G, Glielmo L (2014) Use of model predictive control for experimental microgrid optimization. Appl Energy 115:37–46
62. Pecas-Lopes JA, Hatziargyriou N, Mutale J, Djapic P, Jenkins N (2007) Integrating distributed generation into electric power systems: a review of drivers, challenges and opportunities. Electric Power Syst Res 77(9):1189–1203
63. Petrollese M (2015) Optimal generation scheduling for renewable microgrids using hydrogen storage systems. PhD thesis, University of Cagliari
64. Richardson P, Flynn D, Keane A (2012) Optimal charging of electric vehicles in low-voltage distribution systems. IEEE Trans Power Syst 27(1):268–279
65. Rodriguez J, Pontt J, Silva CA, Correa P, Lezana P, Cortés P, Ammann U (2007) Predictive current control of a voltage source inverter. IEEE Trans Ind Electron 54(1):495–503
66. Rouzbehi K, Miranian A, Candela JI, Luna A, Rodriguez P (2014) Intelligent voltage control in a DC micro-grid containing PV generation and energy storage. In: 2014 IEEE PES T&D conference and exposition, pp 1–5, April 2014
67. Scattolini R (2009) Architectures for distributed and hierarchical model predictive control a review. J Process Control 19(5):723–731
68. Schaltz E, Rasmussen PO (2008) Design and comparison of power systems for a fuel cell hybrid electric vehicle. In: 2008 IEEE Industry applications society annual meeting, IAS'08. IEEE, pp 1–8
69. Thounthong P, Chunkag V, Sethakul P, Sikkabut S, Pierfederici S, Davat B (2011) Energy management of fuel cell/solar cell/supercapacitor hybrid power source. J Power Sour 196(1):313–324
70. Unamuno E, Barrena JA, Review and classification of control strategies, (2015) Hybrid AC/DC microgrids. Part ii. Renew Sustain Energy Rev 52:1123–1134
71. Ustun TS, Ozansoy C, Zayegh A (2011) Recent developments in microgrids and example cases around the world: a review. Renew Sustain Energy Rev 15(8):4030–4041
72. Vahidi A, Greenwell W (2007) A decentralized model predictive control approach to power management of a fuel cell-ultracapacitor hybrid. In: 2007 American Control Conference, ACC'07. IEEE, pp 5431–5437
73. Valverde L, Bordons C, Rosa F (2016) Integration of fuel cell technologies in renewable-energy-based microgrids optimizing operational costs and durability. IEEE Trans Ind Electron 63(1):167–177
74. Valverde L, Rosa F, Bordons C (2013) Design, planning and management of a hydrogen-based microgrid. IEEE Trans Ind Inf 9(3):1398–1404
75. Valverde L, Rosa F, Del Real A, Arce A, Bordons C (2013) Modeling, simulation and experimental set-up of a renewable hydrogen-based domestic microgrid. Int J Hydrog Energy 38(27):11672–11684
76. Vasquez JC (2009) Decentralized control techniques applied to electric power distributed generation in microgrids. PhD thesis, Polytechnic University of Catalonia
77. Vasquez JC, Guerrero JM, Miret J, Castilla M, Garcia de Vicuña L (2010) Hierarchical control of intelligent microgrids. IEEE Ind Electron Mag 4(4):23–29
78. Vergara-Dietrich JD, Morato MM, Mendes PRC, Cani AA, Normey-Rico JE, Bordons C (2017) Advanced chance-constrained predictive control for the efficient energy management of renewable power systems. J Process Control
79. Wang C, Nehrir M (2008) Power management of a stand-alone wind/photovoltaic/fuel cell energy system. IEEE Trans Energy Conversion 23(3):957–967
80. Wu J, Guan X (2013) Coordinated multi-microgrids optimal control algorithm for smart distribution management system. IEEE Trans Smart Grid 4(4):2174–2181
81. Zhang Y, Jiang Z, Yu X (2008) Control strategies for battery/supercapacitor hybrid energy storage systems. In: 2008 IEEE Energy 2030 Conference, ENERGY 2008. IEEE, pp 1–6
82. Ziogou C, Ipsakis D, Elmasides C, Stergiopoulos F, Papadopoulou S, Seferlis P, Voutetakis S (2011) Automation infrastructure and operation control strategy in a stand-alone power system based on renewable energy sources. J Power Sour 196(22):9488–9499

Chapter 2
Model Predictive Control Fundamentals

Abstract This book is focused on Model Predictive Control (MPC) techniques, which will be used to solve different control issues in microgrids. Although there are many techniques that can be used for the control of microgrids, MPC provides a general framework to solve most of the issues using some common ideas in an integrated way. MPC replaces offline determination of a control law by online solution of an optimal control problem that provides the current control action. This chapter presents the fundamentals of this technique. The main ideas and formulations are described here as well as some of the most representative techniques. MPC based on state-space models is detailed, since it will be extensively used along the book. Other techniques such as finite state MPC and MPC for hybrid systems are described too. The chapter also tackles two important issues for the application of MPC in microgrids: disturbances and constraints. Based on the methods presented in this chapter, the most relevant topics related to the control of microgrids will be addressed along the rest of the book.

2.1 Model Predictive Control and Microgrids

Apart from the well-known properties of MPC that have given this methodology its success in the process industry [22], the different types or *flavors* of MPC can provide solutions to many problems found in microgrids:

- The coordinated operation of different RESs and ESSs in the microgrid is a difficult task. The multivariable nature of MPC provides an optimal control solution that can manage the operation of the microgrid units in a coordinated way in order to achieve the objectives.
- The intermittence and variability of renewable generation, as well as demand, can be included in the optimization problem by considering stochastic variables, leading to a control action that can cope with randomness.

C. Bordons et al., *Model Predictive Control of Microgrids*,
Advances in Industrial Control, https://doi.org/10.1007/978-3-030-24570-2_2

- MPC can be used when binary/logical variables must be considered in the optimization. This is the case of the connection/disconnection of units (storage devices, electric vehicles, loads, etc.) or the consideration of changing situations, as is the case of different price of energy for purchasing or selling.
- When sudden changes in the microgrid appear, such as the disconnection or malfunctioning of a certain unit, MPC can adapt to this new situation by changing its structure and therefore allow normal operation of the microgrid, provided that there are degrees of freedom available.
- In case that several agents participate in the problem, as is the case of a network of microgrids or microgrids that are geographically distributed, the problem can be solved in a distributed way. MPC can provide a distributed solution, so that complex problems can be addressed.

2.2 The Model Predictive Control Paradigm

MPC provides an intuitive approach to the optimal control of systems subject to constraints [5]. This fact explains why MPC is the advanced control strategy that has the greatest acceptance in the industry. The term Model Predictive Control does not designate a specific control strategy but rather a wide family of control methods which make explicit use of a model of the system to calculate the control signal by minimizing a cost function. The MPC paradigm is based on the choice of the best amongst all feasible input sequences over a future horizon according to some criteria. Using the concept of receding horizon, the first input of this sequence is applied to the system and the scheme is repeated at the next sampling time, as new state information is available. This way, MPC solves a constrained dynamic optimal control problem by means of a repeated online optimization of the open-loop problem instead of difficult offline computation of control law.

The richness of the field—with wide ranges of choice in model structures, prediction horizons and optimization criteria—allows control designers to customize MPC to their applications, with range from high-speed computational requirements in microgrid power quality issues to the integration of complex power networks composed of several microgrids with different optimization criteria.

MPC can handle real-time state and input constraints in a natural way, enabling systems to operate more closely to their limits. MPC involves the solution at each sampling instant of a finite horizon optimal control problem subject to the system dynamics, and state and input constraints. The MPC methodology is appropriate for multivariable control problems and for the design of hierarchical multilayer systems composed of several control algorithms working at different timescales. MPC is especially powerful through its stochastic formulation, giving the controller the possibility to optimize different scenario conditions (such as failures) at the same time. Its distributed formulation can optimize at the same time subsystems integrated into global systems (such as the case of network of microgrids) optimizing the global cost function but respecting the subsystem cost function.

Several features of MPC makes it a suitable methodology to be used in microgrids. Besides its intuitive formulation, the method is easy to understand and it can include constraints and nonlinearities and manage multivariable as well as distributed cases. However, since an optimization problem is solved at each sampling instant, the computational cost is high compared to traditional control schemes.

The ideas, appearing to some extent in the predictive control family, are essentially [5]:

- Explicit use of a model to predict the system output at future time instants.
- Computation of a control sequence minimizing an objective function.
- Use of a receding strategy, so that at each instant the horizon is displaced toward the future, which involves the application of the first control signal of the sequence calculated at each step.

The different MPC algorithms only differ amongst themselves in the model used to represent the system, the cost function to be minimized and the manner that the optimization is performed. MPC presents a series of advantages over other methods, amongst which the following stand out:

- The concepts are very intuitive and, at the same time, the tuning is relatively easy.
- It can be used to control a great variety of systems, from those with simple dynamics to more complex ones, including nonlinear or unstable systems.
- Different performance criteria considering operational constraints can be fulfilled by an appropriate choice of the cost function during the design process.
- Compensation of measurable disturbances is easily derived.
- The resulting controller (once computed) is an easy-to-implement control law.
- Its extension to the multivariable case is conceptually straightforward.
- Future references (such as scheduled demands) can be easily integrated into the formulation.

However, it also has its drawbacks. The most important is that its derivation is more complex than that of classical controllers. Since an optimization problem is being solved at every sampling time, the price to be paid is the large amount of calculations required, especially in the constrained case or when using long horizons. Another important issue is the availability of an appropriate system model. The algorithm is based on prior knowledge of the model and is independent of it, but it is evident that the performance will be affected by the mismatch between the real system and the model used.

The methodology of all the controllers belonging to the MPC family is characterized by the following strategy [5], represented in Fig. 2.1:

1. The future outputs for a determined horizon N_p, called the prediction horizon, are predicted at each sampling instant t using the dynamic model of the system. These predicted outputs $y(t+k \mid t)$[1] for $k = 1 \ldots N_p$ depend on the known values up to instant t (past inputs and outputs and current state) and on the future control signals

[1] The notation indicates the value of the variable at the instant $t+k$ calculated at the current instant t.

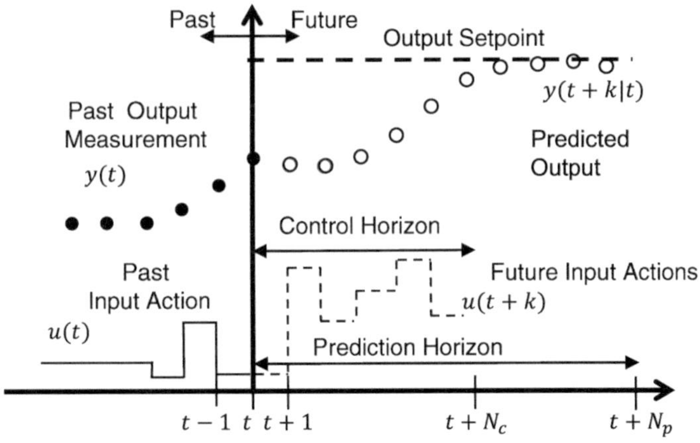

Fig. 2.1 MPC Strategy

$u(t+k \mid t)$, $k = 0 \ldots N_p - 1$, which are those to be computed and sent to the system.

2. The sequence of future control signals is calculated by optimizing a determined criterion which, in general, will try to keep the output as close as possible to the reference trajectory (which can be the setpoint itself or a close approximation to it). This criterion usually takes the form of a quadratic function of the errors between the predicted output signal and the predicted reference trajectory and it can include the necessary control effort. Although the Euclidean norm is the most used, also, the 1-norm or the infinity norm can be considered in the cost function.
3. The control signal $u(t \mid t)$ is sent to the process while the next control signals calculated are discarded, because at the next sampling instant, $y(t+1)$ is already known (feedback action). Step 1 is repeated with this new value and all the sequences are brought up to date. Thus, the signal $u(t+1 \mid t+1)$ is calculated (which may be different from $u(t+1 \mid t)$ because of the new information available) using the receding horizon concept.

This strategy is implemented using the basic structure shown in Fig. 2.2. A dynamical model is used to predict the future system output, based on past and current values and on the proposed optimal future control actions. These actions are calculated by the optimizer taking into account the cost function as well as the constraints.

Notice that the MPC strategy is very similar to the control strategy used when driving a car (see Fig. 2.3). The driver knows the desired reference trajectory for a finite control horizon and, by taking into account the car characteristics (mental model of the car), decides which control actions (accelerator, brakes, steering) must be taken in order to track the desired trajectory. Only the first control action is taken at each instant, and the procedure is repeated for the next control decision in a receding horizon fashion. The behavior of the car may depend on the chosen optimization

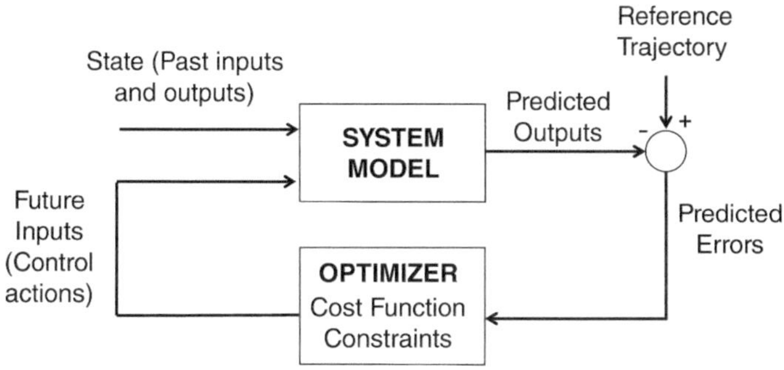

Fig. 2.2 Basic structure of MPC

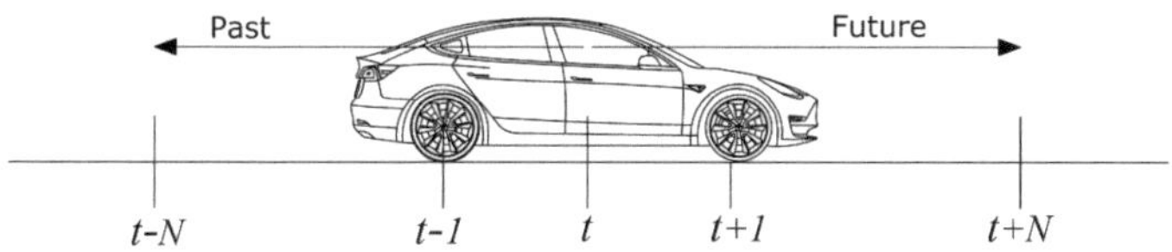

Fig. 2.3 MPC analogy

criterion. For example, if the driver's main objective is the shortest duration of the trip, the action will be more brisk than if fuel consumption is to be minimized.

MPC techniques will be formulated in detail along the book, using the most appropriate one to address the different control issues that appear in microgrids.

2.3 Methodology

MPC is a family of methods that differ amongst themselves in the type of model, the cost function and the solving method. Different formulations of MPC can be used for microgrid control. Since storage is an important component of microgrids, the dynamic models of microgrids are generally formulated as state-space equations where the state variable $x(t)$ coincides with the state of charge of the energy storage units. Therefore, state-space MPC is a good candidate to control microgrids and thus state-space models can be used to formulate the predictive control problem. Besides, this formulation can easily deal with multivariable systems, which is the common case in microgrids. The following equations are used in the linear case to capture system dynamics:

$$\begin{aligned} x(t+1) &= Ax(t) + Bu(t) \\ y(t) &= Cx(t) \end{aligned} \tag{2.1}$$

The state vector is $x(t)$, $y(t)$ and $u(t)$ are scalars in the Single-Input Single-Output (SISO) case, but in Multiple-Input Multiple-Output (MIMO) systems, the input vector $u(t)$ is of dimension m and $y(t)$ of dimension n. In microgrids, usually the output $y(t)$ coincides with the state $x(t)$, so the process is MIMO and matrix C equals identity.

In order to achieve offset-free control, the model can be expressed in incremental form, using the control increment $\Delta u(t)$ as the input, instead of the control signal $u(t)$. This model can be written in the general state-space form considering that $\Delta u(t) = u(t) - u(t-1)$. Then the following representation can be obtained if this expression is included in (2.1):

$$\begin{bmatrix} x(t+1) \\ u(t) \end{bmatrix} = \begin{bmatrix} A & B \\ 0 & I \end{bmatrix} \begin{bmatrix} x(t) \\ u(t-1) \end{bmatrix} + \begin{bmatrix} B \\ I \end{bmatrix} \Delta u(t)$$
$$y(t) = \begin{bmatrix} C & 0 \end{bmatrix} \begin{bmatrix} x(t) \\ u(t-1) \end{bmatrix}$$

If a new state vector $\tilde{x}(t) = [x(t) \;\; u(t-1)]^T$ is introduced, the incremental model takes the general form (2.1):

$$\begin{aligned} \tilde{x}(t+1) &= M\tilde{x}(t) + N \, \Delta u(t) \\ y(t) &= Q\tilde{x}(t) \end{aligned} \tag{2.2}$$

where the values of M, N, and Q can be expressed as functions of A, B, and C comparing Eqs. (2.1) and (2.2). Notice that the general notation $x(t)$ for the state vector will be used from now on, although the incremental form is used, in order to consider any type of state-space model for the MPC derivation.

Once a dynamic model is available, it can be included in the cost function and proceed to its minimization. The various MPC algorithms use different cost functions for computing the control law. Typically, the main goal is that the future output $y(t)$ tracks a certain reference signal $w(t)$ along the horizon while penalizing the control effort $\Delta u(t)$ necessary for doing so. The general expression for such an objective function in the SISO case will be:

$$J(N_p, N_c) = \sum_{j=1}^{N_p} \delta(j)[\hat{y}(t+j \mid t) - w(t+j)]^2 + \sum_{j=1}^{N_c} \lambda(j)[\Delta u(t+j-1)]^2 \tag{2.3}$$

where $\hat{y}$ is the prediction of the output.

An additional term penalizing the control signal (not its increment) can be included. N_p is the prediction horizon and $N_c \leq N_p$ is the control horizon, which does not necessarily have to take the same value. The value N_p sets the limit of the time instants in which it is desirable for the output to track the reference. The control horizon concept (N_c) consists of considering that after a certain interval $N_c < N_p$ the proposed control signals will be kept constant, that is, $u(t+j)$ does not change after $j = N_c$:

$$\Delta u(t+j-1)=0 \quad j>N_c$$

This can significantly reduce the number of decision variables and, therefore, the complexity of the problem. The coefficients $\delta(j)$ and $\lambda(j)$ are sequences that consider the relative weight of error and control effort along the horizon; usually constant values or exponential sequences are considered. In some situations, the state $x(t)$ can be included in the cost function instead of the output $y(t)$. Then, the derivation done below in this section must be slightly changed by making matrices C in (2.1) or Q in (2.2) equal to the identity matrix.

In the case of MIMO processes, inputs and outputs are vectors and therefore the costs are computed using quadratic functions, where R and P are positive definite weighting matrices which are usually diagonal:

$$J(N_p, N_c) = \sum_{j=1}^{N_p} \|\hat{y}(t+j \mid t) - w(t+j)\|_R^2 + \sum_{j=1}^{N_c} \| \Delta u(t+j-1)\|_P^2 \quad (2.4)$$

being $\|.\|_R^2$ the 2-norm.[2]

Due to the predictive nature of MPC, if the future evolution of the reference $r(t+k)$ is known a priori, the controller can react before the change has effectively been made. This happens in many applications; for instance, the power to be exchanged with the grid can be computed by an upper level scheduling process. An improvement in performance can be obtained even though the reference is constant by simply knowing the instant when the value changes and getting ahead of this circumstance. Usually, a reference trajectory $w(t+k)$ is used, which is a smooth approximation from the current output value to the known reference value $r(t+k)$ using a first-order filter:

$$w(t) = y(t) \quad w(t+k) = \alpha w(t+k-1) + (1-\alpha) r(t+k) \quad k = 1 \ldots N_p \quad (2.5)$$

where α is an adjustable parameter between 0 and 1 (the closer to 1, the smoother the approximation).

The output predictions to be used in the objective function (2.3), can be computed using (2.1) or (2.2) if an incremental model is used. In this case, the predictions are given by

$$\hat{y}(t+j) = QM^j x(t) + \sum_{i=0}^{j-1} QM^{j-i-1} N \Delta u(t+i)$$

Notice that $x(t)$ must be calculated using an observer in case the state vector is not accessible. Then, the predictions along the horizon are given by

[2] Defined as $\|x\|_R^2 = x^T R x$.

$$\mathbf{y} = \begin{bmatrix} \hat{y}(t+1|t) \\ \hat{y}(t+2|t) \\ \vdots \\ \hat{y}(t+N_p|t) \end{bmatrix} = \begin{bmatrix} QMx(t) + QN \triangle u(t) \\ QM^2x(t) + \sum_{i=0}^{1} QM^{1-i}N \triangle u(t+i) \\ \vdots \\ QM^{N_p}x(t) + \sum_{i=0}^{N_p-1} QM^{N_p-1-i}N \triangle u(t+i) \end{bmatrix}$$

where boldface lower letters are used to indicate vectors composed of elements along the horizon and boldface upper case letters denote matrices composed of other matrices and vectors. Then, the last equation can be written in vector form as

$$\mathbf{y} = \mathbf{F}x(t) + \mathbf{H}\mathbf{u} \tag{2.6}$$

where vector $\mathbf{u} = [\Delta u(t) \; \triangle u(t+1) \; \ldots \; \triangle u(t+N_c-1)]^T$ contains the future control increments, $\mathbf{H}$ is a block lower triangular matrix with its non-null elements defined by $\mathbf{H}_{ij} = QM^{i-j}N$ and matrix $\mathbf{F}$ is defined as

$$\mathbf{F} = \begin{bmatrix} QM \\ QM^2 \\ \vdots \\ QM^{N_p} \end{bmatrix}$$

Notice that the prediction (2.6) has two terms: the first is the free response of the system $\mathbf{f} = \mathbf{F}x(t)$, which depends on the current state and therefore is known at instant t. The second term depends on the future control sequence $\mathbf{u}$, which is calculated minimizing the objective function (2.3), that (in the case of $\delta(j) = 1$ and $\lambda(j) = \lambda$) can be written as

$$J = (\mathbf{Hu} + \mathbf{F}x(t) - \mathbf{w})^T(\mathbf{Hu} + \mathbf{F}x(t) - \mathbf{w}) + \lambda\mathbf{u}^T\mathbf{u}$$

If there are no constraints, the analytical solution that provides the optimum can be calculated by imposing that the derivative of J must equal 0, giving:

$$\mathbf{u} = (\mathbf{H}^T\mathbf{H} + \lambda\mathbf{I})^{-1}\mathbf{H}^T(\mathbf{w} - \mathbf{F}x(t))$$

As stated in the previous section, the receding horizon implies that only the first element of the control sequence, $\Delta u(t)$, is used and all the computation is repeated at the next sampling time.

In case that the control and the prediction horizons approach infinity and there are no constraints, the predictive controller becomes the well-known Linear Quadratic Regulator (LQR) problem, as shown in [5]. This equivalence allows the use of results coming from the optimal control field to analyze theoretical issues of MPC, as stability (see Sect. 2.9).

If the non-incremental model of Eq. (2.1) is used, the predictions are computed in a slightly different way, as shown in [15]. In this case:

$$\mathbf{y} = \begin{bmatrix} CA \\ CA^2 \\ \vdots \\ CA^{N_p} \end{bmatrix} x(t) + \begin{bmatrix} CB \\ CA^2B \\ \vdots \\ \sum_{i=0}^{N_p-1} CA^iB \end{bmatrix} u(t-1)$$

$$+ \begin{bmatrix} B & \dots & 0 \\ C(AB+B) & \dots & 0 \\ \vdots & \ddots & \vdots \\ \sum_{i=0}^{N_p-1} CA^iB & \dots & \sum_{i=0}^{N_p-N_c} CA^iB \end{bmatrix} \mathbf{u}$$

This equation can be written in vector form as

$$\mathbf{y} = \mathbf{F_1}x(t) + \mathbf{F_2}u(t-1) + \mathbf{H_1}\mathbf{u}$$

A new term that depends on $u(t-1)$ appears, which does not depend on the decision variable $\mathbf{u}$ and therefore does not affect the optimization since. Then, the control action is calculated as

$$\mathbf{u} = (\mathbf{H_1}^T\mathbf{H_1} + \lambda I)^{-1}\mathbf{H_1}^T(\mathbf{w} - \mathbf{F_1}x(t) - \mathbf{F_2}u(t-1))$$

For both types of models (incremental or not), the control law is always a static state feedback law. In the constrained case the solution is obtained solving a Quadratic Programming (QP) algorithm, as will be studied in Sect. 2.5.

2.4 Disturbances

Microgrids, as any system, are subject to disturbances during their normal operation. There are two clear sources of disturbances in microgrids: the power generated by the RESs and the demanded power. Both are external inputs to the system that cannot be manipulated by the controller. Challenges arise from the natural intermittency of renewable energy sources and the requirements to satisfy variable energy demand. Since renewable sources are used for generation, their time-varying nature, difficulty of prediction, and lack of manipulating capability make them a problem to be solved by the control system. The original formulation of MPC does not include disturbances although some MPC schemes have been proposed to ensure stability and compliance with constraints in this case [4]. Notice that MPC, like any other controller, can reject disturbances by the feedback mechanism. However, if disturbances can be measured (or estimated) their influence on the output can be included in the dynamic model and therefore the controller can anticipate their effect on the output. This way, MPC can inherently include feedforward effect.

Therefore, the effect of these disturbances $d(t)$ can be added to the MPC formulation. Then, the dynamic model of the system can be written as

$$\begin{aligned} x(t+1) &= Ax(t) + Bu(t) + B_d d(t) \\ y(t) &= Cx(t) \end{aligned} \tag{2.7}$$

where B_d is the matrix that quantifies the effect of disturbances on the states. Since generation and demand have the same effect on the energy balance (one positive and the other negative), these disturbances can be grouped in only one variable: the net effect of generation and demand: $d(t) = P_{gen}(t) - P_{dem}(t)$. Notice that generation and load can be measured, so they are measurable disturbances.

There are several ways of including disturbances in the system model, providing feedforward effect. Two different approaches are depicted here: (i) add the effect of the disturbance to the output prediction, (ii) include the disturbance vector $d(t)$ as a component of the state vector $x(t)$.

The first approach uses Eq. (2.7) to compute the prediction. If the incremental model is to be used, the discrete model is given by

$$\begin{aligned} \tilde{x}(t+1) &= M\tilde{x}(t) + N \bigtriangleup u(t) + N_d d(t) \\ y(t) &= Q\tilde{x}(t) \end{aligned} \tag{2.8}$$

where matrix N_d is B_d with m (number of inputs) additional rows of zeroes. Now, the prediction includes the values of the disturbance along the horizon, which can be estimated (in case of RESs, they can be supplied by weather forecasts) or can be considered constant and equal to the current value $d(t)$. Then, predictions are given by

$$\mathbf{y} = \begin{bmatrix} \hat{y}(t+1|t) \\ \hat{y}(t+2|t) \\ \vdots \\ \hat{y}(t+N_p|t) \end{bmatrix} = \begin{bmatrix} QMx(t) + QN \bigtriangleup u(t) \\ QM^2x(t) + \sum_{i=0}^{1} QM^{1-i} N \bigtriangleup u(t+i) \\ \vdots \\ QM^{N_p}x(t) + \sum_{i=0}^{N_p-1} QM^{N_p-1-i} N \bigtriangleup u(t+i) \end{bmatrix} + \begin{bmatrix} QN_d d(t) \\ \sum_{i=0}^{1} QM^{1-i} N_d d(t+i) \\ \vdots \\ \sum_{i=0}^{N_p-1} QM^{N_p-1-i} N_d d(t+i) \end{bmatrix}$$

which can be expressed in vector form as

$$\mathbf{y} = \mathbf{F}x(t) + \mathbf{H}\mathbf{u} + \mathbf{H_d}\mathbf{d} \tag{2.9}$$

where $\mathbf{d} = [d(t)\ d(t+1)\ \ldots\ d(t+N_p)]^T$ is the vector of future disturbances and $\mathbf{H_d}$ is a block lower triangular matrix with its non-null elements defined by $\mathbf{H_d}_{ij} = QM^{i-j}N_d$. Then, the form of the prediction is the same as in the undisturbed case, being the free response given by $\mathbf{f} = \mathbf{F}x(t) + \mathbf{H_d}\mathbf{d}$. Now the controller has knowledge of the effect of disturbances on the output along the horizon, which makes it possible to reject its effect, thus providing the feedforward effect.

The second approach, used for instance in [23], modifies the state vector of (2.2) by adding the disturbance variable: $\tilde{x}(t) = [x(t)\ \ d(t)]^T$. Notice that this modification can be done indistinctly for the incremental and the non-incremental state-space models. In the first case, the matrix description is given by

$$\begin{bmatrix} \tilde{x}(t+1) \\ d(t+1) \end{bmatrix} = \begin{bmatrix} M & N_d \\ 0 & I \end{bmatrix} \begin{bmatrix} \tilde{x}(t) \\ d(t) \end{bmatrix} + \begin{bmatrix} N \\ 0 \end{bmatrix} \triangle u(t)$$

$$y(t) = \begin{bmatrix} Q & 0 \end{bmatrix} \begin{bmatrix} \tilde{x}(t) \\ d(t) \end{bmatrix}$$

Therefore, the prediction can be computed as in Eq. (2.6) and the solution is obtained as in the previous subsection.

For the sake of simplicity, $x(t)$ will be henceforth used as the state vector, independently of being incremental or not and including disturbances or not. The accompanying matrices must be defined accordingly. No matter which of the models is used, disturbances are included in the prediction and therefore the controller can anticipate its effect on the microgrid output along the horizon.

Uncertainties in the disturbances (basically generation and demand profiles) have been mainly addressed indirectly in the dispatch problem by using the MPC approach [20]. But these uncertainties can be considered in the optimization problem; one approach is stochastic MPC, which will be addressed in Chap. 7.

2.5 Constraints

In practice, all systems are subject to constraints. The generators and storage units have a limited field of action and a determined power rate. Constructive, safety, regulation or environmental reasons can cause limits in the system variables such as storage levels in ESSs, power flows in lines, or maximum temperatures and pressures. Furthermore, the best operational condition is usually defined by the intersection of certain constraints for basically economic reasons, so that the system will operate close to the boundaries. All of this makes the introduction of constraints in the optimization problem necessary. Normally, bounds in the amplitude and in the ramp rate of the control signal and limits in the output will be considered

$$
\begin{aligned}
u_{\min} \leq & \quad u(t) \quad \leq u_{\max} \quad \forall t \\
\Delta u_{\min} \leq & \; u(t) - u(t-1) \leq \Delta u_{\max} \quad \forall t \\
y_{\min} \leq & \quad y(t) \quad \leq y_{\max} \quad \forall t
\end{aligned}
$$

Notice that constraints in the states $x(t)$ can also be included using the same inequalities as used for output constraints by making matrices C in (2.1) or Q in (2.2) equal to the identity matrix. For an m-input n-output system, the constraints acting over a receding horizon N_p can be expressed as

$$
\begin{aligned}
&\mathbf{1}\, u_{\min} \leq T\mathbf{u} + u(t-1)\, \mathbf{1} \leq \mathbf{1}\, u_{\max} \\
&\mathbf{1}\ \triangle\, u_{\min} \leq \mathbf{u} \leq \mathbf{1}\ \triangle\, u_{\max} \\
&\mathbf{1}\, y_{\min} \leq \mathbf{H}\mathbf{u} + \mathbf{F}x(t) \leq \mathbf{1}\, y_{\max}
\end{aligned}
$$

where $\mathbf{l}$ is an $(N_p \times n) \times m$ matrix formed by N_p $m \times m$ identity matrices and T is a lower triangular block matrix whose non-null block entries are $m \times m$ identity matrices. The following matrix form can be used to express all the constraints:

$$
\mathbf{R}\,\mathbf{u} \leq \mathbf{c}
$$

with:

$$
R = \begin{bmatrix} I_{N_p \times N_p} \\ -I_{N_p \times N_p} \\ T \\ -T \\ \mathbf{H} \\ -\mathbf{H} \end{bmatrix} \qquad
\mathbf{c} = \begin{bmatrix} \mathbf{l}\ \triangle\, u_{\max} \\ -\mathbf{l}\ \triangle\, u_{\min} \\ \mathbf{l}\, u_{\max} - \mathbf{l}u(t-1) \\ -\mathbf{l}\, u_{\min} + \mathbf{l}u(t-1) \\ \mathbf{l}\, y_{\max} - \mathbf{F}x(t) \\ -\mathbf{l}\, y_{\min} + \mathbf{F}x(t) \end{bmatrix}
$$

The introduction of these constraints in the objective function makes the minimization more complex, so that the solution cannot be obtained explicitly as in the unconstrained case. The problem to be solved is the minimization of a quadratic cost function with linear constraints:

$$
\begin{aligned}
\min \quad & \frac{1}{2}\mathbf{u}^T \mathbf{A}\mathbf{u} + \mathbf{b}^T \mathbf{u} \\
\text{subject to:} \quad & \mathbf{R}\mathbf{u} \leq \mathbf{c}
\end{aligned}
$$

which is a Quadratic Programming (QP) problem. QP is a well-known problem for which many robust solvers exist. The three main types of algorithms employed for solving this problem are: active set, interior point, and gradient projection methods. Descriptions of these methods (especially the fast gradient method, which is a kind of projection method) as well as analysis of their performance can be found in [19].

2.6 Other MPC Techniques

There are many MPC techniques based on the same ideas. Comparative studies can be found in [21] and in [22], where a *family tree* for the most significant algorithms, illustrating their connections, is presented. In this section, two of the most important methods: DMC [8] and GPC [7] are reviewed in order to demonstrate their most distinguished features. A description of other methods considered to be representative, like MAC, PFC, and EPSAC, can be found in [5].

2.6.1 *Dynamic Matrix Control*

Dynamic Matrix Control [8] can be considered as the first successful implementation of MPC in industry, due to the easy way of obtaining the process model and its capability to deal with multivariable processes and constraints. DMC uses the system step response to model the process, only taking into account the first N terms, thus assuming the process to be stable and without integrators. As regards the unmeasured disturbance $n(t)$, its value is considered to be constant along the horizon and equal to the one at instant t, that is, to be equal to the measured value of the output $y_m(t)$ minus the one estimated by the model $\hat{y}(t \mid t)$.

$$\hat{n}(t+k \mid t) = \hat{n}(t \mid t) = y_m(t) - \hat{y}(t \mid t)$$

and consequently, the predicted value of the output will be

$$\hat{y}(t+k \mid t) = \sum_{i=1}^{k} g_i \triangle u(t+k-i) + \sum_{i=k+1}^{N} g_i \triangle u(t+k-i) + \hat{n}(t+k \mid t)$$

where the first term contains past values of the control actions and is therefore known, the second contains the future control actions to be calculated and the last represents the disturbances. The computation of the prediction along the horizon can be written as a function of the system *dynamic matrix*, which is composed of the elements of the step response and gives its name to this algorithm. The cost function may consider future errors only, or it can also include the control effort, so that it presents the generic form (2.3), and constraints can be included in the formulation. The inconveniences of this method are, on one hand, the size of the process model required and, on the other hand, the inability to work with unstable processes.

2.6.2 *Generalized Predictive Control*

This method has its roots in minimum variance and adaptive control. The output predictions of the Generalized Predictive Controller [7] are based on using a CARIMA

(Controlled Autoregressive Integrated Moving Average) model:

$$A(z^{-1})y(t) = B(z^{-1})z^{-d}\,u(t-1) + C(z^{-1})\frac{e(t)}{\triangle}$$

where z^{-1} is the backward shift operator and the measurable disturbance is given by a white noise colored by $C(z^{-1})$. As its true value is difficult to know, this polynomial can be used for optimal disturbance rejection or robustness enhancement. The derivation of the optimal prediction is done by solving a Diophantine equation whose solution can be found by an efficient recursive algorithm. GPC uses a quadratic cost function of the form:

$$J(N_p, N_c) = \sum_{j=1}^{N_p} \delta(j)[\hat{y}(t+j \mid t) - w(t+j)]^2 + \sum_{j=1}^{N_c} \lambda(j)[\triangle u(t+j-1)]^2$$

where the weighting sequences $\delta(j)$ and $\lambda(j)$ are usually chosen constant or exponentially increasing and the reference trajectory $w(t+j)$ can be generated by a simple recursion which starts at the current output and tends exponentially to the setpoint. In the case of constant weights, the current control action is computed as

$$\triangle u(t) = K(\mathbf{w} - \mathbf{f})$$

where K is a constant vector which can be calculated beforehand based on the values of the CARIMA model and the Diophantine equation, as shown in [5]. This has a clear meaning: if there are no future predicted errors, that is, if $\mathbf{w} - \mathbf{f} = \mathbf{0}$, then there is no control move, since the objective will be fulfilled with the free evolution of the process. Otherwise, there will be an increment in the control action proportional (with a factor K) to that future error. Notice that the action is taken with respect to *future* errors, not *past* errors, as is the case in conventional PID controllers.

The theoretical basis of the GPC algorithm has been widely studied, and it has been shown [6] that, for limiting cases of parameter choices, this algorithm is stable and also that well-known controllers such as mean level and deadbeat control are inherent in the GPC structure. This technique is described in detail in Chap. 9, in an application to control an electronic power converter.

2.7 MPC with Logic and Continuous Variables

In most systems, there are not only continuous variables but also variables of discrete nature. Many operations linked to the management of microgrids have to be modeled considering continuous/discrete dynamics and switching between different operating conditions. Binary values are needed to describe logical (boolean) variables (which can take the values ON/OFF, 0/1 or connected/disconnected) and represent decision variables and system states. In control engineering, hybrid systems are dynamic systems with continuous, discrete and event variables. That is, the system has time-driven and event-driven dynamics and therefore the controller must deal with continuous and discrete signals.

There are several techniques for modeling hybrid systems, which have to be descriptive enough to capture the behavior of the various elements of the system and to consider interconnections between logic and continuous dynamics. Additionally, the model has to be simple enough to be used for control purposes. Some of the most extended methods are discrete hybrid automaton, Piecewise Affine (PWA) systems, and Mixed Logical Dynamical (MLD) systems. A detailed description and equivalence among the main models of hybrid systems can be found in [11].

2.7.1 Hybrid Model

In order to capture both continuous/discrete dynamics and to switch between different operating conditions, a model of the microgrid within the framework of MLD [3] will be used along this book. MLD is a versatile framework to model a variety of systems. For a detailed description of such capabilities, the reader is referred to [2, 3, 5]. Notice that in microgrids, continuous dynamics (generators, storage units, loads, etc.) coexist with switching operating conditions (charge/discharge of ESSs, sell/purchase from the grid, etc.). MLD models can be very useful in order to transform hybrid dynamic optimization problems into mixed-integer linear and quadratic programs, which can be solved via branch and bound techniques [18].

The development of the MLD form of a hybrid system involves three steps, as described in [9]. The first one is to associate a logical statement S (that can be true or false) with a binary variable $\delta \in \{0, 1\}$ that is 1 if and only if the statement holds true. Then, any statement, which is a combination of elementary statements $S_1, \ldots, S_q$ using the boolean operators AND ($\wedge$), OR ($\vee$) and NOT ($\sim$) can be represented as linear inequalities over the corresponding binary variables δ_i, $i = 1, \ldots, q$. The correspondence between relationships that form a statement and the associated inequalities is reported in Table 2.1, where m and M represent the lower and upper bounds of $a^T x$, where x is the continuous variable and a is a vector of parameters, and $\epsilon > 0$ is the smallest tolerance of the device [3, 9]. As an example, consider P1, which establishes that the statement $S_1 \wedge S_2$ holds true if and only if δ_1 and δ_2 are both equal to 1.

The second step is to represent the product between linear functions and logic variables by introducing an auxiliary variable $z = \delta a^T x$, and the third step is to form a Linear Time-Invariant (LTI) discrete-time dynamic model that includes binary and auxiliary variables. This model describes the evolution of the continuous and logic components of the system, with the general form [3, 9]:

Table 2.1 Conversion of logic relations into mixed-integer inequalities

	Relation	Logic	MLD inequalities
P1	AND ($\wedge$)	$S_3 \Leftrightarrow (S_1 \wedge S_2)$	$-\infty \le -\delta_1 + \delta_3 \le 0$ $-\infty \le -\delta_2 + \delta_3 \le 0$ $-\infty \le \delta_1 + \delta_2 - \delta_3 \le 1$
P2	OR ($\vee$)	$S_1 \vee S_2$	$1 \le \delta_1 + \delta_2 \le 2$
P3	NOT ($\sim$)	$S2 \equiv \sim S_1$	$1 \le \delta_1 + \delta_2 \le 1$
P4	IMPLY ($\Rightarrow$)	$[a^T x \le 0] \Rightarrow [\delta = 1]$	$\varepsilon \le a^T x - (m - \varepsilon)\delta \le \infty$
P5	IMPLY ($\Rightarrow$)	$[\delta = 1] \Rightarrow [a^T x \le 0]$	$-\infty \le a^T x + M\delta \le M$
P6	IFF($\Leftrightarrow$)	$[a^T x \le 0] \Leftrightarrow [\delta = 1]$	$\varepsilon \le a^T x - (m - \varepsilon)\delta \le \infty$ $-\infty \le a^T x + M\delta \le M$
P7	Mixed product	$z = \delta \cdot a^T x$	$-\infty \le z - M\delta \le 0$ $0 \le z - m\delta \le \infty$ $-\infty \le z - a^T x + m(1-\delta) \le 0$ $0 \le z - a^T x + M(1-\delta) \le \infty$

$$x(t+1) = Ax(t) + B_1 u(t) + B_2 \delta(t) + B_3 z(t) \tag{2.10}$$

$$y(t) = Cx(t) + D_1 u(t) + D_2 \delta(t) + D_3 z(t) \tag{2.11}$$

$$E_2 \delta(t) + E_3 z(t) \le E_1 u(t) + E_4 x(t) + E_5 \tag{2.12}$$

where $x = [x_c^T \; x_l^T]^T \in R^{n_c} \times \{0, 1\}^{n_l}$ are the continuous and binary states, $u = [u_c^T \; u_l^T]^T \in R^{m_c} \times \{0, 1\}^{m_l}$ represents the inputs, $y = [y_c^T \; y_l^T]^T \in R^{p_c} \times \{0, 1\}^{p_l}$ the outputs, $\delta \in \{0, 1\}^{r_l}$ represents the binary variables, $z \in R^{r_c}$ represents auxiliary binary and continuous variables. Equation (2.12) gathers the constraints on the states, the inputs, the z and δ variables. Then, this description converts the hybrid dynamics in a set of constraints to be integrated into the MPC formulation. This type of model will be used in Chap. 5 to include different operating conditions (as charge or discharge of batteries) and economic scenarios (as different electricity tariffs for sale and purchase).

2.7.2 *MPC of MLD Systems*

The formulation of the MPC problem for an MLD system is basically the same as in continuous systems: minimization of a cost function with linear constraints. But for the optimization, the decision variables are continuous and binary. These types of optimization problems are known as Mixed Integer Programming (MIP) problems. In case that the cost function is quadratic, the problem is known as Mixed-Integer Quadratic Programming (MIQP) and if the cost function is linear then it is a Mixed-Integer Linear Programming (MILP) problem. Notice that MIP is a much more difficult problem to solve than LP or QP, see [10], since for each of the possible combinations of the binary decision variables, a QP problem (with the remaining continuous decision variables) must be solved. The number of QP problems to be solved would be 2^{n_b} when n_b binary variables exist. However, there are more efficient ways to solve the problem than evaluating all possible combinations. Branch and bound methods are usually used to solve the MIQP problem, because they solve only a portion of all possible QP problems.

In case of binary variables, the optimization problems to be solved can be enumerated in a binary tree, with two branches per node. Then, the concept of branch and bound can be used. The main idea is to partition the feasible set of problems into smaller subsets and, depending on their cost, eliminate several subsets from further consideration, reducing this way the number of QPs to be solved. A detailed description and an illustrative example of the use of branch and bound in hybrid MPC can be found in [5].

2.8 Finite Control Set MPC

In the control of power electronics converters existing in the microgrid, it is typical that the decision variable is constrained to be in a finite set. Therefore, the MPC optimization problem can be simplified by computing a reduced set of predictions, one for each value of the control actions. Then, at each sampling instant, the cost function is evaluated for every admissible value of the control action and the one which provides the optimal value is selected. This approach is known as Finite Control Set MPC (FCS-MPC) [13], since the possible control actions (states of the converter switches) are finite. This method has been successfully applied to a wide range of power converters and drive applications [24, 26]. The methodology is simple, intuitive and easy to implement, provided the horizon is small.

But its complexity increases exponentially with the length of the horizon, so this approach is currently limited to short horizons (usually 1 or 2) due to the fact that the solution is obtained evaluating a cost function in a kind of exhaustive search. In the context of MPC, long horizons are better for stability: the infinite-horizon case, in general, ensures closed-loop stability provided that a solution with a finite cost exists [23]. For FCS-MPC, the stability analysis problem is difficult to address since this MPC strategy, in general, does not provide an explicit solution. The paper [1] provides necessary and sufficient conditions for optimality of quantized MPC with horizon equal to 1. A review of FCS-MPC strategies that achieve long prediction intervals has been published in [12]. This methodology will be used in Chap. 9 for quality control in the microgrid.

2.9 Stability of MPC

There are several works about the main features of MPC. A review of the main techniques and applications can be found in [5] and the main theoretical properties are described in detail in [23]. Some of the main features are briefly described below, but the interested reader is invited to acquire a deeper knowledge through the mentioned books.

Stability of MPC is an important issue, especially when uncertainties (model error, disturbances or estimation errors) or nonlinearities appear. Due to the intrinsic nature of MPC, that solves the constrained optimal control problem online, the horizon must be a finite number and therefore the good properties of infinite-horizon approaches such as LQG (Linear Quadratic Gaussian) cannot be inherited. Even is the process is linear, MPC of constrained systems is nonlinear, needing the use of Lyapunov stability theory. In the infinite-horizon case, the cost function can be shown to be monotonically decreasing and can be interpreted as a Lyapunov function, which guarantees stability. That is the reason why most successful industrial applications use a sufficiently long horizon that tries to resemble the infinite-horizon case, although closed-loop stability is not guaranteed.

The survey paper by Mayne et al. [17] compiled the existing formulations related to stability and presented sufficient conditions for nominal stability of MPC for linear or nonlinear systems with constraints. The key ingredients of the stabilizing MPC are a terminal constraint set and a terminal cost. The state predicted at the end of the horizon (terminal state) is forced to reach a terminal set that contains the steady state. Its associated cost (terminal cost) is added to the cost function.

It has been proved that time-invariant systems can be stabilized by solving the constrained optimization problem in a receding horizon way that includes the following:

- A terminal constraint set: it is a region Ω (terminal region) such that for all $x(t) \in \Omega$, then $K_N(x(t)) \in U$ (set of admissible control actions) and the state of the closed-loop system at the next sample time $x(t+1) \in \Omega$. This region must be an admissible invariant set of the system. That is, there exists a local control law $u = h(x)$ which stabilizes the system in Ω and the control actions are admissible.
- A terminal cost: an additional cost term $F(x(t+N_p))$ of the terminal state that is added to the objective function, where F is a control Lyapunov function.

If the terminal cost and terminal set satisfy certain assumptions (see [17]), then the optimal cost is a Lyapunov function and the MPC control law stabilizes asymptotically the system in the set of states where the optimization problem is feasible. Therefore, if the initial state is such that the optimization problem has a solution, then the system is driven to the steady state in an asymptotic way while constraints are fulfilled along its evolution.

The condition imposed on the terminal set Ω ensures constraint fulfillment and the condition on the terminal cost $F(x(t+N_p))$ ensures that the optimal cost is a Lyapunov function. Thus, it is necessary for the asymptotic convergence of the system to the steady state. Moreover, the terminal cost is an upper bound of the optimal cost of the terminal state.

In some situations, the optimal solution may be very difficult to obtain in the sampling time and only a suboptimal solution may be available. In this case, it suffices to consider any feasible solution with an associated cost strictly lower than the one of the previous sample time to guarantee asymptotic stability of the controller, as proved in [25]. Certainly, any feasible solution ensures feasibility, and thus constraint satisfaction, and the strictly decreasing cost guarantees asymptotic stability. Although suboptimality is not desirable since it implies a loss of performance, it can be useful in practice.

The computational burden of the optimization problem can be reduced by removing the terminal constraint. The use of an implicit terminal constraint by choosing the set of permissible initial states and horizon so that the explicit terminal constraint is automatically satisfied is done in [14], where it is proved that, if weighting of the terminal cost is considered, the domain of attraction of the MPC controller without terminal constraint is enlarged, reaching the same domain of attraction of the MPC with terminal constraint. The details about the computation of terminal regions and costs are out of the scope of this book and can be found in [16].

References

1. Aguilera RP, Lezana P, Quevedo DE (2013) Finite-control-set model predictive control with improved steady-state performance. IEEE Trans Ind Inf 9(2):658–667
2. Bemporad A, Ferrari-Trecate G, Morari M (2000) Observability and controllability of piecewise affine and hybrid systems. IEEE Trans Autom control 45(10):1864–1876
3. Bemporad A, Morari M (1999) Control of systems integrating logic, dynamics, and constraints. Automatica 35(3):407–427
4. Bernardini D, Bemporad A (2009) Scenario-based model predictive control of stochastic constrained linear systems. In: Proceedings of the 48th IEEE conference on decision and control (CDC) held jointly with 2009 28th Chinese control conference, December 2009, pp 6333–6338
5. Camacho EF, Bordons C (2007) Model predictive control, 2nd edn. Springer, London
6. Clarke DW, Mohtadi C (1989) Properties of generalized predictive control. Automatica 25(6):859–875
7. Clarke DW, Mohtadi C, Tuffs PS (1987) Generalized predictive control. Part I. The basic algorithm. Automatica 23(2):137–148
8. Cutler CR, Ramaker BC (1980) Dynamic matrix control- a computer control algorithm. In: Automatic control conference, San Francisco
9. Ferrari-Trecate G, Gallestey E, Letizia P, Spedicato M, Morari M, Antoine M (2004) Modeling and control of co-generation power plants: a hybrid system approach. IEEE Trans Control Syst Technol 12(5):694–705
10. Floudas CA (1995) Non-linear and mixed integer optimization. Oxford Academic Press
11. Heemels WPMH, De Schutter B, Bemporad A (2001) Equivalence of hybrid dynamical models. Automatica, pp 1085–1091
12. Karamamakos P, Geyer T, Oikonomou N, Kieferndork FD, Manias S (2014) Direct model predictive control: a review of strategies that achieve long prediction intervals for power electronics. Ind Electron Mag 8(1):32–43
13. Kouro S, Cortés P, Vargas R, Ammann U, Rodriguez J (2009) Model predictive control-a simple and powerful method to control power converters. IEEE Trans Ind Electron 56(6):1826–1838
14. Limon D, Alamo T, Salas F, Camacho EF (2006) On the stability of constrained MPC without terminal constraint. IEEE Trans Autom Control 51(5):832–836
15. Maciejowski JM (2001) Predictive control with constraints. Prentice Hall, Harlow
16. Mayne DQ (2014) Model predictive control: recent developments and future promise. Automatica 50(12):2967–2986

17. Mayne DQ, Rawlings JB, Rao CV, Scokaert POM (2000) Constrained model predictive control: stability and optimality. Automatica 36:789–814
18. Nemhauser GL, Wolsey LA (1988) Integer programming and combinatorial optimization. Wiley, Chichester; Nemhauser GL, Savelsbergh MWP, Sigismondi GS (1992) Constraint classification for mixed integer programming formulations. COAL Bull 20(8–12)
19. Nesterov Y (2013) Introductory lectures on convex optimization: a basic course, vol 87. Springer Science & Business Media, New York
20. Olivares DE, Mehrizi-Sani A, Etemadi AH, Cañizares CA, Iravani R, Kazerani M, Hajimiragha AH, Gomis-Bellmunt O, Saeedifard A, Palma-Behnke R, Jimenez-Estevez GA, Hatziargyriou ND (2014) Trends in microgrid control. IEEE Trans Smart Grid 5(4):1905–1919
21. Qin SJ, Badgwell TA (1997) An overview of industrial model predictive control technology. In: Kantor JC, Garcia CE, Carnahan B (eds) Chemical process control: assessment and new directions for research. AIChE Symposium Series, vol 93, No 316, pp 232–256
22. Qin SJ, Badgwell TA (2003) A survey of industrial model predictive control technology. Control Eng Pract 11:733–764
23. Rawlings J, Mayne DQ (2009) Model predictive control: theory and design. Nob Hill Publishing
24. Rodriguez J, Pontt J, Silva CA, Correa P, Lezana P, Cortés P, Ammann U (2007) Predictive current control of a voltage source inverter. IEEE Trans Ind Electron 54(1):495–503
25. Scokaert POM, Mayne DQ, Rawlings JB (1999) Suboptimal model predictive control (feasibility implies stability). IEEE Trans Autom Control 44(3):648–654
26. Vargas R, Cortés P, Ammann U, Rodriguez J, Pontt J (2007) Predictive control of a three-phase neutral-point-clamped inverter. IEEE Trans Ind Electron 54(5):2697–2705

Chapter 3
Dynamical Models of the Microgrid Components

Abstract This chapter describes the main components of a microgrid, focusing on their dynamical behavior, a key concept in control engineering and particularly in MPC. Mathematical models of renewable generation devices (photovoltaic panels or wind turbines), and also energy storage systems with high penetration in microgrids (batteries, ultracapacitors, and hydrogen-based systems) are presented in detail in the chapter. These models are the base for the development of the software included in the companion toolbox $Sim\mu grid$. Brief descriptions of alternative storage systems, such as flywheels or compressed air, are also included. Operational issues in energy storage systems to avoid non-adequate use, prevent the degradation of the devices, and improve their performance, reliability, and lifespan are also addressed. These concepts are of considerable importance for the design of MPC controllers, and they will be widely used throughout the book.

3.1 Mathematical Models

In control engineering, a model can be defined as an abstract representation of systems or processes, focusing the interest in the dynamical behavior of its characteristic variables. Models are typically a set of differential and algebraic equations giving rise to a simplification of the real system. Model design is not a trivial task and two opposing issues have to be considered: On one side, it needs to be accurate, but on the other side, it needs to be simple enough to avoid computational burden when it is solved numerically. This fact leads to a crucial issue in model design: What is the objective of the desired model? The key idea is to build the simplest model that meets objectives.

Typical objectives of modeling in control engineering are the analysis of system behavior, control design or simulation, where a more accurate (and more complex) model is usually developed. Model design takes a relevant role in Model Predictive Control, but as mentioned in the previous chapter, the model will be incorporated in an optimization problem, requiring a simple formulation, mostly linear.

The dynamical modeling of systems has to take into account the influence of two types of external inputs, manipulated or control signals (e.g., control of fuel inlet

C. Bordons et al., *Model Predictive Control of Microgrids*,
Advances in Industrial Control, https://doi.org/10.1007/978-3-030-24570-2_3

in a gas turbine) and uncontrollable disturbances (e.g., randomness of wind speed affecting a wind turbine or sunlight fluctuation over a solar panel). Obviously, the dynamical model has to provide the outputs of the system, defined as the variables of interest according to the model design objective.

Focusing on microgrids, models will be different depending on the timescale of the control level of the microgrid because of the different design objectives. In the tertiary control level, where the sample period is higher, aggregated models of the microgrid or simplified models of the main components have to be considered. In the secondary control level a more detailed component model will be needed, while in primary control, the own interface of the component given by its associated power electronic devices will be modeled.

The objective of this chapter is twofold. First, a physical and functional description of the main components of a microgrid is needed to understand how they are controlled and operated in the next chapters. Second, a simple simulation model of each component is presented. On one hand, the mathematical language of component equations can help in understanding its behavior and operation issues, but on the other hand, implementation of these simulation models can be used for testing the MPC algorithms included in this book. These models are provided for the reader in the companion toolbox $Sim\mu grid$ developed for Simulink©. The provided models can be used both in tertiary and secondary levels.

3.2 Distributed Energy Resources

A microgrid will integrate several alternatives to produce energy, given that depleting fossil fuel reserves will encourage renewable technologies to have an increasingly important role in the new energy model. In order to give a comprehensive overview of the different technologies, this section provides the characteristics of the main systems of Distributed Generation (DG) in microgrids through the description and mathematical models of each technology.

3.2.1 Fossil Fuels Generators

Small capacity combustion turbines are one of the nonrenewable alternatives for distributed power generation, which operate with fossil fuels as natural gas, petrol, propane, etc. As a result of a relative low-priced natural gas and also, low installation and maintenance costs, microturbines are one of the most extended DG energy sources today [15, 23, 37]. The controllability and autonomy of this kind of system have been its major advantage, but the increasing price of fossil fuels and the environmental impact will cause this technology to reduce its role in the future.

A gas turbine is basically comprised of a compressor, a combustion chamber and a turbine which is joined to an electric generator through a power shaft. These

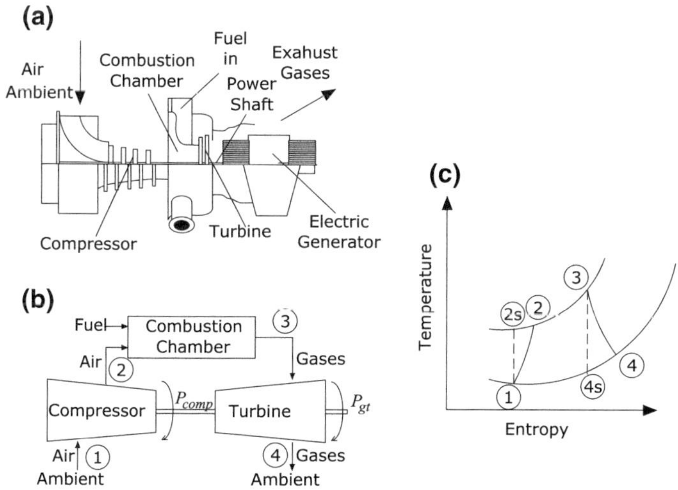

Fig. 3.1 Gas turbine. **a** Main components **b** Working schema **c** Thermodynamic cycle

components work together to produce electric power. A general view of a gas turbine can be observed in Fig. 3.1a. Unlike traditional combustion turbines, microturbines operate at lower temperatures and pressure, and faster speed (up to 150,000 rpm). Their small size is a great advantage of these systems thanks to the use of high-speed turbines with air-foil bearings. The main elements of a microturbine are as follows:

- **Compressor**: Compressors used in gas turbines are rotating type. At the inlet of the compressor, the atmospheric air is filtered to remove the dust, and then, the rotary blades push the air to increase pressure and temperature.
- **Combustion chamber**: Air with high static pressure from the compressor is conducted to the combustion chamber, where fuel is burnt to increase the temperature at nearly constant pressure, and then, the heated air is directed to the gas turbine. In order to increase the efficiency, a regenerator can be inserted between compressor and combustor. A regenerator is basically a heat exchanger to recover heat from the turbine outlet.
- **Turbine**: High pressure and temperature gases are expanded in the turbine, transforming the gas energy into mechanical energy, giving rise to the rotation of the blades. As mentioned above, usually after this stage, the exhaust gases with high temperature are conducted to a regenerator.
- **Electric generator/alternator**: An electric generator or alternator is directly coupled with the turbine through a power shaft. Alternator transforms the mechanical energy of the turbine into electrical energy, which is supplied to the microgrid.

Gas microturbines are combustion engines that share the same physical principle of conventional turbines but with simpler mechanical elements. As regards Fig. 3.1b and c, the microturbine is a Brayton engine, which ideally consists of the following four processes:

1. Isentropic process: atmospheric air is conducted into the compressor, where it is pressurized (1–2s).
2. Isobaric process: the compressed air is drawn through a combustion chamber to increase temperature (2s–3).
3. Isentropic process: Air at high pressure and temperature is expanded, and the energy is discharged to the mechanical system (3–4s).
4. Isobaric process: heat rejection (4s–1).

In real systems, processes in turbines and compressors are non-isentropic, given the real Brayton cycle 1–2–3–4 in Fig. 3.1c. The ideal and real cycles differences define the compressor and turbine efficiencies, η_c and η_t, respectively [8]:

$$\eta_c \approx \frac{T_{2s} - T_1}{T_2 - T_1} \tag{3.1}$$

$$\eta_t \approx \frac{T_3 - T_4}{T_3 - T_{4s}} \tag{3.2}$$

The physical model of the microturbine is obtained through the mathematical formulation of the thermodynamic cycle performed by the fluid inside the turbine. The generated mechanical power P_{gt} that produces the complete cycle is given by the following expressions:

$$P_{gt} = \dot{m}_a[C_{ph}(T_3 - T_4) - C_{pc}(T_2 - T_1)] \tag{3.3}$$

where $\dot{m}_a$ is the airflow, and C_{ph} and C_{pc} are the specific heat of air at constant pressure when air is at hot (T_3) and cold (T_1) temperatures, respectively. The pressure ratio of the cycle is defined as

$$PR = \frac{p_2}{p_1} = \frac{p_3}{p_4} \tag{3.4}$$

If γ_h and γ_c denote the ratio of specific heats at hot and cold temperatures, the relation between pressures and temperatures in the isentropic processes can be expressed as follows:

$$\frac{T_{2s}}{T_1} = PR^{\frac{\gamma_c - 1}{\gamma_c}} = x_c \tag{3.5}$$

$$\frac{T_3}{T_{4s}} = PR^{\frac{\gamma_h - 1}{\gamma_h}} = x_h \tag{3.6}$$

Then, temperatures can be obtained as follows [8]:

$$T_2 = T_1 \left(\frac{x_c - 1}{\gamma_c} + 1 \right) \tag{3.7}$$

$$T_3 = T_2 + \eta_{comb} \frac{\dot{m}_f}{\dot{m}_a} \frac{H}{C_{ph}} \tag{3.8}$$

$$T_4 = T_3 \left[1 - \left(1 - \frac{1}{x_h} \right) \eta_t \right] \tag{3.9}$$

where η_{comb} is the combustor efficiency, $\dot{m}_f$ is the fuel flow and H is the lower heating value of fuel.

3.2.2 Photovoltaic Panels

Photovoltaic (PV) cells are electronic devices that convert solar energy into electrical energy. When solar light falls on a semiconductor device, normally made of silicon and composed of two layers, a voltage difference between them is produced. When this voltage is able to conduct a current through the circuit, useful work is produced. The energy converted by a cell depends on the material properties, on the temperature, and on the solar radiation. Individual cells are connected to form arrays, panels or set of photovoltaic panels.

Electrical data provided by manufacturers about their panels under standard measurement conditions (STC—Standard Test Conditions as defined by IEC-60904-3) are nominal irradiance $G_n = 1000\,\mathrm{W/m^2}$ and ambient temperature $T_n = 25\,°\mathrm{C}$. Some of these technical specifications are:

- **Short-circuit current** (I_{sc}): It is the current between the terminals of a solar panel when there is no resistance. In other words, it is the maximum intensity that can be applied between both terminals.
- **Open circuit voltage** (U_{oc}): It is the maximum voltage value at the ends of the cell, when it is not connected to any load.
- **Peak power**: It is the product of the value of voltage and intensity for which the power delivered to the load is maximum.

Photovoltaic cells are formed by p-type and n-type silicon (see Fig. 3.2). When a photon reaches the cell, it ionizes the atoms in the silicon and separates an electron (negative charge) that creates at the same time a gap (positive charge). The gaps move toward the positive layer or the p-layer and the electrons move toward the negative or the n-layer, thus a potential difference occurs. In Fig. 3.2, the main components of a PV panel are detailed showing also its equivalent circuit. Basically, a photovoltaic cell can be modeled by

- **Diode current** ($I_{d,cell}$): Diode current when it is directly polarized.
- **Photo-generated current** ($I_{L,cell}$): It is the generated current due to the incidence of the sunlight over the photovoltaic cell. This is proportional to the received irradiance over the cell.
- **Generated current** ($I_{pv,cell}$): Current at the output of the photovoltaic cell.
- **Generated voltage** ($U_{pv,cell}$): Voltage at the output of the photovoltaic cell.

The behavior of a cell when it functions as a current generator can be explained as the difference between the photo generated current and the diode current. Figure 3.2

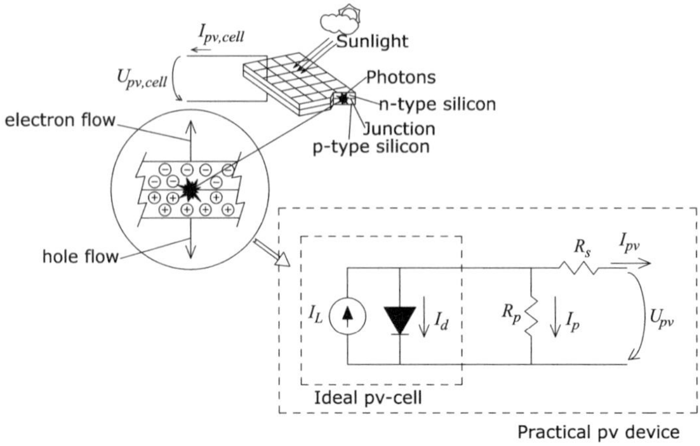

Fig. 3.2 Photovoltaic panel

also presents the equivalent circuit of an ideal PV cell. The mathematical equation that models the current–voltage characteristic (I–V) of the ideal PV cell is [46, 56]:

$$I_{pv,cell} = I_{L,cell} - I_{d,cell} \tag{3.10}$$

$$I_{d,cell} = I_{0,cell}\left[\exp\left(\frac{q \cdot U_t}{a \cdot k \cdot T_{pv,cell}}\right) - 1\right] \tag{3.11}$$

where $I_{L,cell}$ is the photo-generated current by a PV cell, $I_{0,cell}$ is the saturation current of the diode, a is the ideal factor of the diode, U_t is the thermal voltage, k is the Boltzmann's constant, $T_{pv,cell}$ is the temperature of the cell and q is the electron charge.

A photovoltaic panel is formed by a set of photovoltaic cells connected in series and in parallel. As detailed in Fig. 3.2, when the different cells are connected in the panel a serial and a parallel resistor has to be included in the model. For practical PV devices, Eq. (3.10) has to be reformulated as follows:

$$\begin{aligned} I_{pv} &= I_L - I_d - I_p = \\ &= N_p \cdot I_{L,cell} - N_p \cdot I_{0,cell} \cdot \left[\exp\left(\frac{q \cdot (U_{pv} + R_s \cdot I_{pv})}{N_s \cdot k \cdot T_{pv} \cdot a}\right) - 1\right] - \frac{U_{pv} + R_s \cdot I_{pv}}{R_p} \end{aligned} \tag{3.12}$$

where I_{pv} is the photovoltaic current and I_o is the saturation currents of the array of PV cells with N_s cells connected in series and N_p cells connected in parallel, and R_p and R_s are the equivalent parallel and series resistances of the array of cells. The photo-generated current (I_L) of the panel can be represented as a linear function of the solar irradiance, being affected by temperature as expressed in the following equation [17, 21, 56]:

$$I_L = (I_{L,n} + K_I \Delta T_{pv}) \frac{G}{G_n} \tag{3.13}$$

where $I_{L,n}$ is the value of I_L at nominal condition (usually 25 °C and 1000 W/m^2), $\Delta T_{pv} = T_{pv} - T_{pv,n}$ (T_{pv} and $T_{pv,n}$ are the actual and the nominal temperature), G is the present irradiance on the surface of the PV, and G_n is the nominal irradiance. The diode saturation current I_0 depends on the temperature according to [17, 41, 56]:

$$I_0 = I_{0,n} \left(\frac{T_{pv,n}}{T_{pv}}\right)^3 \exp\left[\frac{q \cdot E_g}{a \cdot k}\left(\frac{1}{T_{pv,n}} - \frac{1}{T_{pv}}\right)\right] \tag{3.14}$$

being E_g the band gap energy of the semiconductor and $I_{0,n}$ the nominal value of the saturation current, that can be expressed as

$$I_{0,n} = \frac{I_{sc,n}}{\exp\left(\frac{U_{oc,n}}{aU_{t,n}}\right) - 1} \tag{3.15}$$

where $U_{t,n}$ is the thermal voltage of N_s cells connected in series at nominal temperature. The model can be improved if in Eq. (3.15) the voltage and current correction factors K_U and K_I are included.

$$I_0 = \frac{I_{sc,n} + K_I \Delta T}{\exp\left((U_{oc,n} + K_U \Delta T)/aU_{t,n}\right) - 1} \tag{3.16}$$

Power output of solar plants varies in a deterministic way due to the change of the sun incidence angle on a diurnal and seasonal basis, but also, it can vary in a stochastic way, as a result of cloud movements and temperature variations. Stochastic changes are not easily predictable, and forecasts play a significant role in helping grid operators manage variability. Weather forecast is also important in MPC approaches, where predictions of the variables of the system are needed. A review of the different forecast models in the literature applied to power generation with solar panels can be found in [18].

The photovoltaic panel is interfaced with the microgrid with a power inverter or a DC/DC converter. The power electronics associated to photovoltaic panels work with an algorithm to track the optimal generation point, the so-called Maximum Power Point Tracking (MPPT) algorithm. A review of different MPPTs associated to photovoltaic inverters can be found in [24].

Simμgrid[1] adopts a simplified model of the above equations, looking for a reduction of the number of parameters and a quick simulation even with a long simulation period. The model of a photovoltaic cell consists of the following equations:

Thermal voltage (U_t) is defined as

[1] Simμgrid is the companion software that can be downloaded from http://institucional.us.es/agerar/.

$$U_t = \frac{k \cdot T_{pv}}{q} \tag{3.17}$$

The characteristic equation of a panel can be simplified as a function of the parameters shortcut current (I_{sc}) and open circuit voltage (U_{oc}), normally supplied by the manufacturer:

$$I_{pv} = I_{sc,pv}\left(1 - \exp\left(\frac{V_{pv} - V_{oc} + I_{sc}R_s}{V_t}\right)\right) \tag{3.18}$$

$$I_{sc} = \frac{I_{sc,n} \cdot G}{1000} \tag{3.19}$$

3.2.3 *Wind Turbines*

Wind energy can be transformed into electric energy using wind turbines. The main components of a wind turbine [31, 58] can be found in Fig. 3.3, which are described below:

- **Tower**: The tower raises the generator and the blades to the necessary altitude to generate energy.
- **Rotor and blades**: When the wind spins the blades, they transform the kinetic energy into a torque along the rotor axis.
- **Nacelle**: Is the space where all the mechanical and electrical elements are placed.
- **Gearbox**: Its main function is to adapt the speed of the main axis to the generator.
- **Generator**: This system converts the mechanical energy into electricity.
- **Brake system**: If wind speed is too low to produce a minimal power, the system produces loses. To avoid this negative effect, the brake system is used to block the turbine.
- **Control System**: This system manages the information about the state of the wind turbine to correctly operate the turbine.

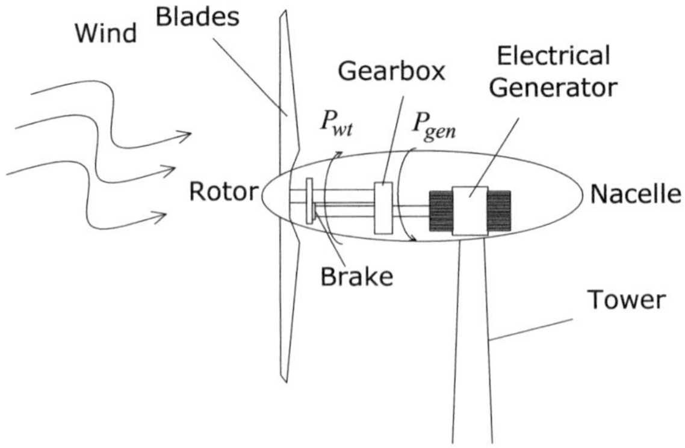

Fig. 3.3 Wind turbine

The wind turbine transforms power from wind into mechanical power. The relation between aerodynamic power and the wind speed is expressed in the following equation [47, 58]:

$$P_{wt} = 0.5 \cdot \pi \cdot \rho \cdot R_{blade}^2 \cdot v_w^3 \cdot C_p(\theta, \lambda) \tag{3.20}$$

where ρ is the density of air, R_{blade} represents the turbine radius, i.e., the blade length, v_w is the wind speed, and C_p is the power coefficient. This coefficient can be obtained as a function of the pitch angle of the rotor θ and the tip speed ratio λ. In order to simplify the model, it will be assumed that C_p has a constant value. The aerodynamic torque T_{wt} can be calculated as the ratio between the power extracted from wind P_{wt} and the rotor speed w_{wt}:

$$T_{wt} = \frac{P_{wt}}{w_{wt}} \tag{3.21}$$

Energy transmission to the generator through the gearbox can be expressed with the following relationships:

$$\frac{d\omega_{gen}}{dt} = \frac{T - T_{wt}}{J_{eq}} - \frac{B_m}{J_{eq}} w_{gen} \tag{3.22}$$

where ω_{gen} is the angular speed of the generator, B_m is the damping coefficient, T is the mechanical torque, T_{wt} is the aerodynamic torque and J_{eq} is the equivalent rotational inertia of the generator.

$$J_{eq} = J_{gen} + \frac{J_{wt}}{n_g^2} \tag{3.23}$$

where J_{gen} and J_{wt} are the rotational inertias corresponding to generator and rotor, and n_g is the gear ratio. Further description of this model can be found in [31, 47].

A detailed model of the generator could also be included, but is out of the scope of the book. Finally, as in the photovoltaic case, an MPPT control algorithm applied to the wind turbine must be chosen. There are different methods to acquire the MPPT value for the wind turbine. A review of these methods can be found in [48].

From the point of view of the control of the microgrid, the forecast of wind speed has a fundamental importance. Basically, wind speed forecast systems use either a physical or a statistical method, or a combination of both [2, 50]. An overall review on several wind power forecasting techniques can be found in [35]. Main characteristics of both approaches are:

- Physical methods simulate large-scale wind flows starting from Numeric Weather Predictions (NWP) and further predict local wind power generation using physical equations. This method offers insight into the physical processes, allowing a solid theoretical basis for predictions.

- Statistical methods mostly start from NWP, and further uses statistic analysis, or even alternative techniques as neural networks or fuzzy logic to calculate local wind power generation. These techniques are based on previous experimental information and need to be trained with a large set of data, but extreme values of wind power will be more difficult to foresee.

3.3 Distributed Energy Storage Systems

The ability to integrate the different characteristics of different kinds of distributed storage technologies and renewable energy sources is one of the most important functionalities of microgrids. One of the main challenges of controlling hybrid storage is how to take advantage of the technical characteristics of the different alternatives. There are systems, as ultracapacitors, with a quick time response to operational changes, but with a limited capacity to store energy. These are systems with a high specific power. Other technologies, as batteries, provide high specific energy, that is, an important capacity of energy storage, but a reduced time response (see Table 3.1). The hybrid ESS control system must consider, in real time, the limitations of each ESS, taking into account time response, load cycling degradation and power and energy constraints.

A comparison of technical features of different ESS technologies is shown in Table 3.2. According to this, different fields of application and implementation of the different energy storage systems can be seen. Systems with very low storage periods (seconds) and high specific power, as ultracapacitors, are used for grid stabilization in issues related to power quality. At the other end of the spectrum, ESSs with the capacity to store important amounts of energy can be found, e.g., hydrogen systems. They can be used to compensate the variability of renewable generation of energy, and smooth peaks in demand power demand. The central zone can be used to ensure uninterruptible power supply, black start and spinning reserve [20].

The installation of ESSs gives the opportunity for a better management of the economic dispatch of renewable energies. There are several ways to store energy, but each ESS has its own advantages and disadvantages considering economic cost, energy and power rates, degradation and lifetime. The control strategy must have the ability to decide which ESS to use on real time, depending on the operational conditions.

3.3.1 Batteries

A Battery Energy Storage System (BESS) is an electrochemical device that stores electrical energy, and it is composed of one or more voltaic cells. Each cell consists of two half-cells connected by a conducting electrolyte containing anions and cations. One part contains the negative electrode while the other part contains the positive

Table 3.1 Comparison of technical characteristics of EES systems (I)[a]

Systems	Power rating and discharge time		Storage duration		Capital cost		
	Power rating	Discharge time	Self-disch./day	Suitable storage	$/kW/cycle	$/kWh	cents/kWh
Pumped hydroelectric	100–5000 MW	1–24 h+	Very small	Hours-months	600–2000	5–100	0.1–1.4
Compressed air	5–300 MW	1–24 h+	Small	Hours-months	400–800	2–50	2–4
Flywheel	0–250 kW	1e-3s–15 min	100%	Seconds-minutes	250–350	1000–5000	3–25
Capacitor	0–50 kW	1e–3s–60 min	40%	Seconds-hours	200–400	500–1000	–
Ultracapacitor	0–300 kW	1e–3s–60 min	20–40%	Seconds-hours	100–300	300–2000	2–20
Superconducting magnetic	0.1–10 MW	1e–3s–8s	10–15%	Minutes-hours	200–300	1000–10,000	–
Lead–acid battery	0–20 MW	Second-hours	0.1–0.3%	Minutes-days	300–600	200–400	20–100
Ni–Cd battery	0–40 MW	Second-hours	0.2–0.6%	Minutes-days	500–1500	800–1500	20–100
Li-ion battery	0–100 kW	Minutes-hours	0.1–0.3%	Minutes-days	1200–4000	600–2500	15–100
Redox flow battery	30 kW–3 MW	Seconds-10 h	Small	Hours-Month	600–1500	150–1000	5–80
Hydrogen system[b]	0–50 MW	Seconds-24 h+	Almost zero	Hours-months	12,000+	–	3–5[c]

[a]Reprinted from *Progress in Natural Science*, **19**(3), Haisheng Chen, Thang Ngoc Cong, Wei Yang, Chunqing Tan, Yongliang Li, Yulong Ding, Progress in electrical energy storage system: A critical review, 291–312, Copyright (2009), with permission from Elsevier

[b]HESS formed by electrolyzer and fuel cell

[c]Cost in cents per hour

Table 3.2 Comparison of technical characteristics of EES systems (II)[a]

Systems	Energy and power density				Lifetime and cycles		Influence on the environment	
	Wh/kg	W/kg	Wh/L	W/L	Years	Cycles	Influence	Description
Pumped hydroelectric	0.5–1.5	–	–	0.5–1.5	40–60	–	Negative	Destruction of green land
Compressed air	30–60	–	3–6	0.5–2	20–40	–	Negative	Emissions from combustion of natural gas
Capacitor	0.05–5	~1e5	2–10	1e5+	~5	5e4+	Small	Little amount of remains
Ultracapacitor	2.5–15	500–5000	2–10	1e5+	20+	1e6+	Small	Little amount of remains
Superconducting magnetic	0.5–5	500–2000	0.2–2–5	1000-	20+	1e3–4e3	Negative	Strong magnetic fields
Lead–Acid battery	30–50	75–300	50–80	10–400	5–15	500–1000	Negative	Toxic remains
Ni–Cd battery	50–75	150–300	60–150	–	10–20	2000–2500	Negative	Toxic remains
Li-ion battery	75–200	150–315	200–500	–	5–15	1e3–10e3	Negative	Toxic remains
Redox flow battery	10–30	–	16–33	–	5–10	12e3	Negative	Toxic remains
Hydrogen system[b]	800–10e3	500+	500–3000	500+	5–15	1000+	Nearly null	Remains of component

[a]Reprinted from *Progress in Natural Science*, **19**(3), Haisheng Chen, Thang Ngoc Cong, Wei Yang, Chunqing Tan, Yongliang Li, Yulong Ding, Progress in electrical energy storage system: A critical review, 291–312, Copyright (2009), with permission from Elsevier

[b]HESS formed by electrolyzer and fuel cell with renewable energy system

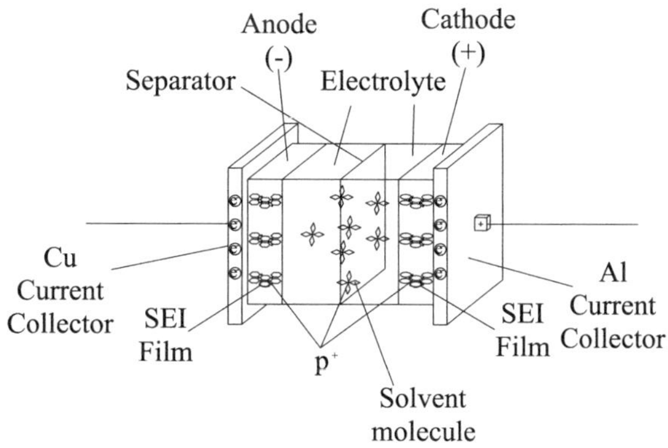

Fig. 3.4 Batteries

electrode. A redox reaction produces the charge of the battery, where the cations are reduced because electrons are added, and the anions are oxidized because the electrons are removed. During the discharge, this process is reversed.

Consequently, an electromotive force (Emf) is produced at each half-cell, being the Emf of the cell, the difference between the Emf of it half-cells. The voltage between the terminals of the cell is known as *voltage difference*. The *open circuit voltage* is defined as the voltage difference when the cell is not being charged or discharged, and it is equal to the Emf of the cell. Due to the internal resistance, the voltage of a cell being discharged is less in magnitude than the open circuit voltage and conversely, the voltage of a cell being charged exceeds the open circuit voltage.

A battery is mainly composed of the following components (Fig. 3.4):

- **Electrodes**: The cathode stores protons and releases them when the battery is charging. The anode stores protons and releases them when the battery is discharging.
- **Electrolyte**: It is a gel or liquid material, that from one side must permit the transportation of ions produced in the chemical reactions that take place in anode and cathode, and on the other side, must restrict the flow of electrons via the electrolyte, making easier the flow of electrons through the external circuit.
- **Separator**: The anode and cathode cannot be in contact to avoid a short-circuit in the battery. This is the objective of the separator, a porous substance that does not react chemically with anode, cathode or electrolyte, but it must permit the flow of ions between cathode and anode. It also avoids the flow of electrons inside the battery structure.
- **Current Collectors**: This element collects current from the individual cells, and it permits channeling the electrical current through the load.
- **SEI layer**: The Solid Electrolyte Interphase (SEI) layer is a microscopically thin interface between the liquid electrolyte and solid electrode inside batteries. SEI layer typically forms on the graphite surface, produced by chemical reactions with the electrolyte solvent. It has important implications for the performance and degradation of the lithium battery, including capacity fade, cycle life limitations,

and safety. The growth of this layer depends on several factors such as temperature, operational conditions, characteristic of the graphite, etc., with the drawback that is not easy to control, being an active research topic.

Battery Types

This section briefly describes the most interesting batteries technologies for microgrid applications.

Lead–acid battery: This type of batteries is a mature technology, and consequently, they present a very important commercial development and a relatively low cost for stationary applications. The lead–acid batteries are composed of a cathode consisting of a lead plate coated with lead dioxide PbO_2, and a spongy lead anode. The electrolyte is a solution of sulfuric acid in water.

Lithium-ion battery: Lithium-ion is the most used battery technology in consumer electronics and also in electric vehicles. But they have also a significant use in stationary applications, residential, commercial or industrial buildings, to be integrated with renewable generation. They are composed of a lithium metal oxide cathode, a carbon material anode, and an electrolyte formed by substances that also contain lithium.

Redox Flow Batteries: Redox Flow Batteries (RFB) are composed of a stack and two external tanks where electrolyte is stored. The stack is a number of connected cells formed by two electrodes and an ion-selective membrane. One of the main advantages of this technology is the decoupling of power (depends on the number of cells) and capacity (a function of the tank electrolyte volume). Also, the number of deep cycles is very high (around 10,000 in a Vanadium RFB) and it is easy to obtain the real SOC at any moment [25, 29].

Operational Issues

Lithium-ion batteries are the most widespread technology in microgrids, and this section presents some topics related to an adequate operation of these batteries to prevent the degradation of their cells, and as a consequence, the increase in the operational costs. Additionally, this type of batteries demands complex protection system to prevent damages. With these objectives in mind, this storage system usually includes a Battery Management System (BMS), an electronic device responsible for proper and secure operation of the batteries. Typical functions of BMS are: (i) cell protection, to guarantee that cells operate within their design limits, (ii) cell balancing, to equalize the charge and avoid that some cells can be overstressed that could be cause of a premature failure, (iii) charge and discharge control, (iv) estimation of the SOC and (v) determination of the State of Health (SOH). To achieve these objectives, a BMS typically needs measures of voltages, currents, and temperatures [29].

Degradation of lithium-ion batteries is an active research field, which includes theoretical results about causes and effects, but also the development of mathematical aging models to estimate the consequences of degradation. In Chap. 5 of this book, battery degradation will be included in the cost function of MPC when economic

Table 3.3 Main features of batteries

Advantages	Disadvantages
Intermediate specific power	Limited number of life cycles
Intermediate specific energy	Fast charging and discharging processes lead to degradation
Intermediate cycle efficiency	
Mature technology	Current ripple produces degradation
Decreasing capital costs	Toxic remains
Low maintenance costs	Hazardous materials
Modular systems	

issues are taken into account. The aging degradation of batteries is manifested in two ways: a reduction in the capability of delivering power and of storing energy. The first one is the power fade, due to an increase of the internal impedance, and typically measured in volts, and the later one is the capacity fade, measured in amperes-hours. Many factors affect to these losses, but mainly, temperature, charge, and discharge currents and the operating SOC have a considerable impact on them [29, 51].

One important degradation phenomenon is gassing, typically resulting of the electrolyte decomposition, that can be accentuated in overcharging or overheating situations, resulting in a shorter life of the battery, and could even lead to the explosion of the cell [29, 44]. Another problem is the formation of permanent oxides when overcharging at high SOC, producing a loss of the active material of the electrode, and an increase in the resistance caused by the growth of the surface film.

However, this problem can be reduced if an optimized current profile is used [6, 49, 51]. This film growth can be expressed with the following expression [6]:

$$\frac{d\iota}{dt} = \kappa \cdot |I_{bat}(t)| \tag{3.24}$$

where ι is the thickness of the film, κ is a battery-dependent parameter and I_{bat} is the current.

Finally, AC ripple can be also the cause of degradation, because of the battery heating, being the inverter design an important issue to the reduction of this problem.

The main features of a typical Li-ion battery as energy storage system are summarized in Table 3.3.

Dynamical Model

The voltage of the battery U_{bat} can be expressed as a function of the capacity of the battery $C_{\max,bat}$ and the current of the battery I_{bat} [55]:

$$\begin{aligned}U_{bat}(t) = U_{o,bat} &- K_{1,bat}\frac{C_{\max,bat}}{C_{\max,bat} - C_{bat}(t)}I_{bat,ch}(t)\\&+ K_{1,bat}\frac{C_{\max,bat}\cdot(\delta_{bat,ch} - \delta_{bat,dis})}{C_{\max,bat} - C_{bat}(t)}C_{bat}(t)\\&- K_{1,bat}\frac{C_{\max,bat}}{C_{\max,bat} + 0.1C_{bat}(t)}I_{bat,dis}(t)\\&- K_{2,bat}\cdot e^{(K_{3,bat}\cdot C_{bat}(t))} + R_{ohm,bat}\cdot I_{bat}(t)\end{aligned} \tag{3.25}$$

where $U_{o,bat}$ is the battery internal voltage (V), $I_{bat,ch}$ is the charge current, $I_{bat,dis}$ is the discharge current, $C_{\max,bat}$ is the maximum battery capacity (Ah), C_{bat} is the capacity of the battery (Ah), $K_{i,bat}$ are internal parameters of the battery and $R_{ohm,bat}$ is the internal ohmic resistor of the battery. The model differs for the charge and discharge modes of the battery, so the binary variables of the state of the battery are introduced $\delta_{bat,dis}$ and $\delta_{bat,ch}$. The capacity (Ah) of a battery can be expressed by the following expression

$$C_{bat}(t) = \int_0^t I_{bat}(t)dt \tag{3.26}$$

Finally, the state of charge of the battery is related to the capacity in the following equation:

$$\mathrm{SOC}_{bat}(t) = \frac{C_{bat}(t)}{C_{\max,bat}} \tag{3.27}$$

3.3.2 *Ultracapacitors*

Capacitors can be considered the most direct way to store electrical energy. A capacitor consists of two metal plates separated by a dielectric nonconducting layer. When one plate is electrically charged, the other plate will induce an opposite charge [1, 13, 20, 40]. The application of a potential difference between the two plates produces a static electric field across the dielectric. As energy is stored directly as an electrostatic field, capacitors can be charged and discharged faster than batteries and can complete tens of thousands of cycles without any appreciable loss of efficiency. On the other side, the main drawback of capacitors is their low specific energy.

The energy stored in a capacitor is given by the known expression:

$$E = \frac{1}{2}CU^2 \tag{3.28}$$

where C is the capacitance (in Farads) and U is the voltage between terminals. The stored charge Q in the capacitor is obtained by the product of the capacity and the

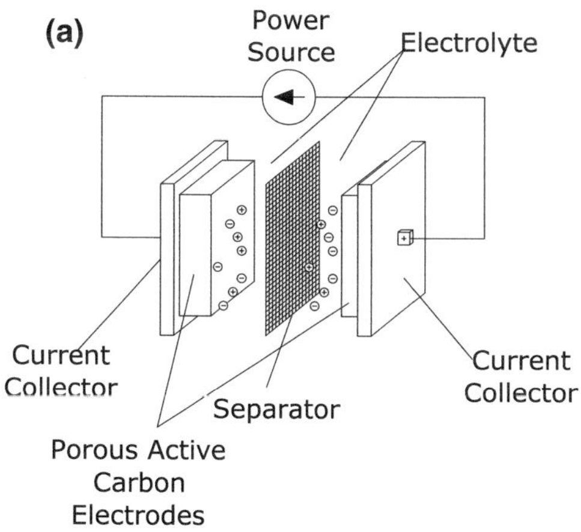

Fig. 3.5 Ultracapacitor (**a** Discharged **b** Charged and equivalent circuit)

voltage. The breakdown characteristics of the dielectric determine the maximum voltage. The capacitance of the dielectric can be modeled as follows [12]:

$$C = \varepsilon_d \frac{A}{d} \tag{3.29}$$

where d is the distance between the plates, ε_d the dielectric constant of the gap between the plates, and finally, A is the area of the plates.

There are some differences between conventional capacitors and ultracapacitors. There is not a conventional dielectric in an ultracapacitor, both plates or current collectors (see Fig. 3.5) are soaked in an electrolyte and separated by a thin insulator layer known as separator. When the current collectors are charged, an opposite charge is produced on either side of the separator, creating what is called an electric double layer. This is the reason why ultracapacitors are known as double-layer capacitors. The principle of construction is based on the following components:

- **Current collectors or plates**: Metal contacts are used as current collectors.
- **Electrodes**: Highly activated carbon is used with high porosity. This fact allows to provide a high specific surface area. This kind of electrodes has good conductivity. This material supports high temperature and has low corrosion. It also provides chemical stability. This material is the basis of the increase of capacitance in the ultracapacitor versus classical capacitors.
- **Electrolyte**: It is the substance, typically a liquid, that provides the ions to form the double layer.
- **Separator**: It provides insulation between the electrodes, allowing the movement of ions from one electrode to another.

A schematic diagram of an ultracapacitor is shown in Fig. 3.5. The capacitance depends mainly on the porous material of the surface of electrodes, and also on the distribution of the pore size. In Fig. 3.5-(b), it is shown the schematic view of the ultracapacitor when it is charged.

Operational Issues

Operating conditions is a key issue in relation to performance, reliability, degradation and lifetime of ultracapacitors. For example, ripple voltage or ripple current from the charging system can induce an overheat in the capacitor. This effect can occur with switch mode Pulse Width Modulation (PWM) converters used for charging ultracapacitors or when using ultracapacitors as energy storage buffers for downstream PWM converters [11].

In order to operate with a high-voltage DC bus, many capacitors must be stacked in series, and it is necessary the use of adequate balancing and charging systems to avoid overcharge of individual capacitor in the series. Charging of ultracapacitors must be performed at rated voltage because a generation of gas can occur in the capacitor if a high value of this voltage is maintained, with the consequent reduction of its lifetime. Temperature is also a factor to be considered during the charging process and it must be taken into account that temperature rises along with charging voltage, which must be reduced when ambient temperature increases. According to [11], an increment in the ambient temperature of 10 °C could reduce the ultracapacitor lifetime by a factor of two.

As mentioned previously, ultracapacitors can be charged substantially faster than conventional batteries and can have a higher number of life cycles. But they have low specific energy and behave like a short-circuit when they are exposed to low levels of SOC. High states of charge can also damage this technology [12].

The main features of the ultracapacitors as energy storage systems are detailed in Table 3.4.

Dynamical Model

A comparison between different dynamical models of ultracapacitors can be found in [27]. In order to give a comprehensive overview of this technology, only a simplified model of the ultracapacitor is going to be developed. The total capacitance of an ultracapacitor $C_{uc}(t)$ depends on voltage and can be modeled as follows [14, 27]:

$$C_{uc}(t) = C_{uc,0} + k_{uc} U_{uc}(t) \tag{3.30}$$

Table 3.4 Main characteristics of ultracapacitors

Advantages	Disadvantages
High specific power	Low specific energy
Fast charging and discharging processes	High self-discharge rate
High cycle efficiency	High cost
Low toxicity	Behavior as short-circuit with low SOC
Modular systems	Overvoltage produces loss of capacitance

where $C_{uc,0}$ is the initial capacitance (electrostatic capacitance) of the capacitor, and k_{uc} is a constant that models the linear dependence with voltage.

$$I_{uc}(t) = \frac{dQ_{uc}(t)}{dt} = \frac{d(C_{uc}(t)U_{uc}(t))}{dt} \tag{3.31}$$

$$I_{uc}(t) = \left(C_{uc,0} + 2k_{uc}U_{uc}(t)\right)\frac{dU_{uc}(t)}{dt} \tag{3.32}$$

The energy stored in an ultracapacitor can be modeled

$$E_{uc}(t) = \frac{1}{2}\left(C_{uc,0} + \frac{4}{3}k_{uc}U_{uc}(t)\right)U_{uc}^2(t) \tag{3.33}$$

and the SOC is given by the ratio between the current stored energy and its maximum value:

$$SOC_{uc}(t) = \frac{\left(C_{uc,0} + \frac{4}{3}k_{uc}U_{uc}(t)\right)U_{uc}^2(t)}{\left(C_{uc,0} + \frac{4}{3}k_{uc}U_{uc}^{max}(t)\right)(U_{uc}^{max}(t))^2} \tag{3.34}$$

This model has been simplified in *Simµgrid*, considering constant capacitance, and consequently, the model of a conventional capacitor has been used.

3.3.3 Hydrogen

Hydrogen is the most common element in the universe, but rarely is found in nature in its free state. However, it can be found in a combined state in water, hydrocarbon, fats, etc. In spite of this, hydrogen can be considered as a promising alternative to be used as an energy storage system, in particular when hydrogen is produced with renewable energies. A complete hydrogen energy storage system is composed of a hydrogen production system, a hydrogen storage system, and another system to transform hydrogen to energy, i.e., a fuel cell or a hydrogen engine [7, 22, 60].

Hydrogen typically is produced from water or fossil fuels using renewable energy sources by processes such as electrolysis, natural gas reforming or biological processes. Nevertheless, the most interesting alternative to be used in microgrids is the production of hydrogen by coupling electrolyzers to renewable sources [7, 22].

There are several ways to store hydrogen, being metal hydride and compressed hydrogen the most conventional and mature technologies.

Hydrogen can be transformed into energy using an Internal Combustion Engine (ICE), transforming chemical energy in the hydrogen into mechanical energy or a Fuel Cell (FC), obtaining electricity. ICE has several advantages as opposed to the

FC because it is a mature technology with lower costs and a higher life cycle. But the efficiency of ICEs is restricted by the Carnot limit, which implies poor efficiency for usual operating temperatures. This is its main disadvantage versus FCs, that can easily double the conversion efficiency of ICEs.

Electrolyzers

Electrolyzers are electrochemical devices which are able to separate hydrogen and oxygen from H_2O molecules when a direct current is applied. In an electrolyzer, water is supplied by the channels of the anode and the cathode of the electrolysis cells, then, when DC voltage is supplied, the catalysis action of the platinum produces the conduction of protons through the membrane which separates the anode and the cathode. This electrochemical reaction allows the decomposition of the water molecule into oxygen through the anode and hydrogen through the cathode. These reactions are produced in the stack of electrolysis cells. The output of the anode is driven to the oxygen separator, and in a similar way, the output of the cathode is linked with the hydrogen separator. To eliminate the residual water in oxygen and hydrogen flows, a condenser is situated in the output of each one [9, 26]. The different components of the electrolyzer Balance of Plant (BOP) are shown in Fig. 3.6.

- **Electrolyte**: In the electrolysis, an electrolyte, typically a salt solution, is added to increased the efficiency of the process.
- **Electrodes**: They are electrical conductors providing an interface between the electrolyte and the electrical circuit.
- **Bipolar Plates**: Their main tasks consists of conducting the electrical current among cells, distribute the water and remove heat.
- **Stack**: A single electrolyzer cell consists of the electrolyte placed between the two electrodes and two bipolar plates. The stack is composed of a number of individual cells to achieve a higher voltage.
- **Separators**: The bubbling gases produced in the electrolysis are collected in the separators, two recipients with electrolyte inside, where the separation of gas and liquid is produced [19].
- **Condenser**: The quantity of water vapor in the hydrogen gas from the separator, is reduced in a heat exchanger (condenser). This process can be also done for oxygen. After the cooling, the gases are conducted through coalescing filters to eliminate the remaining water droplets.
- **Heaters/Coolers**: In order to maximize the efficiency of the electrolysis reaction and to increase the durability of the system, the temperature has to be maintained over a range. The heat management process can be carried out with just air, water or coolant liquids.

Electrolyzers have different degradation mechanisms. Fluctuations in the current applied to electrolyzers produce: (i) variations in the differential pressure, producing reverse conduction of hydrogen to the oxygen separator or vice versa (ii) small concentrations of hydrogen in oxygen or oxygen in hydrogen ($>4\%$) which can produce highly explosive reactions (iii) losses in the mechanical wear of the membrane, (iv) chemical degradation of the membrane via radical attack. On the other hand,

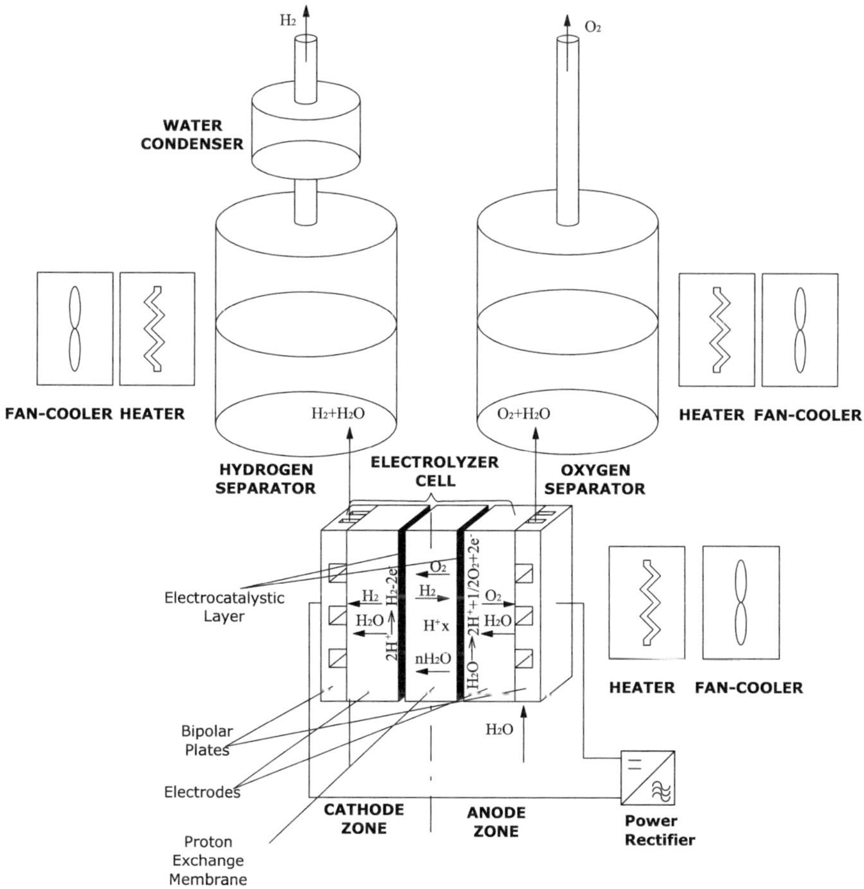

Fig. 3.6 Schematic view of a balance of plant of an electrolyzer

ON/OFF cycling of electrolyzers produces: (i) chemical and mechanical degradation due to the temperature and pressure cycling, (ii) chemical degradation caused by uncontrolled stack polarity, (iii) production of hydrogen peroxide at cathode after turn off with following membrane and carbon carrier oxidation.

In the same way, electrolyzers need to be maintained over an optimal-working temperature. The Electrolyzer Management System (ELMS) is in charge of maintaining the correct water level in the separator in order to avoid drying conditions in the membrane and feeding the stack with enough water to produce the electrolysis reaction. Before the electrolyzer is in the state of hydrogen production, the ELMS conducts the stages of water filling of the separators, nitrogen purge in all the gas circuit and a preheating to the working temperature of the stack. Further description about Proton Exchange Membrane (PEM) electrolyzers can be found in [9, 26]. Finally, electrolyzers have similar current ripple degradation issues than the explained for fuel cells in the next section.

Next, a mathematical model of the electrolyzer will be developed. It will be modeled following the expressions given in [55], providing mathematical equations of the polarization curve of the electrolyzer.

The stack voltage of the electrolyzer $U_{elz}(t)$ (V) can be expressed as the product of the number of electrolysis cells N_{elz}^{cell} and the voltage of a single cell U_{elz}^{cell}.

$$U_{elz}(t) = N_{elz}^{cell} \cdot U_{elz}^{cell}(t) \tag{3.35}$$

The voltage of a single cell can be expressed by the following equation:

$$U_{elz}^{cell}(t) = U_{elz,o}^{cell}(t) + U_{elz,act}^{cell}(t) + U_{elz,ohm}^{cell}(t) + U_{elz,conc}^{cell}(t) \tag{3.36}$$

where $U_{elz,o}^{cell}$ is the reversible potential or "Nernst" voltage, $U_{elz,act}^{cell}$ is the activation overpotential, $U_{elz,act}^{cell}$ is the ohmic overvoltage and $U_{elz,conc}^{cell}$ provides the losses due to concentration mass. Hence, the voltage drop is the sum of the following terms:

$$\begin{aligned} U_{elz,o}^{cell}(t) &= E_{elz}^{o} + \frac{\Delta S_{elz}^{o}}{2F}\left(T_{elz}(t) - T_{elz}^{o}\right) \\ &+ \frac{2.3 \cdot R \cdot T_{elz}(t)}{2F} \ln\left(\frac{p_{H_2}(t) \cdot p_{O_2}^{1/2}(t)}{p_{H_2O}(t)}\right) \end{aligned} \tag{3.37}$$

$$\begin{aligned} U_{elz,act}^{cell}(t) &= \frac{R \cdot T_{elz}(t)}{F}\left[\sinh^{-1}\left(\frac{I_{elz}(t)}{2 \cdot A_{elz} \cdot i_{a_o,elz}}\right)\right. \\ &\left. + \sinh^{-1}\left(\frac{I_{elz}(t)}{2 \cdot A_{elz} \cdot i_{c_o,elz}}\right)\right] \end{aligned} \tag{3.38}$$

$$U_{elz,ohm}^{cell}(t) = R_{ohm} \cdot I_{elz}(t) \tag{3.39}$$

$$U_{elz,conc}^{cell}(t) = K_{1,elz}^{conc} \cdot e^{\left(K_{2,elz}^{conc} \cdot I_{elz}(t)\right)} \tag{3.40}$$

where ΔS_{elz}^{o} is the entropy change, T_{elz} is the temperature of the electrolyzer, T_{elz}^{o} is the temperature in standard conditions, R and F are the known ideal gas and Faraday's constants, p_{O_2}, is the oxygen partial pressure, p_{H_2} is the hydrogen partial pressure, I_{elz} is the electrolyzer current, $i_{a_o,elz}$ and $i_{c_o,elz}$ are the anode and cathode current densities and $K_{1,elz}^{conc}$ and $K_{2,elz}^{conc}$ are the concentration-losses factors of the electrolyzer.

Taking into account the reaction produced in the electrolysis stack the hydrogen mass flow can be modeled with the next expression:

$$W_{elz}^{H_2,pro}(t) = N_{elz}^{cell} \frac{I_{elz}(t)}{F} \tag{3.41}$$

Storage

Basically, the two most commonly used hydrogen storage technologies used in micro-grids are compressed hydrogen and metal hydrides. A very simple model of compressed hydrogen systems has been assumed, consisting of a mass balance equation and the perfect gas equation:

$$\frac{dm_{H_2}}{dt} = \dot{m}_{H_2,elz} - \dot{m}_{H_2,fc} \tag{3.42}$$

$$P_{H_2} \cdot V = N_{H_2} \cdot R \cdot T \tag{3.43}$$

where $\dot{m}_{H_2,elz}$ is the hydrogen flow produced by the electrolyzer, $\dot{m}_{H_2,fc}$ is the hydrogen consumption (typically the fuel cell demand) and N_{H_2} is the number of moles in the tank.

Concerning metal hydrides, certain metals (M), particularly aluminum, titanium, iron, copper, nickel, etc., may react with hydrogen to produce a metal hydride compound through an easily controllable reversible reaction. With this technology, hydrogen is stored at moderate pressures, typically around 2 bar. The general equation is as follows:

$$M + H_2 \longleftrightarrow MH_2 \tag{3.44}$$

Different types of metal alloys can be chosen for the hydride, so that the reaction can take place within a wide range of temperatures and pressures. The reaction where the metal hydride is formed is slightly exothermic, and consequently, a supply of air cooling must be provided. When all the metal has reacted with hydrogen, the pressure will increase.

When hydrogen is needed, the metal hydride container is connected to the fuel cell. Then the reaction proceeds to the left, and hydrogen is released. If the pressure increases above atmospheric the reaction will be slow. Now the reaction is endothermic, so a small amount of energy (heat) must be supplied.

Once the reaction is complete and all the hydrogen has been released, the process can be repeated. Normally, several hundred cycles of charge and discharge can be completed. Although hydrogen is not stored under pressure, the container must be able to handle reasonably high pressures. The mathematical model of the metal hydride is complex, and out of the scope of this book, but the interested reader can find a complete description in [38].

Fuel Cells

Fuel cells are electrochemical devices whose function is to produce electricity from flows of hydrogen and oxygen. A fuel cell consists of two electrodes (anode and cathode), membrane and electrolyte sandwiched between the electrodes. These elements form the Membrane Electrode Assembly (MEA). To increase the electrical power supply, several individual cells are connected in series forming a stack fuel cell. The nature of the electrolyte blocks the movement of electrons through it, but permits protons to pass. In one of the electrodes, the anode, the molecules of hydrogen gas is separated into protons and electrons, using a catalyst for the reaction [45] (see Fig. 3.7):

$$2H_2 \longrightarrow 4H^+ + 4e^- \tag{3.45}$$

Those protons move toward the cathode, the other electrode, through the electrolyte, but electrons are forced to flow through the electrical circuit, generating electricity. Finally, water is produced due to the combination of hydrogen protons and oxygen in the cathode [45]:

$$O_2 + 4H^+ + 4e^- \longrightarrow 2H_2O \tag{3.46}$$

The overall reaction of the fuel cell is therefore:

$$2H_2 + O_2 + 4H^+ + 4e^- \longrightarrow 2H_2O \tag{3.47}$$

The dynamics of the fuel cell, characterized by mass and heat balances, results in a slow transient response compared to batteries or ultracapacitors [51, 53].

There are different types of fuel cells according to the nature of electrolyte, the operation temperature or the used fuel (hydrogen, fossil fuels, etc.). The most popular fuel cell type for stationary and vehicles applications is Proton-Exchange Membrane Fuel Cell (PEMFC), because they operate at relatively low temperature and present a quicker time response. They use a solid polymer membrane as electrolyte, and typically, platinum as catalyst.

A PEM fuel cell system needs also ancillary services, as the membrane humidifier, whose function is to maintain correct humidity in the membrane, the water separator whose function is to separate water in the output oxygen flow, the cooling and heating system, which is used to control the operating temperature (typically around 55 °C) and the compressor, needed to inject the correct quantity of air to produce the desired output power.

There are two configurations in PEM fuel cell systems: open and closed cathode. Open-cathode PEMFCs have cathode channels exposed to the atmosphere, the air is supplied by a small fan or even by convection, and consequently, they work close to the atmospheric pressure. On the other hand, in closed-cathode PEMFCs, the air is supplied by a compressor at higher pressures, typically less than 6 bars.

Open-cathode PEMFCs have the advantage of a simpler design, due to the fact that some of the auxiliary systems, as compressor or cooling system are not needed [32]. As a consequence of this compact design, this kind of fuel cells is more adequate for applications where portability is an issue. There are two different configurations of open-cathode PEMFC: air-breathing, where the air is obtained directly from the atmosphere [10, 32], without any mechanical device to provide the airflow; and air-forced systems, needing a blower or a fan to provide the airflow. Air-breathing systems are limited to low power applications, such as laptops or mobile phone chargers, for the reason that the produced water can only be removed by evaporation. Air-forced configuration permits higher power because the fuel cell requires more air for the cathodic reaction, and also, the air-forced flow can dissipate the heat generated in the system.

On the other side, the operation at higher pressures in closed-cathode PEMFs is more efficient, with a better performance and enables higher current densities [36]. As a drawback, increasing power has the effect of an increase of the parasitic loads, since compressors, cooling or humidification systems, etc., have to be powered by the fuel cell. Also, the operational pressure in closed-cathode PEMFCs requires the regulation of the pressure of the cathode [32, 34].

The operation of fuel cells has also some challenging topics in control. First, the regulation of the airflow through the cathode controlling the compressor or fan, and taking into account the heat dissipation inside the cathode channels. The flow of hydrogen has to be controlled also using the valve placed at the anode entry. Finally, the purge of the excess of hydrogen or nitrogen must be controlled efficiently with the purge valve to minimize hydrogen losses and guarantee a proper fuel cell operation [32].

As a summary, the BOP of a fuel cell system is composed of the following devices, as can be observed in Fig. 3.7:

- **Electrolyte**: It has several important functions, such as ionic conductor, electronic insulation and separator of the anodic and cathodic reactants.
- **Electrodes**: They must allow the contact between the reacting gas and the electrolyte or membrane These electrodes are divided into two layers, the first one is the catalyst and the second is a layer formed by the porous medium.
- **MEA**: Anode, electrolyte, and cathode are sealed together to form a single Membrane Electrolyte Assembly (MEA).
- **Bipolar plates**: Bipolar plates connect the cell with the electrical circuit and separate the cells forming the stack. They have to distribute hydrogen and oxygen within the cell and collaborate in heat dissipation and water elimination [19].
- **Stack**: A single fuel cell consists of two bipolar plates delivering about 0.5 and 1 V voltage. The stack is composed of individual connected cells to achieve a higher voltage and power.
- **Humidifier**: Humidification management of these systems is critical to maintain an adequate humidification in the membrane and to prevent steam condensation. A too dry membrane produces a reduction of the protons transport and the decreasing of the oxygen reduction reaction at the cathode with the result of a poor fuel cell performance or even failure.
- **Compressor**: It injects the correct proportion of air to produce the desired output power.
- **Radiator**: Fuel cells require to be maintained around an operational temperature which is essential for its performance and durability. For this reason, heat exchangers, radiators or fans are included in their balance of plant.

In a similar way to the electrolyzer, startup and shutdown cycles produce degradation effects on the catalyst layer. A fluctuating operation of the fuel cell can produce the loss of the correct humidity in the membrane, but also a more critical process known as starvation, where the airflow at the cathode input is not enough to satisfy the output power demanded to the fuel cell, producing a serious degradation. This situation can occur when operating with low airflow, producing a temporarily low

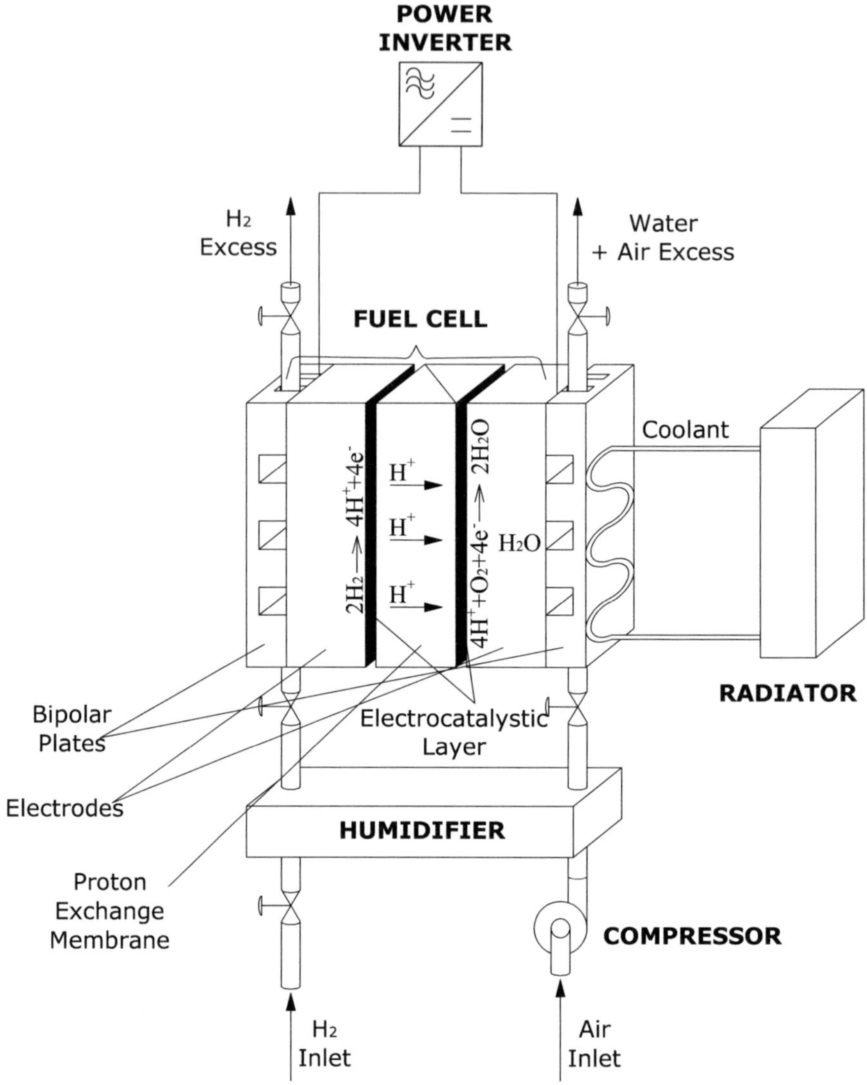

Fig. 3.7 Schematic view of a balance of plant of a fuel cell

oxidant stoichiometry with the consequence of a cell voltage drop and an increment of temperature. Additionally, an insufficient gas flow may produce an accumulation of water. For this reason, fuel cells have to operate with an excess of oxygen and hydrogen flows into the stack [53]. Unsuitable operation conditions can be the cause of irreversible degradation, and to avoid these problems, an adequate control of the fuel cell is indispensable, consisting mainly in the control of reactants supply, temperature, and power management.

A review of the causal degradation of fuel cell is exposed in [51], showing the effects of the load profile, operating temperature, and nature of the membrane on the degradation rate. On one hand, fluctuant load cycling of fuel cells produces: (i) catalyzer (platinum) particles dissolving in the cathode due to the potential cycling, (ii) mechanical wear of the membrane due to thermal and humidity cycling, (iii) chemical degradation of the membrane via radical attack due to the time at open circuit voltage, (iv) starvation phenomena when the current drawn from a fuel cell is not in the correct proportion to the reacted oxygen and (v) inappropriate humidity in the membrane. On the other hand, Start/Stop cycling produces carbon corrosion in the electrodes and deactivation of the catalysis layer [16, 51, 53].

Next, a model of the polarization curve (relationship between current and voltage) of the fuel cell stack will be developed. A more detailed model and description of each component of the fuel cell can be found in [45]. The stack voltage of the fuel cell $U_{fc}(t)$ can be expressed as the product of the number of cells N_{fc}^{cell} and the voltage of a single cell [45]:

$$U_{fc}(t) = N_{fc}^{cell} \cdot U_{fc}^{cell}(t) \tag{3.48}$$

The voltage of a single cell can be represented as follows:

$$U_{fc}^{cell}(t) = U_{fc,o}^{cell}(t) - U_{fc,act}^{cell}(t) - U_{fc,ohm}^{cell}(t) - U_{fc,conc}^{cell}(t) \tag{3.49}$$

where $U_{fc,o}^{cell}$ is the reversible potential or "Nernst" voltage, $U_{fc,act}^{cell}$ is the activation overpotential, $U_{fc,ohm}^{cell}$ is the ohmic overvoltage and $U_{fc,conc}^{cell}$ are the losses due to concentration mass. Hence, the voltage drop is a sum of four terms that can be expressed by the following expressions:

$$U_{fc,o}^{cell}(t) = E_{fc}^{o} + \frac{\Delta S^{o}}{2F}\left(T_{fc}(t) - T_{fc}^{o}\right) + \frac{R \cdot T_{fc}(t)}{2F} ln\left(\frac{p_{H_2}(t) \cdot p_{O_2}^{1/2}(t)}{p_{H_2O}(t)}\right) \tag{3.50}$$

with ΔS_{elz}^{o} is the entropy change, R is the ideal gas constant, F is Faraday's constant, T_{fc} is the stack temperature, p_{H_2} is the hydrogen partial pressure and p_{O_2} is the oxygen partial pressure.

The activation losses in the fuel cell can be modeled as a function of two constant coefficients $K_{1,act}$ and $K_{2,act}$ and the current of the stack I_{fc}

$$U_{fc,act}^{cell}(t) = -K_{1,act} \cdot (1 - e^{-I_{fc}/K_{2,act}}) \tag{3.51}$$

The ohmic losses can be modeled as a function of the equivalent ohmic resistor of the cell R_{ohm} and the stack current I_{fc}:

$$U_{fc,ohm}^{cell}(t) = R_{ohm} \cdot I_{fc}(t) \tag{3.52}$$

The concentration losses can be modeled as a function of two constant coefficients $K_{1,fc}^{conc}$, $K_{2,fc}^{conc}$ and the current of the stack.

Table 3.5 Main features of hydrogen ESS

Advantages	Disadvantages
High specific energy	Low specific power
Systems can be fully discharged	Low cycle efficiency
No self-discharge	Electrolyzers and fuel cells require cooling
Materials can be recycled	Time delay in the startup processes
Low toxicity	Limited number of life hours
	ON/OFF cycles and fluctuating operation conditions lead to degradation processes

$$U_{fc,conc}^{cell}(t) = K_{1,fc}^{conc} \cdot e^{K_{2,fc}^{conc} \cdot I_{fc}(t)} \tag{3.53}$$

As a conclusion, the main features of hydrogen as energy storage system are detailed in Table 3.5.

3.3.4 Other Energy Storage Systems

The above review of energy storage systems has not been exhaustive and other alternatives can be considered in a microgrid framework. Also, new emerging technologies are appearing which in the future can be used in microgrids. In this section, an overview of these alternatives EESs is described.

Flywheels

A flywheel storage system is a disk able to spin at high speed and to store a significant amount of energy. Most flywheel systems consist of a disk driven by an electrical machine that can work as a motor or generator. The energy applied to accelerate the rotor is maintained as kinetic energy in the spinning disk. Flywheels release the energy by reversing the process, typically producing electricity through a generator, inducing a deceleration of the flywheel rotor. Modern flywheels use advanced composite materials to reduce the weight, allowing high speeds [5, 13, 20, 39].

The main advantages of flywheel systems are the quick charge and discharge, low maintenance, resistant to temperature changes, and long life span. As can be seen in Table 3.1, flywheels occupy an intermediate position between batteries and ultracapacitors in relation to specific power and energy [40].

Superconducting Magnetic Energy Storage

In Superconducting Magnetic Energy Storage (SMES), energy is stored as a magnetic field, which is generated by DC current circulating in a superconducting coil. A SMES unit consists of a large superconducting coil, typically made of Niobium–Titanium (NbTi), at cryogenic temperature, which has to be maintained by a cryostat. Their overall cost, as a result of the cooling system and the operating temperature, is one of the limitations for the application of this ESS [4, 33, 43]. As can be seen in Table 3.1, SMES has high specific power but with low specific energy. It possesses high efficiency and long lifetime and also fast response. SMESs are highly suitable for power quality applications [4, 20].

Compressed Air

Compressed Air Energy Storage (CAES) can be considered as another promising energy storage technology to be applied in microgrids, including aboveground CAES and Mini-CAES. Their operating principle consists of compressing air with electrical power and store it into underground or aboveground containers. When the stored energy is demanded, the compressed air is passed through turbines to produce electricity. CAES has very appropriate features to be a considerable option to be used in microgrids, i.e., a reduced startup time and a long cycle life. Their energy density can provide large-scale storage capacities [54, 59]. Several studies of application of CAES in microgrids can be found in [30, 59].

Pumped Hydroelectric Storage

Pumped Hydroelectric Storage (PHS) has been traditionally used in electrical grids. PHS consists of two large water reservoirs positioned at different elevations. During low energy demand periods, water is pumped to the upper reservoir using lower cost electricity. In this way, excess energy from the grid is stored in the form of potential energy. When it is necessary to generate electricity, the stored water is released to the lower reservoir through hydraulic turbines, generating electrical power. Its self-discharge is very low, so this ESS is quite appropriate for long-term energy storage. Small PHS systems of less than a megawatt, or even less, are sometimes used in isolated grids [13, 20, 39, 42].

PHS is a technology with a long life cycle, high-efficiency conversion, and low cost. However, it requires large green land spaces which could entail the destruction of trees and associated ecosystems.

3.4 Loads

As mentioned for the case of photovoltaic or wind turbine generation, the forecast of demand in a microgrid is one of the crucial control aspects. As can be seen in the following chapters, one of the first steps in microgrid control is the scheduling of the microgrid, which implies the knowledge of the load during the schedule horizon. There are different models for load forecast applicable in microgrids, such as those based time series, econometric, or Autoregressive Integrated Moving Average (ARIMA) models. Other techniques used for demand prediction are bioinspired methods such as neural networks, genetic algorithms, particle swarm optimization, and also fuzzy logic or support vector regression. A review of the methods for demand forecasting in microgrids can be found [28, 52].

3.5 Grid

The exchange of energy of the microgrid with the main grid is done as an economic transaction where energy is sold to or purchased from the main grid. These economic transactions are based on the prices of energy in the electrical markets. As it will be detailed in Chap. 5, the energy prices are unknown when the schedule of the microgrid is carried out. An approach to know the value of the energy prices has to be done through forecast algorithms. The necessary models are commonly based on stochastic time series, causal models, artificial intelligence based models, etc. A review of some of the most popular electricity price forecasting methods can be found in [3, 57].

References

1. Abbey C, Joos G (2007) Supercapacitor energy storage for wind energy applications. IEEE Trans Ind Appl 43(3):769–776
2. Ackermann T (2005) Wind power in power systems, vol 140. Wiley Online Library
3. Aggarwal SK, Saini LM, Kumar A (2009) Electricity price forecasting in deregulated markets: a review and evaluation. Int J Electric Power Energy Syst 31(1):13–22
4. Ali MH, Wu B, Dougal RA (2010) An overview of SMES applications in power and energy systems. IEEE Trans Sustain Energy 1(1):38–47
5. Arghandeh R, Pipattanasomporn M, Rahman S (2012) Flywheel energy storage systems for ride-through applications in a facility microgrid. IEEE Trans Smart Grid 3(4):1955–1962
6. Arora P, White RE, Doyle M (1998) Capacity fade mechanisms and side reactions in lithium-ion batteries. J Electrochem Soc 145(10):3647–3667
7. Ball M, Wietschel M (2009) The future of hydrogen-opportunities and challenges. Int J Hydrog Energy 34(2):615–627
8. Bank Tavakoli MR, Vahidi B, Gawlik W (2009) An educational guide to extract the parameters of heavy duty gas turbines model in dynamic studies based on operational data. IEEE Trans Power Syst 24(3):1366–1374
9. Barbir F (2005) Pem electrolysis for production of hydrogen from renewable energy sources. Solar Energy 78(5):661–669
10. Barreras F, López AM, Lozano A, Barranco JE (2011) Experimental study of the pressure drop in the cathode side of air-forced open-cathode proton exchange membrane fuel cells. Int J Hydrog Energy 36(13):7612–7620
11. Basu S, Undeland TM (2007) Voltage and current ripple considerations for improving lifetime of ultra-capacitors used for energy buffer applications at converter inputs. In: 2007 IEEE Power Electronics Specialists Conference, PESC 2007. IEEE, pp 1453–1457
12. Burke A (2000) Ultracapacitors: why, how, and where is the technology. J Power Sour 91(1):37–50
13. Chen H, Cong TN, Yang W, Tan C, Li Y, Ding Y (2009) Progress in electrical energy storage system: a critical review. Prog Nat Sci 19(3):291–312
14. Conway BE (2013) Electrochemical supercapacitors: scientific fundamentals and technological applications. Springer
15. Davis MW, Gifford AH, Krupa TJ (1999) Microturbines-an economic and reliability evaluation for commercial, residential, and remote load applications. IEEE Trans Power Syst 14(4):1556–1562
16. De Bruijn FA, Dam VAT, Janssen GJM (2008) Review: durability and degradation issues of pem fuel cell components. Fuel Cells 8(1):3–22

17. De Soto W, Klein SA, Beckman WA (2006) Improvement and validation of a model for photovoltaic array performance. Solar Energy 80(1):78–88
18. Diagne M, David M, Lauret P, Boland J, Schmutz N (2013) Review of solar irradiance forecasting methods and a proposition for small-scale insular grids. Renew Sustain Energy Rev 27(C):65–76
19. Dieguez PM, Ursúa A, Sanchis P, Sopena C, Guelbenzu E, Ganda LM (2008) Thermal performance of a commercial alkaline water electrolyzer: Experimental study and mathematical modeling. Int J Hydrog Energy 33(12):7338–7354
20. Dötsch C (2009) Energy storage. In: Technology guide. Springer, pp 362–367
21. Driesse A, Harrison S, Jain P (2007) Evaluating the effectiveness of maximum power point tracking methods in photovoltaic power systems using array performance models. In: 2007 IEEE Power electronics specialists conference, PESC 2007. IEEE, pp 145–151
22. Dunn S (2002) Hydrogen futures: toward a sustainable energy system. Int J Hydrog Energy 27(3):235–264
23. Etezadi-Amoli M, Choma K (2001) Electrical performance characteristics of a new microturbine generator. In: 2001 Conference Proceedings IEEE power engineering society winter meeting (Cat. No.01CH37194), vol 2, Jan 2001, pp 736–740
24. Faranda R, Leva S, Maugeri V (2008) MPPT techniques for PV systems: energetic and cost comparison. In: 2008 IEEE Power and energy society general meeting-conversion and delivery of electrical energy in the 21st Century. IEEE, pp 1–6
25. Fialho L, Fartaria T, Narvarte L, Collares Pereira M (2016) Implementation and validation of a self-consumption maximization energy management strategy in a vanadium redox flow BIPV demonstrator. Energies 9(7)
26. Görgün H (2006) Dynamic modelling of a proton exchange membrane (PEM) electrolyzer. Int J Hydrog Energy 31(1):29–38
27. Grbovic PJ, Delarue P, Le Moigne P, Bartholomeus P (2011) Modeling and control of the ultracapacitor-based regenerative controlled electric drives. IEEE Trans Ind Electron 58(8):3471–3484
28. Hahn H, Meyer-Nieberg S, Pickl S (2009) Electric load forecasting methods: tools for decision making. Eur J Oper Res 199(3):902–907
29. Hartmann II RL (2008) An aging model for lithium-ion cells. PhD thesis, The University of Akron
30. Ibrahim H, Belmokhtar K, Ghandour M (2015) Investigation of usage of compressed air energy storage for power generation system improving-application in a microgrid integrating wind energy. Energy Proc 73:305–316
31. Iov F, Hansen AD, Sorensen P, Blaabjerg F (2004) Wind turbine blockset in matlab/simulink. Aalborg University, March
32. Ishaku J, Lotfi N, Zomorodi H, Landers RG (2014) Control-oriented modeling for open-cathode fuel cell systems. In: 2014 American control conference (ACC). IEEE, pp 268–273
33. Jin JX, Xu W, Chen XY, Zhou X, Zhang JY, Gong WZ, Ren AL, Xin Y (2012) Developments of SMES devices and potential applications in smart grids. In: 2012 IEEE Innovative smart grid technologies-Asia (ISGT Asia). IEEE, pp 1–6
34. Leksono E, Pradipta J, Tamba TA (2012) Modelling and identification of oxygen excess ratio of self-humidified pem fuel cell system. Mechatron Electric Power Vehicular Technol 3(1):39–48
35. Lydia M, Suresh Kumar S (2010) A comprehensive overview on wind power forecasting. In: 2010 Conference proceedings IPEC, October 2010, pp 268–273
36. Meyer Q, Ronaszegi K, Pei-June G, Curnick O, Ashton S, Reisch T, Adcock P, Shearing PR, Brett DJL (2015) Optimisation of air cooled, open-cathode fuel cells: current of lowest resistance and electro-thermal performance mapping. J Power Sourc 291:261–269
37. Mohamed Y (2008) New control algorithms for the distributed generation interface in grid-connected and micro-grid systems
38. Mohammadshahi SS, Gray EM, Webb CJ (2016) A review of mathematical modelling of metal-hydride systems for hydrogen storage applications. Int J Hydrog Energy 41(5):3470–3484

39. Mohd A, Ortjohann E, Schmelter A, Hamsic N, Morton D (2008) Challenges in integrating distributed energy storage systems into future smart grid. In: 2008 IEEE International Symposium on Industrial Electronics, ISIE 2008. IEEE, pp 1627–1632
40. Molina MG (2010) Dynamic modelling and control design of advanced energy storage for power system applications. In: IntechOpen, pp 49–92
41. Nakanishi F, Ikegami T, Ebihara K, Kuriyama S, Shiota Y (2000) Modeling and operation of a 10 kW photovoltaic power generator using equivalent electric circuit method. In: 2000 Conference Record of the Twenty-Eighth IEEE Photovoltaic specialists conference. IEEE, pp 1703–1706
42. Nanahara T (2002) Pumped hydro storage. J Jpn Inst Energy 81(9; ISSU 905):811–818
43. Padimiti DS, Chowdhury BH (2007) Superconducting magnetic energy storage system (SMES) for improved dynamic system performance. In: 2007 IEEE power engineering society general meeting. IEEE, pp 1–6
44. Plett GL (2004) Extended kalman filtering for battery management systems of LiPB-based HEV battery packs: part 1. Background. J Power Sources 134(2):252–261
45. Pukrushpan JT, Stefanopoulou AG, Peng H (2004) Control of fuel cell power systems: principles, modeling, analysis and feedback design. Springer Verlag, London
46. Rauschenbach HS (1980) Solar cell array design handbook. The principles and technology of photovoltaic energy conversion. Springer, New York
47. Rolan A, Luna A, Vazquez G, Aguilar D, Azevedo G. Modeling of a variable speed wind turbine with a permanent magnet synchronous generator. In: 2009 IEEE international symposium on industrial electronics, July 2009, pp 734–739
48. Shirazi M, Viki AH, Babayi O (2009) A comparative study of maximum power extraction strategies in PMSG wind turbine system. In: 2009 IEEE electrical power & energy conference (EPEC). IEEE, pp 1–6
49. Sikha G, Ramadass P, Haran BS, White RE, Popov BN (2003) Comparison of the capacity fade of Sony US 18650 cells charged with different protocols. J Power Sources 122(1):67–76
50. Soens J (2005) Impact of wind energy in a future power grid. Belgica, Katholieke Universiteit Leuven, Leuven
51. Stevens MB (2008) Hybrid fuel cell vehicle powertrain development considering power source degradation. PhD thesis, University of Waterloo
52. Suganthi L, Samuel AA (2012) Energy models for demand forecasting: a review. Renew Sustain Energy Rev 16(2):1223–1240
53. Suh KW (2006) Modeling, analysis and control of fuel cell hybrid power systems. PhD thesis, University of Michigan
54. Tan X, Li Q, Wang H (2013) Advances and trends of energy storage technology in microgrid. Int J Electric Power Energy Syst 44(1):179–191
55. Valverde L, Rosa F, Del Real A, Arce A, Bordons C (2013) Modeling, simulation and experimental set-up of a renewable hydrogen-based domestic microgrid. Int J Hydrog Energy 38(27):11672–11684
56. Villalva MG, Gazoli JR (2009) Comprehensive approach to modeling and simulation of photovoltaic arrays. IEEE Trans Power Electron 24(5):1198–1208
57. Weron R (2014) Electricity price forecasting: a review of the state-of-the-art with a look into the future. Int J Forecast 30(4):1030–1081
58. Yin M, Li G, Zhou M, Zhao C Modeling of the wind turbine with a permanent magnet synchronous generator for integration. In: 2007 IEEE Power engineering society general meeting. IEEE, pp 1–6
59. Zhang J, Li K-J, Wang M, Lee W-J, Gao H, Zhang C, Li K (2016) A bi-level program for the planning of an islanded microgrid including CAES. IEEE Trans Ind Appl 52(4):2768–2777
60. Züttel A (2004) Hydrogen storage methods. Naturwissenschaften 91(4):157–172

Chapter 4
Basic Energy Management Systems in Microgrids

Abstract This chapter addresses the basic Energy Management System (EMS) for microgrids, which aims to balance generation and demand using storage or the external grid, and corresponds to secondary control, as presented in Chap. 1. This is also known as power sharing or power dispatch, whose purpose is to drive the dispatchable units (Distributed Energy Resources, DERs) to supply local loads in an appropriate way. A basic MPC algorithm is developed in this chapter, which can solve the problem using only continuous variables. In order to illustrate the concept and methodology, the design and implementation of the basic EMS on a pilot-scale microgrid is presented. Simulated and experimental tests are performed under realistic scenarios, showing how MPC can be customized to a particular microgrid. Other issues such as schedule, consideration of degradation and maintenance costs, integration of energy tariffs, or connection to the electrical market will be addressed in Chap. 5.

4.1 Problem Description

The Energy Management System is in charge of achieving the energy balance in the microgrid in the most effective way. The primary goal is, therefore, to ensure stable delivery of electrical power to its local load consumers. This may include just managing the excess/deficit of energy or considering other functionalities with economic or operational criteria. The EMS has to balance the power generation and demand by means of the energy storage, the dispatchable generators, and demand management if possible. At the same time, ideally, the EMS can optimize the system efficiency and minimize the operational cost.

The EMS manages the excess or deficit of energy from the renewable sources; when possible, the electric power from the renewable sources is delivered directly to the loads. Any excess of power is shunted to the storage units or the grid and, if power is unavailable from the renewable sources, it must be supplied by the storage units or the grid. The primary objective of the EMS is, therefore, to effectively balance the power in the microgrid but, additionally, depending on the control algorithm, the

C. Bordons et al., *Model Predictive Control of Microgrids*,
Advances in Industrial Control, https://doi.org/10.1007/978-3-030-24570-2_4

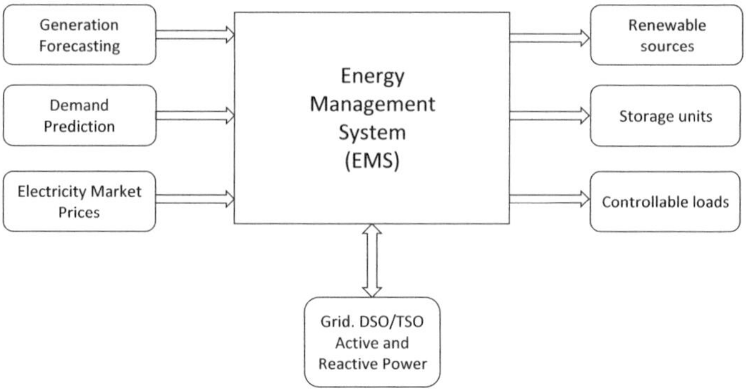

Fig. 4.1 EMS Scheme

EMS may try to optimize production toward assigned objectives. The appropriate amount of energy that must be exchanged among generators, storage units, loads, and external grid will be dictated by the control policy used, which may range from simple heuristic rules to complex optimization algorithms.

The general scheme of a microgrid EMS is shown in Fig. 4.1. Notice that if techno-economic optimization is included in the problem, additional information such as generation and load forecasting and evolution of market prices are needed. The problem addressed in this chapter is the computation of the set points to the dispatchable components of the microgrid taking into account the available information about renewable generation and demand as well as the state of energy storage. This can be done using several methods, which are described in the following section.

4.2 Review of Methods

EMS algorithms range from simple if-then rules to complex multiparametric optimization that can include the forecasting of electricity prices, loads, weather, operational cost, electrical markets, degradation issues, etc. It can be considered that there are two main families of methods: those based on heuristics and those based on the optimization of some criteria. The most representative methods for a centralized solution are described in the following sections. Distributed methods can be of interest in the case of large or geographically distributed microgrids or networks of microgrids; they will be addressed in Chap. 8.

4.2.1 Heuristic Methods

In general, heuristic methods are a set of algorithms that use rules to handle the energy mismatch in the microgrid. Usually, they are characterized for being simple and reliable, which has made them very popular for being implemented in small microgrids [17] and in other energy systems such as hybrid vehicles or industrial plants [30].

Hysteresis Band Control (HBC) is the most used heuristic method [2, 20, 33], where the operation of the ESSs follows a hysteresis band whose thresholds are defined according to the state of charge of energy storage units. If only one ESS is used, such as a battery, the operation is simple: the battery absorbs the unbalance between generation and demand provided its SOC is between the upper and lower thresholds. If the upper threshold is reached, generation must be stopped, or the excess energy sold to the grid (for grid-connected microgrids); if the lower limit is reached, some loads must be disconnected or the lack of energy must be purchased from the grid. In the presence of several energy storage systems (such as batteries, hydrogen, ultracapacitors or flywheels), the criterion is to use several hysteresis bands and always use first the system with higher efficiency. The complexity of the algorithms increases as more rules are added [18].

The rules for switching among different ESSs are usually based on the stored energy. The switching of the electrolyzer and fuel cell in a microgrid that uses hydrogen and batteries as energy buffers is often based on the battery SOC. The basic control scheme is illustrated in Fig. 4.2. It shows the on–off switching thresholds for the electrolyzer and fuel cell. The basic principle of this strategy is the simplicity of operation. The electrolyzer is switched on when the battery SOC is high. On the other hand, the fuel cell is activated when the SOC is very low, according to certain thresholds. Likewise, the switching off of both equipment is defined by upper hysteresis thresholds.

In addition, the battery bank needs to be protected from overcharging (high SOC) or undercharging (low SOC). In this case, the control system transfers energy from the grid in order to save the battery integrity. This methodology is widely used in demonstration microgrids with hybrid storage as can be seen in [34].

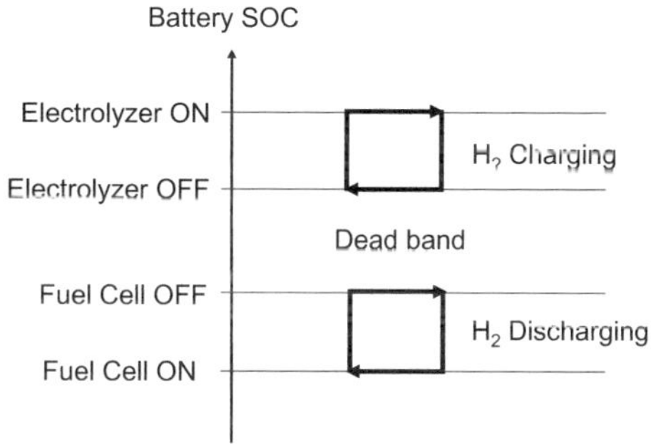

Fig. 4.2 Control strategy by hysteresis band showing working regions of the electrolyzer and the fuel cell

Linguistic rules used in fuzzy logic can simplify the management and control of the microgrid when the addition of heuristic rules becomes difficult to handle, providing a suitable and practical solution. Fuzzy logic is a logical system built on the basis of fuzzy sets theory [40] that deals with sets to which the elements may belong to a certain degree. In classical set theory, the elements belong or not to a set, while in the theory of fuzzy sets, the membership is defined by a function called membership function, which can take multiple values. In fuzzy logic, unlike standard conditional logic, the truth of any statement is a matter of degree, which allows to handle problems with imprecise and incomplete data. Fuzzy Control (FC) establishes the controller directly based on measurements and the knowledge of expert operators, using an inference mechanism based on fuzzy rules. FC is used in several microgrid applications [7], either as the main controller or for tuning or supporting a conventional controller. Fuzzy control can be combined with other methodology, as done in [32] to capture nonlinearity and uncertainty in order to formulate a robust EMS that uses MPC theory as the mathematical framework. The EMS is formulated using a fuzzy prediction interval model as the prediction model. Fuzzy logic is also used for demand and price forecasting in microgrids participating in the electricity market [25].

Although simplicity is the primary feature of rule-based methods, their main drawbacks are that the solution is not optimal and that management of many ESSs is very complicated. In order to overcome this, advanced methods that solve an optimization problem are considered.

4.2.2 Optimization-Based Methods

Some of the optimal criteria that can be considered in the operation of microgrids have been previously presented in Chap. 1. This section briefly describes the methods that can be used to solve the optimization problem. Notice that this is basically an enumeration of some methods of interest in microgrids; the reader who wants to go deeper into the algorithms is referred to these books on the subject: [3, 21].

The optimization problem can be posed as a cost function to be minimized and a set of constraints. The solution will provide the optimal values of the powers that must be managed (supplied or absorbed) by each DER, which are the decision variables. In general, the cost function is composed of a weighted sum of the powers of the DERs, where the corresponding weights are associated to the cost of using a certain unit (cost of purchasing from the grid, operating and maintenance costs, etc.). Then the cost function is linear (a linear combination of the decision variables over a period of time) [27], but in some occasions it can be useful to have a quadratic cost function [12]. This can penalize large values of powers more aggressively than a linear cost and it can also allow the consideration of tracking errors (for instance, a deviation of the SOC from a desired value). Other norms such as 1-norm (absolute value) or ∞-norm (maximum value) can also be used in the cost function.

By using single- or multi-objective mathematical optimization, recent energy management techniques enable to include in the microgrid operation a wide range of cost functions and parameters to encompass optimization techniques and realistic constraints. As previously pointed out, the problem formulation can include operation costs, equipment lifetime, emissions concerns, spot market price, availability of hydrogen, etc.

The constraints are usually limits imposed to the operating range or ramp rates of the DERs, and can be represented by inequalities, that is, the constraints are linear on the decision variables. Therefore, the problem to be solved is the minimization of a linear cost function with linear constraints, which is solved by Linear Programming (LP) or the minimization of a linearly constrained quadratic function, which can be solved by Quadratic Programming (QP). Notice that the model of the microgrid must be included in the minimization as a constraint; if the model or other constraints are nonlinear, the problem has to be solved by Nonlinear Programming (NLP). For example, CO_2 equivalent emissions and cost of the consumed gas of a microturbine can be expressed as a nonlinear function of its power output, as done in [7] to solve day-ahead operational planning, leading to an NLP formulation.

Besides, if apart from the continuous variables, binary (logical) values are considered (for instance, associated to connection or disconnection of units), the problem turns into a mixed-integer problem: MILP, MIQP, or MINLP. Mixed problems are more difficult to solve than those with continuous variables and usually branch and bound techniques are needed. Some approximate solutions exist, as the one presented in [23] that transforms an MIQP into a set of Quadratic Programming (QP) that are easier to solve. Mixed problems can also appear when approximating nonlinear constraints (such as the model of a diesel generator) by piecewise linear models, which leads to an optimization problem that is solved using mixed-integer linear programming (MILP), as done in [26]. Additionally, the optimization problem can be deterministic or stochastic, in this last case, Stochastic Programming (SP) techniques must be used.

Robust and efficient methods exist for LP and QP [6], which can provide solutions even for short execution times. However, the solution of the nonlinear problem can be quite computationally demanding since the problem is nonconvex and there is no guarantee that the global optimum can be found.

Dynamic Programming (DP) can also be used to solve the optimization problem. DP (see [9]) refers to simplifying a complicated problem by breaking it down into simpler subproblems in a recursive manner. It is a methodology that evaluates a large number of possible decisions in a multistep problem. Dynamic programming algorithms are commonly used for solving optimal control problems for dynamic systems, which can be accomplished by forward or backward iteration. As pointed out by [7], a subset of possible decisions is associated with each sequential problem step and a single one must be selected. There is a cost associated to each possible decision, which may be affected by the decision made in the previous step. Transition costs are associated to a decision toward the following step. The objective is to make a decision in each problem step which minimizes the total cost for all the decisions made. The biggest limitation of using DP is the high number of partial solutions (that grow with the dimension of the problem) that must keep track of and that no general formation of DP is available (except for simple cases). Besides, dividing the problem into subproblems and storing intermediate results consume memory. In spite of these drawbacks, DP has also been applied to microgrids. An application to the economic dispatch in a microgrid using forward iteration can be found in [39].

Metaheuristics

In many situations, the nonlinearity of the problem, resulting from the cost function, the dynamic model or the constraints, makes it difficult to find a solution. In addition, the high number of variables makes that the optimization of these systems cannot be done satisfactorily using analytical methods, leading to inadmissible computation times. Although there exist many algorithms that can be used, in some situations metaheuristic methods can provide practical solutions.

Metaheuristics (also known as soft computing or bioinspired methods) are a family of methodologies based on heuristics testing to efficiently find solutions to optimization problems. They can obtain highly accurate approximate solutions to optimization problems quite quickly for a variety of cost functions and constraints. These methods can be an option in the case of nonconvex problems or large number of decision variables. A wide range of metaheuristics have been proposed in several applications, including microgrid control. The most representative ones are introduced here.

Tabu search (TS) [14] is an efficient local search procedure that works in a similar way that the human memory process. The searching is done looking for solutions that move toward the best evaluation function value through the use of a list which memorizes search history (tabu list). It is forbidden to revisit solutions that have already been explored and to move to similar solutions, so TS can escape from local minima to find a global optimal solution. An example of application of this method for determining the operating schedule of an energy network that minimizes emission and energy costs can be found in [31]. There, the transition direction when updating the schedule (e.g., reducing or increasing the output of each generator) is stored in the tabu list.

Genetic Algorithms (GAs) [15] imitate the evolution process of living beings. The main idea is to propose solutions that evolve from one generation to another. Each candidate solution in the population is expressed as a gene. The next generation of solutions is generated by genetic operations, such as selection, crossover, and mutation. The beneficial features of a solution with a good evaluation function value are transmitted to the next generation with high probability. GAs are suitable for global searches because of their multipoint search and genetic manipulation with multiple solutions. Management of a microgrid based on genetic algorithm is presented in [10].

The Particle Swarm Optimization (PSO) method [19] is a heuristic stochastic optimization method that is originated from the concept of swarm intelligence. This concept describes how a swarm of insects (ant colony), a flock of birds, or a school of fish search for food. Following this idea, a PSO method uses a population-based search engine to determine an optimal set of solutions in the objective space. It has been used in [29] to find real-time optimal energy management solutions for an islanded hybrid wind–microturbine energy system. The proposed PSO-based energy management algorithm incorporates many objectives such as minimizing the cost of generated electricity, maximizing microturbine operational efficiency, and reducing environmental emissions.

An Artificial Neural Network (ANN) is a model-free estimator which can replicate a certain input–output relationship without the need of a complex mathematical model; they can be used for many applications (identification, forecasting, pattern recognition, machine learning, etc.) including optimization. The general idea of the ANN is to mimic the neural network in the brain connecting different neurons, also referred to as nodes, in several layers. Each neuron produces an output which is a nonlinear function of the weighted sum of its inputs. The optimal problem is solved offline, training the ANN to replicate the optimization during run-time execution, when the ANN maps all the input values to the corresponding output control values. An example of the use of ANN for optimization in a microgrid is found in [11], where the proposed neural network determines the optimal amount of power over a time horizon of 1 week for wind, solar, battery systems, and an electric car, in order to minimize the power acquired from the utility grid and to maximize the power supplied by the RESs. Other examples of ANN-based control applications for frequency and voltage control in microgrids are found in [16]. ANNs are also used for forecasting tasks in microgrids, which are executed offline; an application for short-term load forecast is shown in [1].

Model Predictive Control

MPC is an optimization-based method that can compute control actions (set points to the different units that integrate the microgrid) in order to fulfill some criteria. In this sense, it is similar to any other optimization-based strategy. But the main advantage of MPC is that the optimization process is embedded in a control scheme which incorporates feedback. This way, MPC can face disturbances and model mismatch, recomputing the necessary control actions in a receding way when fresh information about the microgrid state is available. MPC can incorporate any optimization procedure (QP, LP, NLP, MILP, MIQP, metaheuristics, etc.) depending on the type of model used (linear, nonlinear, hybrid, etc.) and the cost function employed. Therefore, the main feature of MPC is the replacement of an (usually complex) offline determination of the control actions by a repeated online solution of the optimization of an open-loop problem whose solution provides the current control action. MPC can incorporate optimal use of EESs and external grid as well as management of demand response to compensate imbalances generated by the difference between generation (usually non-dispatchable) and demand.

Many MPC schemes have been proposed to optimally manage the power flows inside the microgrid. In general, controllers have a hierarchical structure [12] and the microgrid can operate in islanded [4] or grid-connected mode [35]. Stochastic approaches are also considered [8]. Discrete dynamics (charge/discharge of ESSs and ON/OFF of DER units and loads, buy/sell tariffs for electricity) and the switching between different operating conditions can also be considered. Then, the problem must be solved using MILP techniques, as done in [24] for a household application, or in [27] for a grid-connected microgrid. This latter work shows how the use of the feedback control law provided by MPC improves the operational cost versus the use of MILP optimization alone, due to uncertainties in generation and demand.

4.3 Basic Model Predictive Control Algorithm

The primary goal of the EMS is to ensure stable delivery of electrical power to its local load consumers, trying to encompass performance optimization and preventing equipment damage. A basic MPC controller can perform this task if the cost function and operation constraints are properly defined. The predictive controller calculates online optimal set points which are sent as control signals to the power converters of the generators, loads, storage units, and grid connection. Then, the onboard electronic control units of the different elements of the microgrid (generators, batteries, fuel cells, etc.) determine the best way to reach these set points, according to their own manufacturer's controllers.

The MPC optimization problem supplies a solution that proposes a trajectory of inputs and states into the future that satisfy operational constraints while optimizing some given criteria. That is, at each sampling instant, an optimal plan is formulated based on forecast of demand and generation and on the knowledge of the energy storage level (state of the system). The first element of the sequence is implemented and the horizon is shifted. At the next sampling time, a new optimization problem is solved using the newly available information (state of the system that is measured or estimated). By using this feedback mechanism, the new optimal plan can potentially compensate for disturbances that act on the microgrid.

The EMS can take decisions on unit commitment (when should each generation unit be started and stopped) and economic dispatch (how much should each unit generate to meet the load at minimum cost and how much energy should be purchased from or sold to the grid). Curtailment and Demand-Side Management (DMS) (which controllable loads must be shed/curtailed and when) can also be included. This section will address the baseline case which seeks to operate the microgrid fulfilling operational criteria, but without considering the economic optimization, that will be further dealt with in Chap. 5.

The MPC will be responsible for the reliable operation of the microgrid. This task becomes particularly challenging in microgrids with the presence of still-expensive technologies, such as fuel cells, where the dispatch command should be fast enough to track the sudden load changes while at the same time achieve certain objectives of their lifetime, highly dependent on the load profile. The use of weights in the cost function and operational constraints can help to fulfill this objective.

Specifically, the proposed control aims to fulfill the following objectives:

- To minimize the power flows exchanged among the units to keep the balance.
- To protect the battery bank from deep discharging and overcharging.

- To limit the units' power rates (especially electrolyzer and fuel cell) in order to protect such expensive equipment from intensive use.
- To take into account the energy efficiency in the plant, using the most efficient units when possible. For instance, in a microgrid with hydrogen storage, it is better to use batteries as first energy storage means whenever possible. Since the hydrogen roundtrip efficiency is much lower than batteries efficiency, this path is used only when there is a large imbalance between production and demand.
- To provide flexibility in the operation, guaranteed by establishing weights in the cost function for reference tracking.
- To minimize the energy exchanged with the grid, when a high degree of autonomous operation is desired.

In this multi-objective optimization problem, the goal is to achieve an optimal solution for several competing objectives. In such problems, the satisfaction of the cost function becomes a Pareto optimum where the solution represents a state of trade-off among objectives. Therefore, the microgrid will reach a state of energy resource allocation in which it is impossible to make any solution better without making at least one solution worse.

During normal operation, storage or generation units will have to cope with ripples and/or sudden power changes. In the grid-connected case, the controller can be tuned so that the grid has to cope with the rapid demand changes in order to protect the rest of the equipment from intensive use. The alternative approach, more favorable for the grid operator, is also possible.

4.3.1 Control-Oriented Model

As stated along the book, MPC needs a model of the microgrid to perform predictions. This control-oriented model is a simplified one (different from those presented in Chap. 3) that can be integrated into the optimization procedure. At the EMS level, the dynamics of loads and generators is very fast compared to the characteristic sampling time and it can be neglected. Therefore, the main dynamics to be considered is that of the storage units, which, together with the balance equation of powers in the bus, will constitute the model to be used by MPC. The signal criteria used along this book are that powers injected into the bus are positive and powers extracted from the bus are negative. Therefore, in the case of storage systems, which can inject or extract from the bus, their power is considered positive when discharging and negative when charging. Figure 4.3 illustrates this: power supplied by the grid, solar panels, wind turbine, fuel cell, batteries, and ultracapacitor are positive, while power demanded by the load and the electrolyzer are negative from the point of view of the bus.

The stored units can be modeled by an energy balance equation that determines the increment in the level of energy $x(t)$ as the integral of the charged power $P_{sto}(t)$ which is positive for charging and negative for discharging:

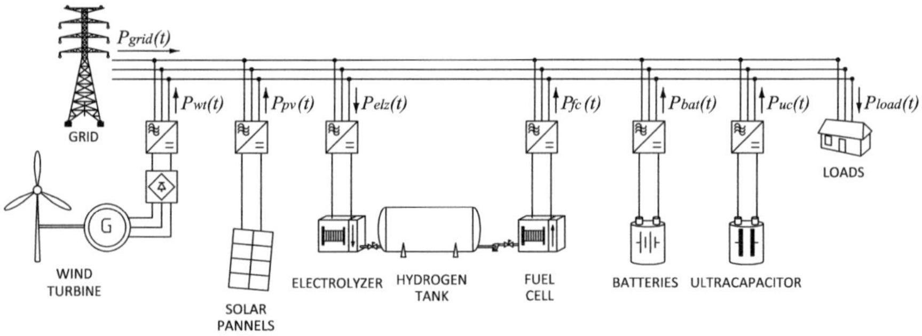

Fig. 4.3 Example of power flows in a microgrid

$$x(t+1) = x(t) - \eta T_s P_{sto}(t) \tag{4.1}$$

where T_s is the sampling time, given in seconds. In general, the influence of the charge/discharge of the storage units on the stored energy levels is not the same, so different efficiencies for charge/discharge are used:

$$\eta = \begin{cases} \eta_{ch}, \text{ if } P_{sto}(t) < 0 \text{ (charging mode)} \\ \frac{1}{\eta_{dis}} \text{ otherwise (discharging mode)} \end{cases} \tag{4.2}$$

η_{ch} and η_{dis} are the charging/discharging efficiencies (which in general take different values). In the common case that the storage unit is a battery, the level of energy is given by the SOC, defined as the ratio between the current capacity $C_{bat}(t)$ and the maximum battery capacity C_{max} (see Chap. 3), then its evolution is given by

$$SOC(t+1) = SOC(t) - \frac{\eta T_s}{C_{max}} P_{bat}(t) \tag{4.3}$$

where C_{max} is the battery capacity expressed in energy units (tipically Wh).

In order to manage the different behaviors in charging and discharging, a binary variable $\delta(t)$ must be considered, which takes value 1 for charging and 0 for discharging. Then, the dynamics of storage can be written as

$$x(t+1) = x(t) - \eta_{ch}\delta(t)T_s P_{sto}(t) + \frac{1}{\eta_{dis}}(1-\delta(t))T_s P_{sto}(t) \tag{4.4}$$

Notice that this equation is nonlinear and includes continuous and binary variables (it is a hybrid model), therefore is not easy to manage. The problem can be simplified if the different efficiencies for charge/discharge are neglected, considering η in Eq. (4.1) as a fixed value. This does not represent accurately the storage behavior due to the different multiplicative factors in charging and discharging, but can be justified in order to simplify the model. This will be done along this chapter that addresses the basic EMS problem. When logical variables are considered, the model is hybrid in the sense that it involves continuous and integer (binary) variables in the equations, and therefore the optimization to be solved by MPC becomes a Mixed-Integer Quadratic Problem (MIQP). This issue as well as other logical conditions will be detailed in Chap. 5, where the framework of Mixed Logical Dynamic (MLD) systems [5] is used to formulate the problem.

Another possibility is to consider two variables for each storage unit, P^s_{ch} which is the power charged into the storage unit and P^s_{dis} which is the one discharged. In this case, they must be complementary variables (only one of them can be different from 0 at each instant) and thus this constraint must be added to the model equations. This is of particular interest in the case of hydrogen storage, where the efficiency of charging (electrolyzer) is certainly different of discharging (fuel cell) and the storage unit is in fact composed of two separate equipment, so the fuel cell can be considered as a generator and the electrolyzer as a load (both manipulable). This assumption is used in [35] for a microgrid that has electrical and hydrogen storage, where one variable is used for the battery (P_{bat}), other for the power delivered by the fuel cell (P_{fc}) and another for the power consumed by the electrolyzer (P_{elz}). In [28] the problem is simplified more, using one variable for the battery (P_{bat}) and only one variable for hydrogen storage (P_{H_2}), which is positive when the fuel cell is used and negative when the electrolyzer is running.

These assumptions allow to address the problem with an MPC methodology where all the variables take continuous values, as the one described in Sect. 2.3.

Then the model of the microgrid is formed by one equation like (4.1) per storage unit and the following energy balance equation of the microgrid, which implies that the net sum of all the energy flows in the bus is zero:

$$\sum_{i=1}^{n_g} P_{gen,i}(t) + \sum_{i=1}^{n_e} P_{ext,i}(t) + \sum_{i=1}^{n_s} P_{sto,i}(t) - \sum_{i=1}^{n_s} P_{load,i}(t) = 0 \tag{4.5}$$

where $P_{gen,i}$ is the power generated by the generation unit i (that can be dispatchable or not), $P_{sto,i}$ is the power exchanged with the storage units (positive when discharging), $P_{ext,i}$ is the power exchanged with the external connections, like main utility grid or other microgrids (positive when importing energy), and $P_{load,i}$ is the power consumed by the loads. In general, one of the storage units (commonly the battery) is used to balance the microgrid, absorbing extra power or supplying the necessary one at each time instant; in this case, the power of this device (k) can be expressed as

$$P_{sto,k}(t) = \sum_{i=1}^{n_g} P_{gen,i}(t) + \sum_{i=1}^{n_e} P_{ext,i}(t) - \sum_{i=1,i\neq k}^{n_s} P_{sto,i}(t) - \sum_{i-1}^{n_s} P_{load,i}(t) \tag{4.6}$$

and substituted in Eq. (4.1).

The model obtained from Eqs. (4.1) and (4.5) only contains continuous variables and it is suitable for a basic EMS, as shown in [35]. In case that switching on/off of loads or generators or different operating conditions are considered, logical variables (that can only take binary values: 0 or 1) must be included. In [12], the connection and disconnection of a fuel cell and a electrolyzer is modeled with logical variables in order to design an EMS that can improve durability of the equipment by minimizing switching and reducing state transitions. Similarly, different prices for selling and buying energy to/from the grid can be included in the cost function using logical variables. A detailed description of a microgrid model with continuous and logical variables can be found in [27].

State-Space Model

When the microgrid is described using only continuous variables, a state-space model of the form (2.7) can be used. The states $x(t)$ are the energy stored in the different ESSs: SOC of battery or ultracapacitor, Level of Hydrogen (LOH) in tanks, kinetic energy in flywheels, etc. Usually, the outputs $y(t)$ will coincide with the states and the manipulated variables $u(t)$ will be the power flows that can be manipulated: power to charge/discharge the battery, power from/to the main grid, and power supplied by the dispatchable generators (such as a microturbine or a diesel engine), P^{gd}. The power generated by the renewable sources, P^{gr}, as well as the demand are considered as disturbances $d(t)$ which, in general, can be measured but not manipulated. So, the following vectors can be defined:

$$
\begin{aligned}
x(t) &= [x_1(t)\ x_2(t) \ldots x_{n_s}(t)]^T && (4.7)\\
u(t) &= [P_{sto,1}(t)\ \ldots P_{sto,n_s}(t)\ P_{ext,1}(t)\ \ldots P_{ext,n_e}(t)\\
&\qquad P_{gd,1}(t)\ \ldots P_{gd,n_{gd}}(t)]^T && (4.8)\\
d(t) &= \sum_{i=1}^{n_{gr}} P_i^{gr}(t) - \sum_{i=1}^{n_l} P_{load,i}(t) && (4.9)\\
y(t) &= x(t) && (4.10)
\end{aligned}
$$

and the dynamics can be written in the general state-space form with appropriate matrices:

$$
\begin{aligned}
x(t+1) &= Ax(t) + Bu(t) + B_d d(t)\\
y(t) &= Cx(t) && (4.11)
\end{aligned}
$$

Notice that matrices A and C are equal to the identity matrix I and that B and B_d are composed of terms that depend on the storage efficiency, which is used to convert the input/output flows of an storage device into its stored energy. An example is provided in Sect. 4.4.2.

Although the loads have been included as part of the disturbances, it may occur that some loads can be manipulated (curtailed or shifted in time). In that case, they must be included in the $u(t)$ vector, as will be done in Chap. 8.

In order to derive the complete model, Eq. (4.5) must be integrated into this matrix equation. This is equivalent to consider that not all the variables in vector $u(t)$ are independent and one of them (that will be chosen depending on the application) can be expressed as a linear combination of the others, as shown in (4.6). This will be illustrated in Sect. 4.4.2.

4.3.2 Problem Formulation

Once the model is available, the formulation of the MPC problem requires the definition of the cost function to be minimized and operational constraints to be imposed. For this basic EMS, the cost function can include costs associated to power exchanged by the units (amplitude and power rate limits) as well as the cost associated to storage reserve, that is, to keep storage at certain levels. It can be written as

$$J = \sum_{j=1}^{N_p} \|\hat{x}(t+j \mid t) - w(t+j)\|_R^2 + \sum_{j=1}^{N_c} \|u(t+j-1)\|_{Q_1}^2 + \sum_{j=1}^{N_c} \| \triangle u(t+j-1)\|_{Q_2}^2 \quad (4.12)$$

Notice that the first term of the summatory is included for setpoint tracking of reference values for storage devices and the second term is related to the cost of using each of the generation and storage units. The weighting matrix Q_1 will in general be diagonal, and their values are related to the priority of using a certain unit, either for efficiency or operating cost. For instance, in a microgrid with hydrogen storage, it is better to use batteries first if possible, since the hydrogen path efficiency is lower. Thus, the term associated to the battery in matrix Q_1 will be smaller than that associated to the fuel cell. The third term of Eq. (4.12) is used to limit the units' power rates (especially electrolyzer and fuel cell) in order to protect equipment from sudden changes in demanded power, which seriously affect their lifetime. In general, the setpoint tracking (first term) is not a tight condition, since the objective is to maintain storage levels around operational values, so the weighting matrix R will in general be smaller than the other weights.

This simple cost function will be augmented in the following chapters in order to consider other criteria: maintenance costs, durability, electrical tariffs, integration into the electrical market, etc. This can be done with the inclusion of logical variables, which turns the problem into a MILP.

Constraints

Additionally, operational constraints must be taken into account during the minimization. There are basically two types of constraints: those associated to physical limits of the units that cannot be trespassed and those related to operational limits that should not be exceeded. The first type includes the limited power that can be supplied by the units (batteries, fuel cells, electrolyzers, dispatchable generators, external grid, etc.). Those are physical thresholds that cannot be trespassed for constructive reasons. An upper limit exists for all the units but it is also common that a lower limit appears, meaning that once the unit is connected, it has to deliver a minimum power (for instance, a gas turbine has to supply at least a minimum amount of power in order not to stall). Therefore, these constraints apply on the power (variable $u(t)$) and also on the capacity of the storage units (maximum energy that can be stored in a battery or an ultracapacitor) and take the form:

$$\begin{aligned} P_{sto}^{\min} &\leq P_{sto}(t) \leq P_{sto}^{\max} & \forall t \\ P_{gen}^{\min} &\leq P_{gen}(t) \leq P_{gen}^{\max} & \forall t \\ P_{ext}^{\min} &\leq P_{ext}(t) \leq P_{ext}^{\max} & \forall t \\ SOC^{\min} &\leq SOC(t) \leq SOC^{\max} & \forall t \end{aligned}$$

Notice that the maximum and minimum values can be exactly the physical limits but a safety band can also be considered, avoiding to work very close to dangerous regions.

The second type of constraints is imposed to avoid sudden changes in the power supplied by the units. These are limits which affect the degradation of the units and will be important in expensive equipment such as fuel cells.

$$\begin{aligned} \triangle P_{sto}^{\min} &\leq \triangle P_{sto}(t) \leq \triangle P_{sto}^{\max} & \forall t \\ \triangle P_{gen}^{\min} &\leq \triangle P_{gen}(t) \leq \triangle P_{gen}^{\max} & \forall t \\ \triangle P_{ext}^{\min} &\leq \triangle P_{ext}(t) \leq \triangle P_{ext}^{\max} & \forall t \\ \triangle SOC^{\min} &\leq \triangle SOC(t) \leq \triangle SOC^{\max} & \forall t \end{aligned}$$

Notice that some of these constraints can be moved to the category of *soft constraints* if the inequalities are substituted by a weighted term in the cost function. That is the case of constraints on energy storage capacity.

During microgrid operation, the balance between energy generation and demand must always be met, so Eq. (4.5) must be added to the formulation as a equality constraint. Notice that this equality can be either considered in the minimization procedure as a constraint or included in the model, in the sense that one of the manipulated variables can be substituted by a linear combination of the others (as suggested in Eq. (4.6)). This will be done in the case study below.

Then the MPC is formulated as the optimization of a quadratic cost function with linear constraints:

$$\begin{aligned} \text{minimize} \quad & \frac{1}{2}\mathbf{u}^T\mathbf{A}\mathbf{u} + \mathbf{b}^T\mathbf{u} \\ \text{subject to:} \quad & \mathbf{R}\mathbf{u} \leq \mathbf{c} \end{aligned}$$

where all the involved variables are continuous, that is, $x(t)$, $u(t)$ and $d(t)$ are vectors of real numbers. Therefore, it can be solved by Quadratic Programming (QP) algorithms, for which fast and robust solvers exist.

4.4 Pilot-Scale Implementation

This section illustrates the design and implementation of the basic EMS presented above on a pilot-scale microgrid. The main concepts are demonstrated on an experimental utility under realistic scenarios, showing how MPC can be customized to a particular microgrid.

4.4.1 Plant Description

The plant under study is an experimental renewable-energy-based microgrid platform installed at the University of Seville, Spain, which is used to test control strategies applied to energy management. The microgrid is based on renewable energy sources and hydrogen storage [33, 34, 37]. A picture of the laboratory plant, called Hylab, is shown in Fig. 4.4, which displays its main components.

During the normal operation of the microgrid, the energy produced does not usually match the demand. Then, the excess of energy from renewable sources can be stored in the battery bank or used to produce hydrogen through electrolysis. When power from renewable sources is not available, a fuel cell may use hydrogen to supplement the lack of generation. Additionally, the microgrid has a connection to the main grid allowing energy purchase and sale. This hybrid storage (electricity and hydrogen) allows operation strategies on two different timescales: the battery can absorb/provide small amounts of energy on fast transients, while hydrogen storage supplements bigger oscillations. To replicate renewable energy systems, the microgrid has a programmable power supply that can emulate the dynamic behavior of photovoltaic field and a wind turbine. It also includes an electronic load to emulate different consumption profiles, a hydrogen storage system comprising a Proton Exchange Membrane (PEM) electrolyzer to produce hydrogen, a metal hydride tank

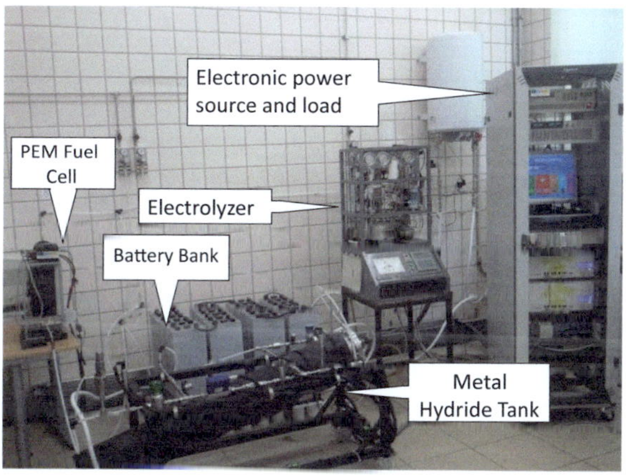

Fig. 4.4 Hylab microgrid

Table 4.1 Microgrid Units

Equipment	Nominal value
Programmable power supply	6 kW
Electronic load	2.5 kW
Electrolyzer	0.23 Nm^3/h at 1 kW
Metal hydride tank	7 Nm^3 at 5 bar
Fuel cell	1.5 kW at 20 Nl
Battery bank	$C_{max} = 17.6$ kWh

to store hydrogen and PEM fuel cell to produce electricity. Therefore, hardware-in-the-loop is used in combination with real electrolyzer, fuel cell, batteries, and hydrogen storage. The technical characteristics of the equipment are summarized in Table 4.1.

All this equipment is connected to a DC current bus with the necessary power electronics. To facilitate the understanding of the microgrid topology, a schematic representation of the system with electric and control signals is shown in Fig. 4.5.

In this figure, it can be seen that the grid power exchange is also electronically emulated. With regard to the control system, the microgrid has a dedicated control system based on a Programmable Logic Controller (PLC). This device performs the required

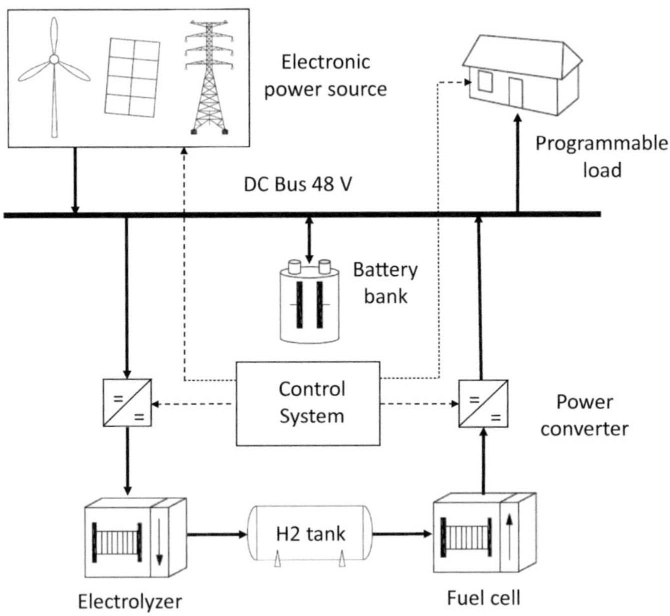

Fig. 4.5 Microgrid layout

calculations and determines the basic control actions (startup, shutdown, communications, etc.). Power supply and electronic load are controlled analogically, whereas the electrolyzer and fuel cell are controlled by means of the power converters and Controller Area Network (CAN) bus communications. As model-based controllers are usually high in computational demands, real-time implementation in a commercial PLC is troublesome. In order to overcome this issue, the MPC control actions are calculated using Simulink Real-Time software on a control computer installed in the plant. Using the MATLAB® OPC library, the computer sends these control commands to the PLC, which executes the orders. The MPC controller receives the plant outputs (SOC and LOH) to compute the optimal sequence of control actions.

The fuel cell and the electrolyzer units have their own local controllers, which execute the commands received from the upper layer. Thus, a compromise between fully centralized and fully decentralized control architectures is achieved by means of the hierarchical control architecture. Two DC/DC converters associated to the electrolyzer and fuel cell allow the DC bus to transfer power. In contrast, the lead–acid battery bank is plugged to the DC bus directly. Hence, bus voltage is held by the battery bank, simplifying the topology. This is a common option in DC microgrids in order to reduce costs and increase reliability, as any unbalance in the system is absorbed by the batteries [20]. A detailed description of the microgrid design and full characterizations of each subsystem can be found in [33].

Many control strategies have been tested on this microgrid. The results of using a heuristic controller are shown in [35], the design and evaluation of an MPC with degradation costs are described in [12], and the connection of electric vehicles to the microgrid is solved in [22] in the MPC framework. Stochastic [38] and Economic MPC [28] have also been tested in this benchmark. A performance comparison of some of these methods can be found in [36].

In this section, the basic EMS using MPC is designed and tested, first in simulation and later in the real plant.

4.4.2 Control-Oriented Model

A control-oriented linear model is used to design the MPC. Using Eq. (4.1) for the battery and the hydrogen storage, a state-space model can be derived. Therefore, the state vector is $x(t) = [SOC(t)\ LOH(t)]^T$, where SOC is the state of charge of the battery and LOH is the level of hydrogen in the hydride tanks. In order to avoid the use of binary variables, a fixed value of the efficiency of the battery will be used here.

$$SOC(t+1) = SOC(t) - \frac{\eta_{bat} T_s}{C_{\max}} P_{bat}(t) \tag{4.13}$$

$$LOH(t+1) = LOH(t) + \frac{\eta_{elz} T_s}{V_{\max}} P_{elz}(t) - \frac{T_s}{\eta_{fc} V_{\max}} P_{fc}(t) \tag{4.14}$$

P_{bat} is the power supplied by the battery and $V_{\max}$ is the maximum volume of H_2 (in Normal cubic meters) that can be stored in the tanks.

The manipulated variables are the power that can be exchanged with the fuel cell (P_{fc}), the electrolyzer (P_{elz}), and the grid (P_{grid}), with the direction of power flows as shown in Fig. 4.3. Notice that, as said above, the battery is directly connected to the DC bus and absorbs the unbalance, so P_{bat} must compensate the rest of powers in the DC bus:

$$P_{bat}(t) = P_{load}(t) + P_{elz}(t) - P_{fc}(t) - P_{grid}(t) - P_{gen}(t) \tag{4.15}$$

So, the storage equations, defining $d(t) = P_{gen}(t) - P_{load}(t)$ as the measurable disturbance, are

$$SOC(t+1) = SOC(t) - \frac{\eta_{bat} T_s}{C_{\max}}(P_{elz}(t) - P_{fc}(t) - P_{grid}(t) - d(t)) \tag{4.16}$$

$$LOH(t+1) = LOH(t) + \frac{\eta_{elz} T_s}{V_{\max}} P_{elz}(t) - \frac{T_s}{\eta_{fc} V_{\max}} P_{fc}(t) \tag{4.17}$$

The values of the conversion from charging power to electrical and hydrogen storage are obtained performing experiments on the plant. A set of tests were carried out for different operating points: SOC and LOH between 10 and 90% and charging and discharging powers between 500 and 1750 W. The mean value obtained for the conversion coefficient of the battery was $K_{bat} = \frac{\eta_{bat}}{C_{\max}} = 1.56 \times 10^{-3}$. In the case of hydrogen, the mean values were $K_{elz} = \frac{\eta_{elz}}{V_{\max}} = 3.216 \times 10^{-3}$ for charging (electrolyzer) and $K_{fc} = \frac{1}{\eta_{fc} V_{\max}} = 8.116 \times 10^{-3}$ for discharging (fuel cell).

4.4.3 Controller Design

The primary goal of this basic EMS is to ensure stable delivery of electrical power to its local load consumers. In addition to this, it encompasses performance optimization and prevents equipment damage. The proposed control aims to fulfill the objectives described in Sect. 4.3. The EMS intends to fulfill the following goals: (i) to extend the battery lifetime by preventing deep discharging and overcharging, (ii) to protect fuel cell and electrolyzer from intensive use by limiting their power rates, (iii) to take into account energy efficiency in the plant by using the most efficient storage when possible, and (iv) to achieve a high degree of autonomy and reduce operation costs by minimizing the energy exchanged with the grid. The priority in the fulfillment of these objectives can be established by their weights in the cost function.

Cost Function

The goal of this multi-objective optimization problem is to accomplish an optimal solution for several objectives, so the result will be a compromise among the objectives. Consequently, the solution is a Pareto front where the microgrid will reach a state where no objective can be improved without sacrificing at least another. Therefore, one of the DERs will have to cope with the sudden power changes in order to protect the rest from intensive use. This role can be assigned by modifying the weights in the cost function, which can be adapted to the operating conditions, as done in [35].

The cost function can include terms that take into account the values of the different powers involved (related to the cost of using each DER) and also the power rates (related to their lifetimes). It can also penalize the deviation of the stored energy from a desired operation point. A cost function of the form (4.12) can be customized to this case study:

$$J = \sum_{k=1}^{N_c} \alpha_1 P_{fc}^2(t+k) + \alpha_2 P_{elz}^2(t+k) + \alpha_3 P_{grid}^2(t+k) + \alpha_4 P_{bat}^2(t+k) + \\ + \beta_1 \Delta P_{fc}^2(t+k) + \beta_2 \Delta P_{elz}^2(t+k) + \beta_3 \Delta P_{grid}^2(t+k) + \beta_4 \Delta P_{bat}^2(t+k) + \\ + \sum_{k=1}^{N_p} \gamma_1 (SOC(t+k) - SOC_{ref})^2 + \gamma_2 (LOH(t+k) - LOH_{ref})^2 \quad (4.18)$$

In this cost function, the first four terms weigh the usage of the manipulated variables, the following four terms penalize their rates and the last ones help to keep the stored energy around an operation point. A quadratic cost function has been chosen but a linear one could also be used. In this microgrid, the battery bank is directly connected to the DC bus, so P_{bat} is not a manipulated variable (it is not part of the u vector). However, it can be expressed as a linear combination of the other manipulated variables P_{grid}, P_{fc}, and P_{elz} and d.

Different plant behaviors will be obtained depending on the choice of the weighting factors. These weights can be computed based on available data of capital and operation and management (O&M) costs (as done in the next chapter) or in a qualitative manner (as done here). This way, if the variable associated to weight α_i must be used prior to the one associated to α_j then α_i must be greater than α_j. If the equipment associated to β_i must be protected from intensive use more than the one associated to β_j then the choice is $\beta_i > \beta_j$. In general, the values of γ_i will be much smaller than the other weights (or even null) since reference tracking for stored energy is not a significant objective. Some simulations with different choices for the weights will be shown below.

Constraints

The microgrid components have physical and operational constraints that must be considered in the optimization problem. Equipment constraints in terms of power limits and power rates are included in order to optimize efficiency, lifetime, and O&M cost. The battery bank must operate in a range of SOC values in order to avoid over and undercharging that remarkably reduce the number of admissible cycles. Although the power exchanged by the battery (P_{bat}) is not a manipulated variable, its constraints can be expressed in terms of the others, as shown in (4.6).

Table 4.2 Constraints

Variable	Power (W)	Power rate (W/s)	State of Charge (%)
Battery	0–2500	Unconstrained	40–75
Fuel Cell	100–1200	20	–
Electrolyzer	100–900	20	–
H_2 storage	–	–	10–90
Grid	0–2500	Unconstrained	–
Generation	0–6000	Unconstrained	–

The electrolyzer is not designed to operate at fluctuating conditions and it must be operated above a minimum power threshold in order to avoid impurities in the gases and hazardous mixtures. In addition, high current densities may produce overvoltage in the electrodes, accelerating its degradation. In the case of fuel cells, severe load cycling leads to performance reduction and membrane deterioration (see [35] for detailed explanation). Regarding the metal hydride tank, although it is not damaged by deep discharge, it must maintain a minimum delivering pressure, which corresponds to a minimum value of LOH. Upper bound is imposed for safety reasons. For the grid power, it is limited by the PCC connection and, in this laboratory plant, the power rate is considered to be unlimited.

These constraints can be quantified as shown in Table 4.2. Note that some of them are physical limits (e.g., power supplied by the generator or the fuel cell), while others are limits imposed for a safe operation (e.g., power rate requested to the fuel cell).

4.4.4 Case Study 1

This case study considers the operation of the microgrid using the lead–acid battery, the electrolyzer, the fuel cell, and the external grid. The EMS can be tuned using the microgrid simulator Simμgrid. This modular simulator can be used to design the control strategy before its implementation on the real plant. The use of a simulator allows the comparison of different controllers in the same scenarios. In this section, the EMS will be tested in different weather conditions (that affect renewable generation) and with different modifications of the MPC, first under simulation and later on the real microgrid.

In this configuration with hydrogen storage, P_{fc} and P_{elz} always take positive values and both variables cannot be different from zero at the same time instant, so they are complementary variables. In order to avoid the consideration of this constraint in the model and for the sake of simplicity, their effect can be condensed in only one variable $P_{H_2} = P_{fc} - P_{elz}$, which is the net hydrogen storage power.

This approximation is used in [28, 38]. So the vector of manipulated variables is $u(t) = [P_{H_2}(t)\ P_{grid}(t)]^T$.

In order to obtain the state-space matrices for this case, the hydrogen conversion efficiency is obtained as the mean value between charging (electrolyzer) and discharging (fuel cell) of Eq. (4.16), $K_{H_2} = -5.66 \times 10^{-3}$. Notice that this is a simplified model: the plant is assumed to be linear, and the same efficiency is considered for charge and discharge of electrical and hydrogen storage. More complex models will be used in the following chapters.

Therefore, considering these simplifications, the following control-oriented model will be used for this case study. If a sampling time T_s of 1 second is used, which allows fast response to sudden disturbances in both demand and renewable production, the model is given by[1]

$$\begin{bmatrix} SOC(t+1) \\ LOH(t+1) \end{bmatrix} = \begin{bmatrix} SOC(t) \\ LOH(t) \end{bmatrix} + \begin{bmatrix} 1.56 \times 10^{-3} & 1.56 \times 10^{-3} \\ -5.66 \times 10^{-3} & 0 \end{bmatrix} \begin{bmatrix} P_{H_2}(t) \\ P_{grid}(t) \end{bmatrix} + \\ + \begin{bmatrix} 1.56 \times 10^{-3} \\ 0 \end{bmatrix} d(t)$$

And consequently the system matrices are given by

$$A = I, \quad B = \begin{bmatrix} 1.56 & 1.56 \\ -5.66 & 0 \end{bmatrix} \times 10^{-3}, \quad B_d = \begin{bmatrix} 1.56 \times 10^{-3} \\ 0 \end{bmatrix}, \quad C = I \qquad (4.19)$$

The cost function is a slight variation of (4.18), where P_{H_2} is used and P_{bat} is neglected:

$$J = \sum_{k=1}^{N_c} \alpha_1 P_{H_2}^2(t+k) + \alpha_2 P_{grid}^2(t+k) + \beta_1 \Delta P_{H_2}^2(t+k) + \beta_2 \Delta P_{grid}^2(t+k) + \\ + \sum_{k=1}^{N_p} \gamma_1 (SOC(t+k) - SOC_{ref})^2 + \gamma_2 (LOH(t+k) - LOH_{ref})^2 \qquad (4.20)$$

Since setpoint tracking is not a substantial issue, very small values ($\gamma_i = 10^{-8}$) have been chosen for their associates weights. The other ones are $\alpha_1 = 5 \times 10^{-3}$, $\alpha_2 = 8 \times 10^{-3}$, $\beta_1 = 4$, and $\beta_2 = 10^{-3}$. The choice of these weights (in particular $\alpha_1 < \alpha_2$) encourages the use of hydrogen (which is produced locally) versus buying energy to the grid. In this example, P_{bat} has not been included in the cost function. The chosen horizons are $N_p = 10$ and $N_c = 2$.

[1] In case that another sampling time is used, the elements of these matrices are computed multiplying these values by the new sampling time (in seconds).

Different Generation Scenarios

In order to illustrate the theoretical background, simulated and experimental tests were carried out to study the controller behavior under different external conditions (weather and demand changes). Two types of renewable sources were considered and studied separately: a photovoltaic (PV) array and a wind turbine. The demand profile was taken from a typical household daily pattern on a weekday [13] and adapted to the power levels of this laboratory microgrid, and the irradiance data were gathered from a local meteorological station. Before testing the controller on the real plant, the performance should be demonstrated in simulations. Different generation profiles can be emulated with the programmable energy source, as is the case of wind data.

- **Sunny day**: This generation profile corresponds to a sunny day, with high irradiance during the central hours of the day, having excess of energy then and deficit during the night. The EMS manages both storage units (battery and hydrogen) in order to supply the demand. It can be observed in Fig. 4.6 that during the night the battery is used to fulfill the demand, until there is an excess of energy. Then the battery starts charging and, since there is still a surplus of energy, it is stored in the form of hydrogen using the electrolyzer and even selling energy to the grid. When PV generation cannot supply the demand, the battery is used again until depleted and later the fuel cell starts providing energy with a modest contribution of the grid. Notice that SOC and LOH evolve almost freely between their operational limits since the weights utilized in the cost function for reference tracking are small.
- **Cloudy day**: In this scenario, the PV generation cannot supply the demand during the major part of the day (the evolution of the net power is below zero most of the time, as shown in Fig. 4.7). This deficit must be supplied by the available energy resources: battery, fuel cell, and grid. The EMS decides to use first the battery and later (around $t = 12$ h) the fuel cell (in a smooth way) supported by the grid. So, during the second half of the day, when the battery has reached its minimum SOC value, the load is fed by the fuel cell and the external grid. Notice that the electrolyzer is always off since there is no excess of energy to store.
- **Wind turbine generation**: In this scenario, the renewable source considered is a wind turbine, which generates an excess of power in the microgrid. Therefore, energy is stored during most of the day: the electrolyzer is working for almost all day and some excess of energy is sold to the grid, as shown in Fig. 4.8. The battery also stores energy but it soon gets filled up (from $t = 3$ h to 19 h), only injecting power to the bus a couple of times in that period of time. The fuel cell is not switched on.

These simulations show how the MPC can adapt to different scenarios, providing a good solution to power sharing among the DERs and taking into account the operational constraints and the optimization of the imposed operational criteria. Different results in terms of power distribution can be obtained for different weights associated to each DER.

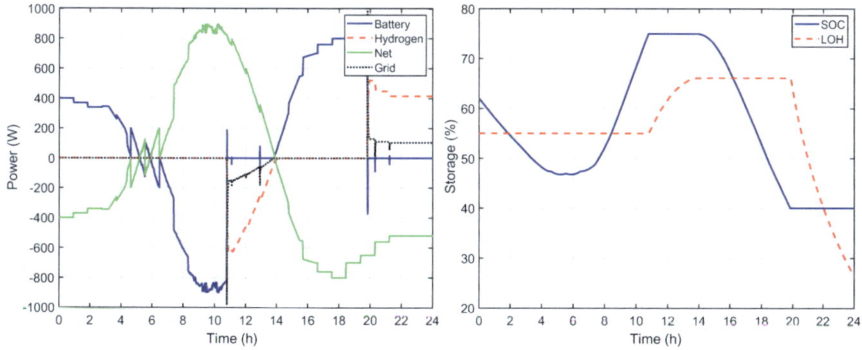

Fig. 4.6 Power flows and storage for a sunny day. The net power is the difference between generation and load

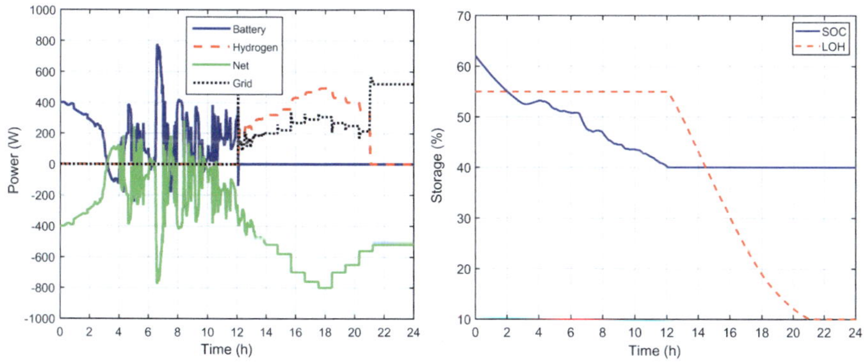

Fig. 4.7 Power flows and storage for a cloudy day

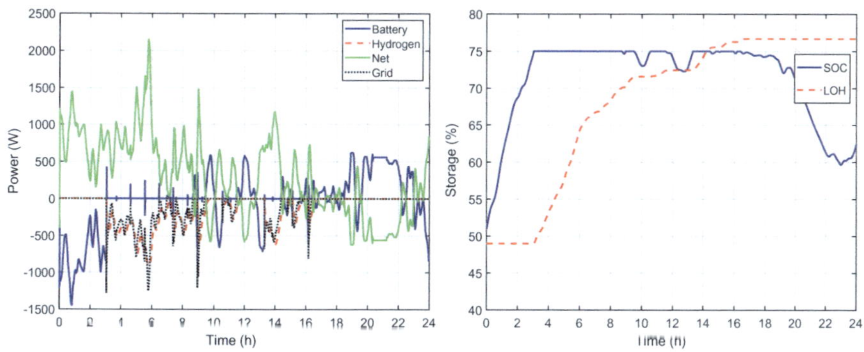

Fig. 4.8 Power flows and storage for a windy day

Experimental Results

Once the MPC has been tuned using Simμgrid, it is tested on the laboratory microgrid, obtaining similar results in the three scenarios.

- **Sunny day**: It can be observed in Fig. 4.9 that the electrolyzer was triggered when the battery SOC reached 75%, due to the excess of energy since the irradiance remained very high. Therefore, the energy surplus had to be stored in form of hydrogen. The electrolyzer power consumption was increased gradually as shown in the figure. Notice that in the first moments of the electrolyzer operation, part of the surplus of energy was exported to the grid and gradually decreased as the electrolyzer consumed more power. The fuel cell operation followed a similar pattern. It was switched on when the battery SOC reached the lower threshold (40%). The grid assumed the transient power required by the load. At the end of the day, the grid and the fuel cell shared the demand according to their weights in the cost function.
  The power sharing among battery, electrolyzer, fuel cell, and external grid is determined by the weights used in the cost function.
- **Cloudy day**: In the experiment results shown in Fig. 4.10, it can be observed an energy deficit during most of the experiment, which was supplied in different ways. In the first stage, when strong power fluctuations were present, the control determined that the cost of using fuel cell power is too expensive in techno-economic terms, due to the high power fluctuations. Thus, the controller used the grid to satisfy the demand. In contrast, over the second stage of the experiment, the fluctuations of the cloudiness disappeared. Then, the control decided to use the fuel cell as the main source to cover the demand. Notice that the electrolyzer was always off since there is no excess of energy to store. It is therefore confirmed that the control worked properly according to the design.
- **Wind turbine generation**: The wind turbine produced a significant power fluctuation as can be observed in Fig. 4.11. In this experiment, a predominant excess power motivated the electrolyzer to operate during most of the day. It should be noted that, despite the high variability of the wind power, the power rate constraints included in the controller design induced a smooth operation of the electrolyzer, whose behavior was therefore very satisfactory. Since there is a surplus of energy during most of the day, there is no need to switch the fuel cell on. Consequently, it is not subject to an intensive use that would highly reduce its lifespan. The MPC controller changed the set points gradually, according to the optimum calculated by the cost function.

These experiments show that MPC is able to effectively manage the energy in the system using the electrolyzer, batteries, and grid power. The results agree with those obtained by simulation, even when the generation profiles are not exactly the same. It is important to point out that the MPC performance has been evaluated in a wide range of operating conditions, varying from low to top level of battery SOC thresholds and different generation profiles, and the performance is successful.

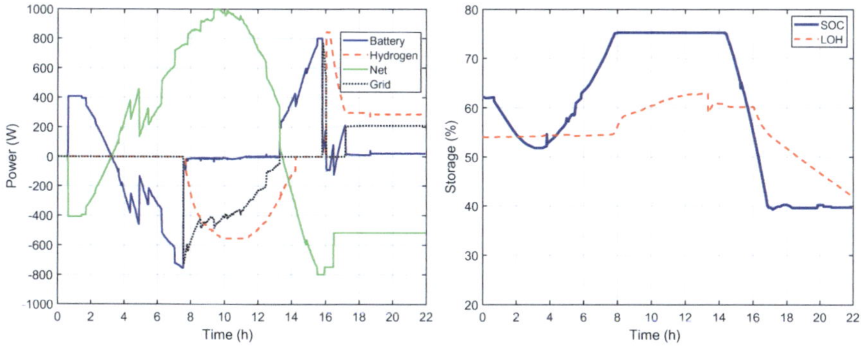

Fig. 4.9 Sunny day: Experimental results

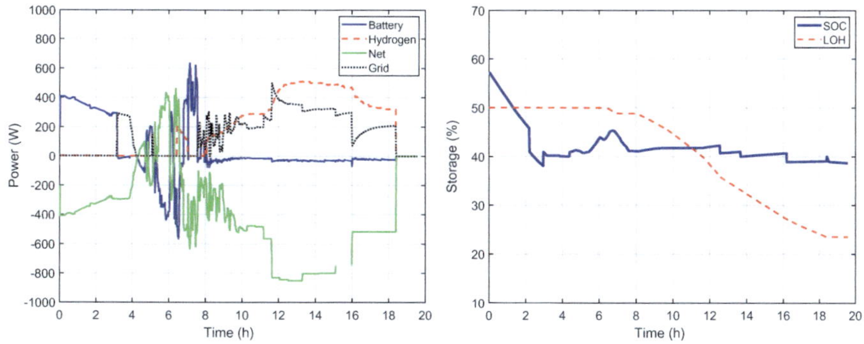

Fig. 4.10 Cloudy day: Experimental results

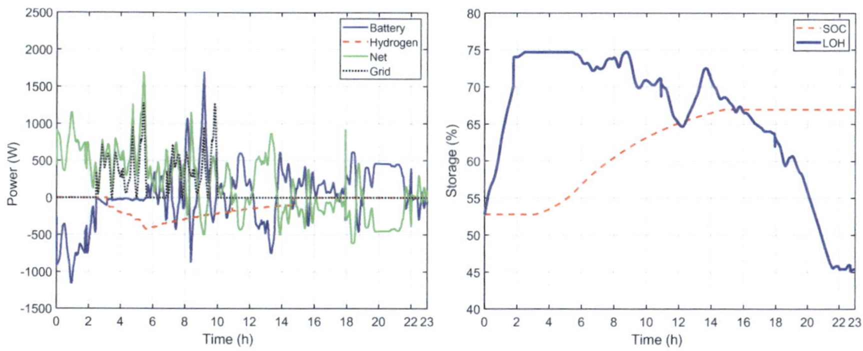

Fig. 4.11 Windy day: Experimental results

The results obtained from the plant operation under the proposed MPC can be compared against the SOC-based Hysteresis Band Control (HBC) technique presented in Sect. 4.2.1, which is a simple strategy widely used in small hydrogen microgrids. A detailed comparison is presented in [35] for the partly cloudy day experiment shown above. That study shows that MPC can improve the behavior obtained with HBC by reducing the number of startups and shutdowns of the fuel cell and electrolyzer. Besides, the smooth power reference computed by MPC avoids that the fuel cell (or electrolyzer) absorbs directly solar (or wind) fluctuations. By preventing the intermittent and jerky operation of the HBC method, MPC can lower the operational cost by 30% with respect to HBC.

This comparison opens the door to consider capital and O&M cost in the optimization problem. If cost can be quantified in terms of the control actions, it can not only be an indicator of the economic benefit, but also it can provide valuable information for system diagnosis and fault detection. In the next chapter, new binary variables will be introduced in the model and the optimization in order to deal with operation cost and degradation issues in a more systematic way.

Effect of Disturbances Prediction

The power generated by the RESs and the demanded power are the main disturbances acting on the microgrid. Although these variables cannot be manipulated by the controller (except in the case of demand response that will be addressed in Chap. 6), the available information (current measurement and future estimation) can be used by MPC to predict the system output along the horizon. This section illustrates how MPC can integrate knowledge of disturbances to anticipate their effect and improve microgrid performance.

Two situations are tested here: (i) the controller has no information about the future evolution of disturbances and (ii) the prediction of the disturbances is perfect (this is an ideal case which provides the best reachable performance, to be used for comparison). The first situation is the most common in MPC: since there is no information about the future, the most reasonable assumption is to consider that the disturbance will be the same along the horizon, as done in the previous examples. But, as detailed in Sect. 2.4, if knowledge of the future evolution of the disturbance is available, it can be integrated into the MPC formulation. In this case study, the disturbance is given by the net power, that is, the difference between generation and demand: $d(t) = P_{gen}(t) - P_{dem}(t)$, which can be measured at the current time instant t. Then Eq. (2.9) can be used for prediction.

In this case, a sampling time of $T_s = 60$ s is used and the model is given by

$$\begin{bmatrix} SOC(t+1) \\ LOH(t+1) \end{bmatrix} = \begin{bmatrix} SOC(t) \\ LOH(t) \end{bmatrix} + \begin{bmatrix} 0.0936 & 0.0936 \\ -0.339 & 0 \end{bmatrix} \begin{bmatrix} P_{H_2}(t) \\ P_{grid}(t) \end{bmatrix} + \begin{bmatrix} 0.0936 \\ 0 \end{bmatrix} d(t)$$

The same cost function as in the previous experiments is used, with the weights given by $\gamma_i = 10^{-8}$, $\alpha_1 = 5 \times 10^{-3}$, $\alpha_2 = 0.02$, $\beta_1 = 40$, and $\beta_2 = 10$. The chosen horizons are $N_p = 60$ (long enough to appreciate the effect of disturbance prediction) and $N_c = 2$.

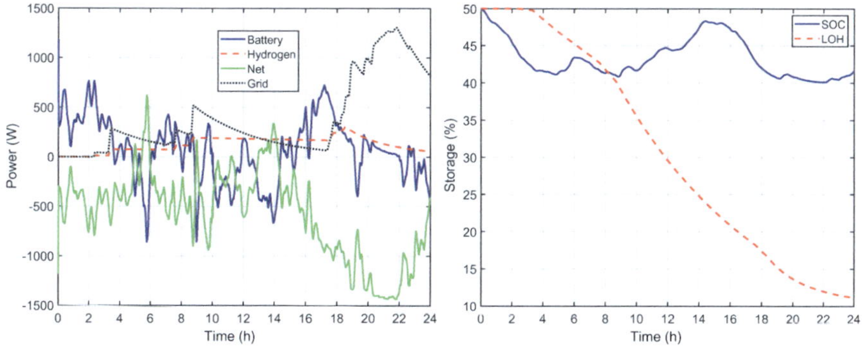

Fig. 4.12 Power flows and storage levels for constant disturbance prediction

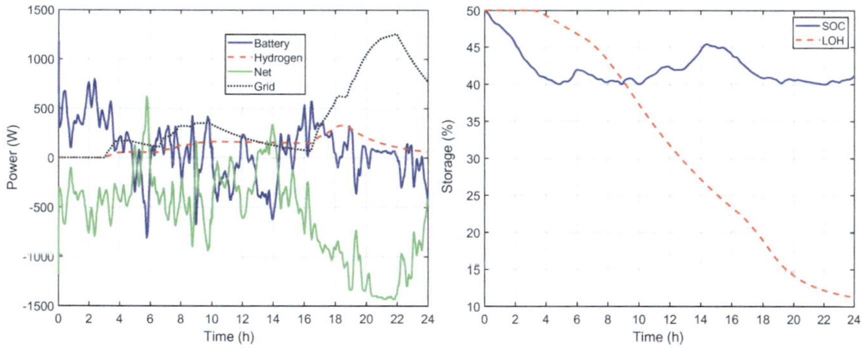

Fig. 4.13 Power flows and storage levels for perfect disturbance prediction

The following tests are done in simulation in order to compare both situations. Figure 4.12 shows the results of using constants predictions along the horizon. The disturbance is measured in the current instant and is kept constant during the minimization process. Good results are obtained, similar to the previous section.

On the other hand, Fig. 4.13 displays the power flows when future disturbances are known and included in the computation of the free response. Although the flows are similar, it can be visualized that the evolution of the manipulated variables P_{H_2} and P_{grid} is smoother, since they anticipate the progression of the disturbance. This is useful to avoid degradation and extend lifetime of the components of the microgrid.

This improvement can be quantified by evaluating the cost function for the whole experiment in both cases. In the first one, the cost function takes the value $J = 9.215$ and in the second case $J = 8.566$, which means a 7% improvement. Similar increases can be obtained in other situations. Therefore, this example has illustrated how the prediction capabilities of MPC can improve the operation of microgrids, provided good forecasting is available.

4.4.5 Case Study 2

Another case study can be considered in order to show other example of the basic EMS in microgrids. In this case, a new lithium–ion battery bank is added to the laboratory microgrid. Therefore, a new MPC must be devised for this configuration. In order to show an example of dispatchable generators, the fuel cell is used as a DG, with a cost associated to the use of hydrogen (which is not generated in the microgrid). The microgrid is composed of a PV plant, two different kinds of batteries and the fuel cell, as shown in Fig. 4.14.

The power exchanged with the DC bus by this Li-ion battery can be adjusted using its DC/DC converter, so a new manipulated variable P_{bat2} appears. A new state variable is included, $SOC_2(t)$, corresponding to the new Li-ion battery, so the state vector is given by $x(t) = [SOC_1(t)\ SOC_2(t)\ LOH(t)]^T$ and the manipulated variables are $u(t) = [P_{fc}(t)\ P_{grid}(t)\ P_{bat2}(t)]^T$. The disturbance is the same as in the previous case: $d(t) = P_{gen}(t) - P_{load}(t)$. The SOC of this battery is constrained between 30 and 80%, with the maximum charge/discharge power limited to 2950 W.

In this situation, the control-oriented model given by (4.16) results in

$$SOC_1(t+1) = SOC_1(t) - \frac{\eta_{bat1} T_s}{C_{1\max}}(-P_{fc}(t) - P_{grid}(t) - d(t)) \quad (4.21)$$

$$LOH(t+1) = LOH(t) - \frac{T_s}{\eta_{fc} V_{\max}} P_{fc}(t) \quad (4.22)$$

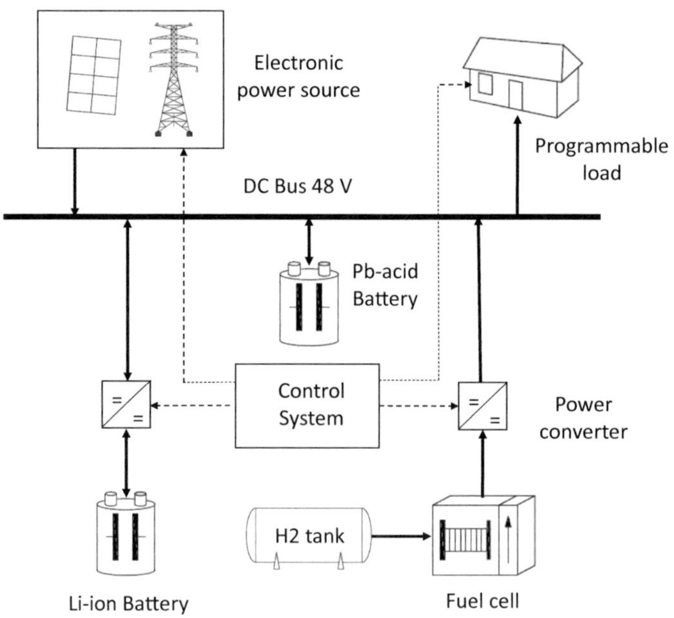

Fig. 4.14 Microgrid layout for case study 2

$$SOC_2(t+1) = SOC_2(t) - \frac{\eta_{bat2} T_s}{C_{2\max}} P_{bat2}(t) \tag{4.23}$$

The conversion coefficient of the Li-ion battery, obtained experimentally, is $K_{bat2} = \frac{\eta_{bat2}}{C_{2\max}} = 1.254 \times 10^{-3}$ and $K_{fc} = \frac{1}{\eta_{fc} V_{\max}} = 7.181 \times 10^{-3}$. Then, for a sampling time of $T_s = 30s$, the model in matrix form is

$$\begin{bmatrix} SOC_1(t+1) \\ LOH(t+1) \\ SOC_2(t+1) \end{bmatrix} = \begin{bmatrix} SOC_1(t) \\ LOH(t) \\ SOC_2(t) \end{bmatrix} + \begin{bmatrix} 0.0468 & 0.0468 & 0.0468 \\ -0.215 & 0 & 0 \\ 0 & 0 & -0.0376 \end{bmatrix} \begin{bmatrix} P_{fc}(t) \\ P_{grid}(t) \\ P_{bat2}(t) \end{bmatrix} + \\ + \begin{bmatrix} 0.0468 \\ 0 \\ 0 \end{bmatrix} d(t) \tag{4.24}$$

The cost function has the form given by Eq. (4.18), with $\gamma_i = 10^{-8}$ (very small values in order to neglect reference tracking for the batteries) $\alpha_1 = 5 \times 10^{-3}$, $\alpha_2 = 0.2$, $\alpha_3 = 5 \times 10^{-4}$, and $\alpha_4 = 1$. This big value of α_4 has been chosen so as to impose that the lead–acid battery is mainly used to keep the DC bus at its operation voltage and does not contribute to supply the demand. This way, the major effort is done by the lithium–ion battery. The increments in power are weighted by $\beta_1 = 0.4$, $\beta_2 = 10^{-3}$, $\beta_3 - 10^{-5}$, and $\beta_4 = 1$. The chosen horizons are $N_p = 10$ and $N_c = 2$.

An experiment is performed with this configuration for a sunny day. The results displayed in Fig. 4.15 indicate that the three DERs operate in a coordinated way to supply the demand along the day. Since the fuel cell consumes hydrogen, it is switched off during most of the day and it only works at nighttime ($t > 18$ h) when the energy stored in the Li-ion batteries is not enough to fulfill the load (notice that it reaches its lower limit that has been set at 30%). Notice that, in this experiment, the lead–acid battery is almost not employed. This could be easily changed by modifying the weights α and β in the cost function.

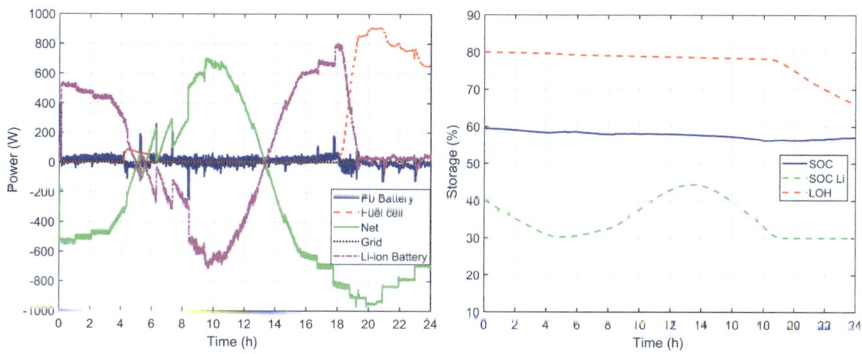

Fig. 4.15 Power flows and storage levels

This pilot-scale implementation has demonstrated how an EMS based on a basic MPC algorithm can be successfully applied to a microgrid. Different scenarios and configurations have been used, showing the flexibility and applicability of MPC. In the next chapter, this basic EMS will be extended with the consideration of electricity tariffs, degradation costs, and participation in the electrical market.

References

1. Amjady N, Keynia F, Zareipour H (2010) Short-term load forecast of microgrids by a new bilevel prediction strategy. IEEE Trans Smart Grid 1(3):286–294
2. Arul PG, Ramachandaramurthy VK, Rajkumar RK (2015) Control strategies for a hybrid renewable energy system: A review. Renew Sustain Energy Rev 42:597–608
3. Bazaraa MS, Sherali HD, Shetty CM (2013) Nonlinear programming: theory and algorithms. Wiley
4. Bella AL, Cominesi SR, Sandroni C, Scattolini R (2017) Hierarchical predictive control of microgrids in islanded operation. IEEE Trans Autom Sci Eng 14(2):536–546
5. Bemporad A, Morari M (1999) Control of systems integrating logic, dynamics, and constraints. Automatica 35(3):407–427
6. Bemporad A, Patrinos P (2012) Simple and certifiable quadratic programming algorithms for embedded linear model predictive control. IFAC Proc 45(17):14–20. 4th IFAC Conference on Nonlinear Model Predictive Control
7. Bevrani H, François B, Ise T (2017) Microgrid dynamics and control. Wiley
8. Cominesi SR, Farina M, Giulioni L, Picasso B, Scattolini R (2018) A two-layer stochastic model predictive control scheme for microgrids. IEEE Trans Control Syst Technol 26(1):1–13
9. Dimitri P (2005) Bertsekas Dynamic programming and optimal control. Athena Scientific Belmont, MA
10. Dufo-López R, Bernal-Agustn JL, Contreras J (2007) Optimization of control strategies for stand-alone renewable energy systems with hydrogen storage. Renew Energy 32(7):1102–1126
11. Gamez Urias ME, Sanchez EN, Ricalde LJ (2015) Electrical microgrid optimization via a new recurrent neural network. IEEE Syst J 9(3):945–953
12. Garcia-Torres F, Valverde L, Bordons C (2016) Optimal load sharing of hydrogen-based microgrids with hybrid storage using model predictive control. Ind Electron IEEE Trans 63(8):4919–4928
13. Ghaemi S, Brauner G (2009) User behavior and patterns of electricity use for energy saving. In: Proceedings 6th international energiewirtschaftstagung TU Wien, Vienna, Austria
14. Glover F, Greenberg HJ (1989) New approaches for heuristic search: a bilateral linkage with artificial intelligence. Eur J Oper Res 39(2):119–130
15. Goldberg DE (1989) Genetic algorithms in search, optimization and machine learning. Addison-Wesley Longman Publishing
16. Hiyama T, Bevrani H, François B, Ise T (2016) Intelligent automatic generation control. CRC Press
17. Ipsakis D, Voutetakis S, Seferlis P, Stergiopoulos F, Papadopoulou S, Elmasides C (2008) The effect of the hysteresis band on power management strategies in a stand-alone power system. Energy 33(10):1537–1550
18. Ipsakis D, Voutetakis S, Seferlis P, Stergiopoulos F, Elmasides C (2009) Power management strategies for a stand-alone power system using renewable energy sources and hydrogen storage. Int J Hydrog Energy 34(16):7081–7095
19. Kennedy J (2011) Particle swarm optimization. In: Encyclopedia of machine learning. Springer, pp 760–766

20. Little M, Thomson M, Infield D (2007) Electrical integration of renewable energy into stand-alone power supplies incorporating hydrogen storage. Int J Hydrog Energy 32(10):1582–1588. EHEC2005
21. Luenberger DG, Ye Y Linear and nonlinear programming. Springer
22. Mendes PRC, Valverde L, Bordons C, Normey-Rico JE (2016) Energy management of an experimental microgrid coupled to a v2g system. J Power Sour 327:702–713
23. Mendes PRC, Maestre JM, Bordons C, Normey-Rico JE (2017) A practical approach for hybrid distributed mpc. J Process Control 55:30–41
24. Negenborn RR, Houwing M, Schutter BD, Hellendoorn J (2009) Model predictive control for residential energy resources using a mixed-logical dynamic model. In: International conference on networking, sensing and control, pp 702–707
25. Nunez-Reyes A, Marcos D, Bordons C, Ridao MA (2017) Optimal scheduling of grid-connected PV plants with energy storage for integration in the electricity market. Solar Energy 144(1):502–516
26. Palma-Behnke R, Benavides C, Lanas F, Severino B, Reyes L, Llanos J, Sáez D (2013) A microgrid energy management system based on the rolling horizon strategy. IEEE Trans Smart Grid 4(2):996–1006
27. Parisio A, Rikos E, Glielmo L (2014) A model predictive control approach to microgrid operation optimization. IEEE Trans Control Syst Technol 22(5):1813–1827
28. Pereira M, Limon D, Muñoz de la Peña D, Valverde L, Alamo T (2015) Periodic economic control of a nonisolated microgrid. IEEE Trans Ind Electron 62(8):5247–5255
29. Pourmousavi SA, Nehrir M, Colson CM, Wang C (2010) Real-time energy management of a stand-alone hybrid wind-microturbine energy system using particle swarm optimization. IEEE Trans Sustain Energy 1(3):193–201
30. Serna A, Tadeo F, Torrijos D (2015) Rule-based control of off-grid desalination powered by renewable energies. J Renew Energy Sustain Dev (RESD), 205–213
31. Takeuchi A, Hayashi T, Nozaki Y, Shimakage T (2012) Optimal scheduling using metaheuristics for energy networks. IEEE Trans Smart Grid 3(2):968–974
32. Valencia F, Collado J, Sáez D, Marín LG (2016) Robust energy management system for a microgrid based on a fuzzy prediction interval model. IEEE Trans Smart Grid 7(3):1486–1494
33. Valverde L, Rosa F, Bordons C (2013) Design, planning and management of a hydrogen-based microgrid. IEEE Trans Ind Inf 9(3):1398–1404
34. Valverde L, Rosa F, Del Real A, Arce A, Bordons C (2013) Modeling, simulation and experimental set-up of a renewable hydrogen-based domestic microgrid. Int J Hydrog Energy 38(27):11672–11684
35. Valverde L, Bordons C, Rosa F (2016) Integration of fuel cell technologies in renewable-energy-based microgrids optimizing operational costs and durability. IEEE Trans Ind Electron 63(1):167–177
36. Valverde L, Rosa F, Bordons C, Guerra J (2016) Energy management strategies in hydrogen smart-grids: a laboratory experience. Int J Hydrog Energy 41(31):13715–13725
37. Valverde L, Bordons C, Rosa F (2012) Power management using model predictive control in a hydrogen-based microgrid. In: Annual conference on IEEE industrial electronics society, pp 5669–5676
38. Velarde P, Valverde L, Maestre JM, Ocampo-Martinez C, Bordons C (2017) On the comparison of stochastic model predictive control strategies applied to a hydrogen-based microgrid. J Power Sour 343:161–173
39. Xiaoping L, Ming D, Jianghong H, Pingping H, Yali P (2010) Dynamic economic dispatch for microgrids including battery energy storage. In: 2010 2nd IEEE international symposium on power electronics for distributed generation systems (PEDG). IEEE, pp 914–917
40. Zadeh LA (1965) Fuzzy sets. Inf Control 8(3):338–353

Chapter 5
Energy Management with Economic and Operation Criteria

Abstract In this chapter, the basic Energy Management System (EMS) presented in the previous chapter is extended to consider operational and degradation costs. The chapter introduces a formulation to integrate the terms related to operational and degradation issues associated to hybrid storage systems in an MPC-based EMS. The participation of microgrids in the different stages of the electrical markets is described. First, a two-step MPC-based algorithm corresponding to the tertiary and secondary control level of the microgrid it is developed. Later on, based on the stages of the electrical markets, MPC controllers are proposed to follow the operation rules in the day-ahead, intraday, and real-time markets interacting with both the market operator and the system operator. The proposed formulation optimizes the final cost of the energy consumption in the microgrid by means of improving its participation in the different stages of the market but also minimizing the degradation issues, as well as the operational costs of hybrid energy storage systems. The formulation requires to deal with both logic and continuous variables. For this reason, the different operation modes in the microgrid are modeled with the Mixed Logic Dynamical (MLD) framework and the MPC controller is formulated as a Mixed-Integer Quadratic Programming (MIQP) problem. Different results based on simulation and experiments are exposed.

5.1 Economic and Operation Issues in EMS of Microgrids

The primary goal of the Energy Management System (EMS) for microgrids is to ensure the delivery of electrical power to its local load consumers. The use of advanced control methods for EMS in microgrids allows to optimize the final cost of the energy which is consumed by local loads. In grid-connected mode, the energy price has an hourly component. The use of Energy Storage Systems (ESSs) can compensate both the intermittent nature of renewable energy systems and the randomness of the consumer behavior, allowing also the optimization of the energy exchange with the main grid. In islanded mode, an appropriate schedule has to be carried out by

C. Bordons et al., *Model Predictive Control of Microgrids*,
Advances in Industrial Control, https://doi.org/10.1007/978-3-030-24570-2_5

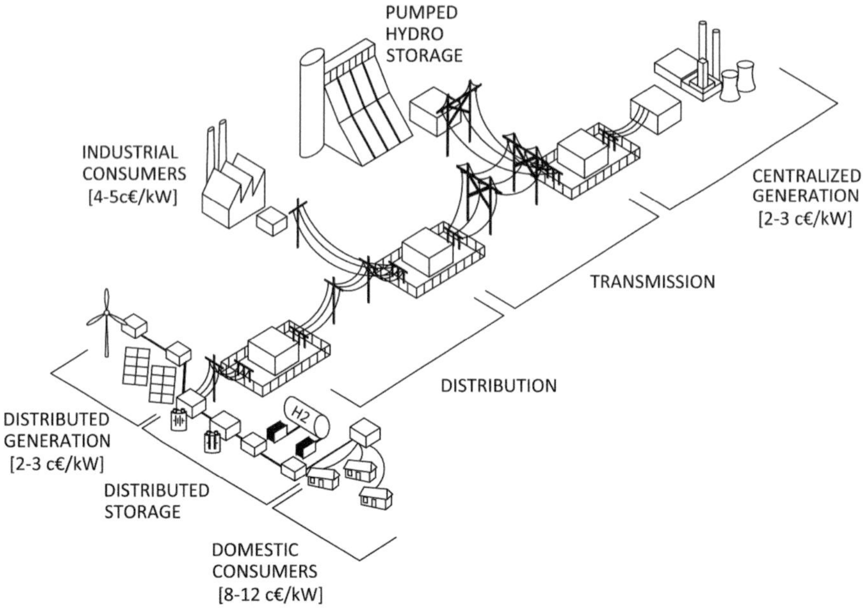

Fig. 5.1 Different conversions steps and energy prices in the electrical power system (Data based on [41])

the EMS in order to guarantee that the supply of energy can be done in all instants, whether there is generation available or not.

Microgrids paradigm breaks with the traditional scheme regarding the final energy price that end consumers have to pay. As can be seen in Fig. 5.1, the different conversion steps in the power systems apply an increasing price at each step since the energy is produced. In this way, industrial consumers, e.g., pay a lower price than domestic consumers since they are connected to Medium Voltage (MV), while domestic consumers are connected to Low Voltage (LV). On the other hand, the final price for selling the energy in the electrical market that the distributed generation receive is similar to that of traditional generation. A cooperative behavior in the LV/MV side of the grid between distributed generators and consumers in microgrids with ESS opens the possibility of an active behavior in demand side. The consumers can become prosumers with the possibility to sell/purchase energy to/from the main grid. Microgrids will allow prosumers not only to decide when to sell or purchase but also to store the energy or to supply the loads through ESSs in order to maximize the economic benefit.

Although the use of ESSs allows the optimization of final cost of energy in microgrids, every ESS has different limitations from the point of view of time autonomy, time response, degradation issues, or acquisition cost, as discussed in Chap. 3. In order to optimize the final cost of the energy consumption in microgrids, not only the aspects related to energy prices for selling or purchasing energy with the main grid have to be taken into account. Operational and degradation issues have to be incorporated in the EMS too. Under this perspective, the control problem of EMS in microgrids has been carried out from different perspectives and control methods. A review of the optimization of microgrids can be found in [10, 31]. The economic costs of each storage system are introduced in [16], where an evaluation of the economics of operating a wind/hydrogen system is explained. Authors in [9] consider a strategy applied to a stand-alone microgrid composed of renewable sources (wind, PV, and hydro), batteries, a fuel cell, an AC generator, and an electrolyzer, all of which optimize the control of the hybrid system while minimizing the total cost throughout their lifetime using genetic algorithms. The work carried out in [18] makes an optimization control algorithm for the combined heat and power of a microgrid with batteries and a fuel cell fed with natural gas neglecting degradation aspects of the ESS. In [23, 51], different strategies in the control of microgrids with hybrid storage using heuristical methods are presented. Advanced scheduling to encourage customers not only to participate in energy generation but also in efficient electricity consumption is proposed in [1]. A self-organizing computing framework, based on self-organizing agents, for solving the fundamental control and monitoring problems of a microgrid is presented in [45]. The study carried out in [21] proposes an energy ecosystem facilitated by microgrid management system with hierarchical agents and optimization, where updated external information and user's task preference are considered to make the control decisions. In [29], a framework for optimizing energy trading operations of a microgrid aggregator in the energy market is suggested.

Several studies and projects have been carried out in reference to microgrids with hybrid ESS. Heuristic methods with classical controllers are applied in [3, 38, 49], managing renewable energy microgrids with hybrid ESSs considering the system dynamics of each ESS. The study presented in [50] proposes a hybrid ESS between hydrogen and ultracapacitor, where the latter absorbs the transients that fuel cell and electrolyzer dynamics cannot absorb. The articles presented in [13, 39] use a control method to manage the hybridization between different ESSs, using frequency filters which generate good results. Authors in [26] propose a strategy based on adaptive control using neural networks. In [43, 44], a hybridization between a fuel cell, a battery, and ultracapacitor is used in an electric vehicle and a microgrid taking into account dynamic and degradation aspects of each ESS. As explained in [36, 42], transient response is a key characteristic feature of ESSs, sometimes more critical than efficiency, due to the importance of accepting rapidly changing, uncertain electric loads. Fast transient response is essential for an ESS in startup processes and fast power response, but concepts such as long-term autonomy in order to have the lower level of dependence with the main grid have to be also considered. An optimal ESS which would be able to respond in the short, medium, and long terms still does not exist. Hybrid ESSs with an optimal algorithm for sharing power, minimizing the usage cost of the whole ESS, and which manages the different timescales in the renewable energy microgrid are required. Advanced controllers are required in order to consider the large number of control variables that must be studied to minimize degradation aspects in hybrid ESSs. The dissertation presented in [42] solves the hybridization control problem with the following objectives: (i) protect the fuel cell system from abnormal load including transient, (ii) maintain state of charge of the battery, (iii) maintain the DC bus voltage, and (iv) regulate the current (from both the fuel cell and battery) to the optimized values if supervisory control demand exists through LQR strategies.

Model Predictive Control seems to be a good way of addressing these issues. In [11], the development of MPC for hybrid cogeneration power plants is carried out introducing the MLD framework. Authors in [25] use an MPC controller to integrate hydrogen ESSs in the electrical market not considering other ESS possibilities, neglecting degradation issues associated to the ESS. In [7, 47], MPC strategies are applied to the field of microgrid optimization. Mixed-Integer Linear Programming (MILP), including operational costs in the ESS, is used in [6, 22, 27, 48]. Mixed-Integer Programming (MIP) techniques are used in [33] in a microgrid with batteries, including the State of Health (SOH) of the battery in the cost function. Microgrid optimal scheduling considering multi-period islanding constraints is solved using MIP in the study carried out in [24]. In [17, 46], MPC techniques are applied to control the load sharing of a hybrid ESS composed of a fuel cell and an ultracapacitor, including some degradation issues. Similar developments have been presented for the hybridization of a fuel cell and a battery in [2, 5]. A Lyapunov-based hybrid MPC is applied to the EMS of microgrids in [30]. An MPC controller for an integrated energy microgrid combining power to heat and hydrogen is presented in [12]. Authors in [37] present an MPC strategy for economic diesel–PV–battery island microgrid operation in a rural area.

5.2 Integration of Operation and Degradation Aspects of ESSs in MPC

The consideration of the operational cost and degradation issues of different ESS technologies integrated into the microgrid allows the controller to optimize the cost of use of the whole ESS, integrating the operational cost of each ESS and considering the degradation mechanisms associated to each technology. As described in Chap. 3, the different ESSs have different operational and degradation issues. A summary of these mechanisms can be found in Table 5.1. The introduction of these criteria in a global cost function of the microgrid allows the controller to decide about the best ESS technology to be used at each control instant. The proposed solution not only optimizes the price of the energy exchange with the main grid but also improves the economic competitiveness of microgrids with hybrid ESS, prolonging the lifetime of the whole ESS and minimizing their operational cost.

The inclusion of degradation and operation issues in EMS implies that a high number of constraints and variables appear in the optimization problem. The addressed problem will be very difficult to be solved using traditional heuristic method such as the hysteresis band. Too many subcases will be found, being difficult to reach the optimal schedule or operation in the microgrid.

Table 5.1 Degradation issues of ESS

Energy storage system	Degradation issues
Ultracapacitors	Overcharge, Undercharge
Batteries	Number of charge/discharge cycles
	Overcharge, Undercharge
	High-stress current ratio
	AC Current Ripple
Electrolyzer	Number of working hours
	Operation and Maintenance Costs
	Fluctuations of current
	Start/Stop Cycles
Fuel Cell	Number of working hours
	Operation and Maintenance Costs
	Fluctuations of current
	Start/Stop Cycles

MPC offers the possibility of the inclusion of a model of the different ESSs integrated into the microgrid which can predict their behavior. Besides, the use of a multi-objective cost function makes the controller able to quantify the operation cost of the ESS according to their number of life cycles/hours, but also considering their degradation mechanisms in order to avoid them. Also, the price of the energy exchange with the main grid can be easily incorporated. Each objective is formulated as a term in a global cost function. A priority order can be easily assigned by the application of weighting factors in this mentioned cost function. All the limitations of the ESSs, as maximum and minimum limits for the level of stored energy, as well as the power limits can be easily included in the controller constraints.

As discussed in Chap. 1, in order to manage the different timescales of the several aspects to be managed in microgrids, it is usual to divide the control problem into three hierarchical levels. In this section, a two-stage MPC-based EMS for microgrids which includes all the aspects related to constraints, operation costs, and degradation issues is introduced. A generic microgrid as the one displayed in Fig. 5.2 (with its associated control block diagram) will be used to illustrate the development of the MPC. As can be seen in the figure, the two stages correspond to the tertiary and secondary control levels. Aspects related to the primary control level will be detailed in Chap. 9.

As shown in Fig. 5.2, the EMS interacts with the microgrid sending reference set points to the different components of the microgrid and acquiring measures from the system through a Local Net Area (LAN) communication. The tertiary controller is devoted to the schedule of the microgrid, where the long-term issues are solved. In this step, the control decisions are taken based on the generation, demand, and prices forecast. The selected schedule horizon (SH) corresponds to the 24 h of the day, discretized with a sample period $T_s = 1$ h.

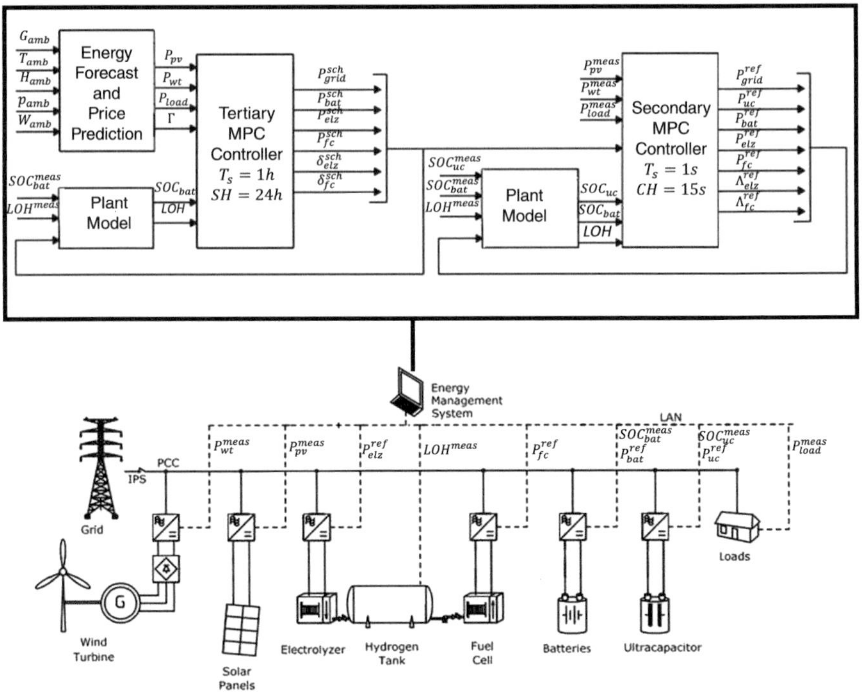

Fig. 5.2 Microgrid and control block diagram

Real operational scenario differs from forecast, which requires a second step where an optimal power sharing among all the components of the microgrid must be carried out. This corresponds to the secondary control level of the microgrid. At this control level, the control decisions are taken with a faster sampling time of $T_s = 1$ s and a shorter horizon $N_p = 15$ s. This control horizon is selected in order to integrate the startup delay in the electrolyzer and the fuel cell into the controller. The main objective of this control stage is the tracking of the schedule established in the long-term control level (tertiary controller). As it will be shown, in both control levels, the different operation and degradation aspects of the ESS are included.

5.2.1 Tertiary Control: Economical Optimization

This control layer manages the long-term schedule of the microgrid [14]. As discussed in Chap. 1, in grid-connected mode, the tertiary controller is in charge of the economical optimization of the microgrid while in islanded mode it considers the energy autonomy of the microgrid. Tertiary control decides the schedule of active and reactive power exchange with the external grid and among the different units of the microgrid. Based on inputs such as forecast, operational costs, or prices, the tertiary controller prepares the sources and storage dispatch schedule, which is communicated to the secondary controller. As can be seen in Fig. 5.2, this control layer is divided into three blocks: (i) the forecasting of energy and price prediction, (ii) the plant model, and (iii) the tertiary MPC Controller. The inputs to the controller are the measurement of the meteorological variables (solar irradiance G_{amb}, ambient temperature T_{amb}, atmospheric pressure p_{amb}, relative humidity H_{amb}, and wind speed W_{amb}, (including the wind direction) and the measurements of the SOC of the batteries SOC_{bat} and the LOH in the hydrogen tanks LOH). The tertiary controller calculates as outputs the schedule given for the energy exchange with the main grid ($P^{sch}_{grid}(t+k)$), the schedule of the batteries power ($P^{sch}_{bat}(t+k)$), the schedule of the electrolyzer, and fuel cell powers ($P^{sch}_{elz}(t+k)$ and $P^{sch}_{fc}(t+k)$), for all the instants belonging to the schedule horizon ($k = 1, ..., SH$). Notice that in the case of the electrolyzer and the fuel cell, their control signals are the power set points but also the on/off (logical) signals $\delta_{elz}(t+k)$ and $\delta_{fc}(t+k)$, which are also calculated in the tertiary controller. The presence of logical and continuous variables makes necessary the use of an MLD formulation and therefore the use of Hybrid MPC techniques, as described in Chap. 2.

Energy Forecast and Price Prediction

A model that provides a forecast of the generation and demand as well as the electricity prices is needed for the optimization. The study of the forecasting and prediction methods for energy prices, renewable generation, and load consumption is out of the scope of this book. One method that can be used is the one presented in [14], which is based on an Artificial Neural Network (ANN) developed with an Autoregressive Integrated Moving Average (ARIMA) model using time series. As discussed, this model uses the meteorological variables to predict the future value of these variables. Based on this forecast, the power production for the photovoltaic array P_{pv} and wind turbine generator P_{wt} can be easily predicted following the models presented in Chap. 3. As detailed in [14], the load consumption and the price prediction are also forecasted based on the meteorological prediction and statistical methods. The result of the model gives the forecast prediction for the power generation by the photovoltaic panels ($\hat{P}_{pv}(t+k|t)$) and the wind turbine ($\hat{P}_{wt}(t+k|t)$), the load consumption ($\hat{P}_{load}(t+k|t)$), as well as the price prediction for the selling/purchasing energy with the main grid $\hat{\Gamma}_{sale}(t+k|t)$) and $\hat{\Gamma}_{pur}(t+k|t)$) for the schedule horizon. Notice that other existing methodologies could be used at this step.

System Model

The system model corresponds to the dynamic linear model of the microgrid state variables, given by the level of the stored energy in the ESSs. Equation (5.1) describes the evolution of the state of charge (SOC_{bat}) of the battery and Eq. (5.12) corresponds to the level of hydrogen in the tank (LOH). Notice that one equation of this form must be introduced for each of the existing ESSs.

Table 5.2 Conversion of logic relations into mixed-integer inequalities

	Relation	Logic	MLD inequalities
P1	AND ($\wedge$)	$S_3 \Leftrightarrow (S_1 \wedge S_2)$	$-\infty \leq -\delta_1 + \delta_3 \leq 0$ $-\infty \leq -\delta_2 + \delta_3 \leq 0$ $-\infty \leq \delta_1 + \delta_2 - \delta_3 \leq 1$
P2	OR ($\vee$)	$S_1 \vee S_2$	$1 \leq \delta_1 + \delta_2 \leq 2$
P3	NOT ($\sim$)	$S2 \equiv\sim S_1$	$1 \leq \delta_1 + \delta_2 \leq 1$
P4	IMPLY ($\Rightarrow$)	$[a^T x \leq 0] \Rightarrow$ $[\delta = 1]$	$\varepsilon \leq a^T x - (m - \varepsilon)\delta \leq \infty$
P5	IMPLY ($\Rightarrow$)	$[\delta = 1] \Rightarrow$ $[a^T x \leq 0]$	$-\infty \leq a^T x + M\delta \leq M$
P6	IFF($\Leftrightarrow$)	$[a^T x \leq 0] \Leftrightarrow$ $[\delta = 1]$	$\varepsilon \leq a^T x - (m - \varepsilon)\delta \leq \infty$ $-\infty \leq a^T x + M\delta \leq M$
P7	Mixed product	$z = \delta \cdot a^T x$	$-\infty \leq z - M\delta \leq 0$ $0 \leq z - m\delta \leq \infty$ $-\infty \leq z - a^T x + m(1 - \delta) \leq 0$ $0 \leq z - a^T x + M(1 - \delta) \leq \infty$

$$SOC_{bat}(t+1) = SOC_{bat}(t) + \frac{\eta_{bat,ch} P_{bat,ch}(t) T_s}{C_{\max}} + \left.\frac{P_{bat,dis}(t) T_s}{\eta_{bat,dis} C_{\max}}\right|_{bat} \tag{5.1}$$

Notice that the efficiencies in the charging and discharging process of the batteries, $\eta_{ch,bat}$ and $\eta_{dis,bat}$, are considered to be different. This is an improvement over the assumptions done in the basic EMS of Chap. 4. Also, as can be seen in Eq. (5.1), they affect in different manner to the charging/discharging process, multiplying or dividing the battery power. This allows a better modeling of the SOC, although this fact requires the battery power P_{bat} to be divided into its negative and positive parts: the charging power of the batteries ($P_{bat,ch}$) and the discharging power ($P_{bat,dis}$). This conversion is carried out using two logic variables δ_{ch} and δ_{dis}, whose value is "0" or "1" according to the following expressions:

$$P_{bat}(t) \leq 0 \Leftrightarrow \delta_{bat,ch}(t) = 1 \tag{5.2}$$

As explained in Chap. 2, this logic relationship can be introduced as a constraint in the controller using the conversions given in [4, 11], obtaining the constraints described below which have to be included in the controller. These conversions are detailed in Table 5.2, where m and M represent the lower and upper bounds of $a^T x$, where x is the continuous variable, a is a vector of parameters, and $\epsilon > 0$ is the smallest tolerance of the device [4, 11].

Using the proposition P6 defined in Table 5.2 this relation can be integrated as a constraint in the controller with the next inequalities:

$$\varepsilon \leq P_{bat}(t) - (P_{bat}^{min} - \varepsilon)\delta_{bat,ch}(t) \leq \infty \quad (5.3)$$

$$-\infty \leq P_{bat}(t) + P_{bat}^{max}\delta_{bat,ch}(t) \leq P_{bat}^{max} \quad (5.4)$$

Using the mixed product, the negative part of $P_{bat}(t)$ can be defined:

$$P_{bat,ch}(t) = -P_{bat}(t)\delta_{bat,ch}(t) \quad (5.5)$$

The introduction of variables related to the positive values of $P_{bat}(t)$: $P_{bat,dis}(t)$ and $\delta_{bat,dis}(t)$, is done through its relationship with $P_{bat,ch}(t)$ and $\delta_{bat,ch}(t)$. These relationships have to be introduced as constraints in the MPC formulation.

$$P_{bat}(t) = P_{bat,dis}(t) - P_{bat,ch}(t) \quad (5.6)$$

$$\delta_{bat,ch}(t) + \delta_{bat,dis}(t) = 1 \quad (5.7)$$

Only the constraints of the transformation of one of the two mixed products given by the definitions of the variables $P_{bat,ch}$ and $P_{bat,dis}$ have to be introduced in the controller since they are related by the constraint given in Eq. (5.6). For simplicity the positive part $P_{bat,dis}$ is selected to be transformed using P7 of Table 5.2.

$$-\infty \leq P_{bat,dis}(t) - P_{bat}^{max}\delta_{bat,dis}(t) \leq 0 \quad (5.8)$$

$$0 \leq P_{bat,dis}(t) - P_{bat}^{min}\delta_{bat,dis}(t) \leq \infty \quad (5.9)$$

$$-\infty \leq P_{bat,dis}(t) - P_{bat}(t) + P_{bat}^{min}(1 - \delta_{bat,dis}(t)) \leq 0 \quad (5.10)$$

$$0 \leq P_{bat,dis}(t) - P_{bat}(t) + P_{bat}^{max}(1 - \delta_{bat,dis}(t)) \leq \infty \quad (5.11)$$

The evolution of the second state variable of the system is given by the changes in the Level of Hydrogen (LOH) in the tank. This evolution depends on the reference power set points given to the electrolyzer and the fuel cell, but also depends on their ON/OFF state. This gives rise to the emergence of mixed products in the evolution of this state variable as exposed in Eq. (5.12)

$$LOH(t+1) = LOH(t) + \frac{\eta_{elz} P_{elz}(t)\delta_{elz}(t)T_s}{V_{max}} - \frac{P_{fc}(t)\delta_{fc}(t)T_s}{\eta_{fc} V_{max}} \quad (5.12)$$

where η_{elz} and η_{fc} correspond to the efficiency of the electrolyzer and the fuel cell, P_{elz} and P_{fc} correspond to power references given by the controller, and V_{max} is the maximum volume of hydrogen which can be stored in the tank. The following mixed products of the power of the electrolyzer and the fuel cell are introduced:

$$z_{elz}(t) = P_{elz}(t) \cdot \delta_{elz}(t) \quad (5.13)$$

$$z_{fc}(t) = P_{fc}(t) \cdot \delta_{fc}(t) \quad (5.14)$$

Using the proposition P7 defined in Table 5.2, the mixed product z_{elz} can be integrated into the controller as the following constraints:

$$-\infty \leq z_{elz}(t) - P_{elz}^{max}\delta_{elz}(t) \leq 0 \quad (5.15)$$

$$0 \leq z_{elz}(t) - P_{elz}^{min}\delta_{elz}(t) \leq \infty \quad (5.16)$$

$$-\infty \leq z_{elz}(t) - P_{elz}(t) - P_{elz}^{min}(1 - \delta_{elz}(t)) \leq 0 \quad (5.17)$$

$$0 \leq z_{elz}(t) - P_{elz}(t) + P_{elz}^{max}(1 - \delta_{elz}(t)) \leq \infty \quad (5.18)$$

Similar procedure has to be followed to introduce the mixed product of the power of the fuel cell $z_{fc}(t)$.

MPC Controller

The cost function to be minimized by MPC can be formulated as a sum of the different cost functions of the components of the microgrid:

$$\min J(t) = \sum_{k=1}^{SH} \left(J_{grid}(t+k \mid t) + J_{bat}(t+k \mid t) + J_{H_2}(t+k \mid t)\right) \quad (5.19)$$

where J_{grid} refers to the cost function of energy exchange with the main grid, and J_{bat} and J_{H_2} are associated to battery and hydrogen, respectively. The minimization is subject to the following constraints along the schedule horizon ($k = 1, ..., SH$):

$$\begin{aligned} \hat{P}_{pv}(t+k|t) + \hat{P}_{wt}(t+k|t) - \hat{P}_{load}(t+k|t) + P_{grid}(t+k)) + \\ + P_{bat}(t+k) - z_{elz}(t+k) + z_{fc}(t+k) = 0 \end{aligned} \quad (5.20)$$

$$0 \leq \delta_i(t+k) \leq 1|_{i=elz,fc} \quad (5.21)$$

$$P_i^{min} \leq P_i(t+k) \leq P^{max}|_{i=grid,bat,elz,fc} \quad (5.22)$$

$$SOC_i^{min} \leq SOC_i(t+k) \leq SOC^{max}|_{i=bat} \quad (5.23)$$

$$LOH_i^{min} \leq LOH(t+k) \leq LOH^{max} \quad (5.24)$$

where the first constraint (5.20) corresponds to the balance of energy in the microgrid, being the constraints given by the relationships (5.21)–(5.24) the physical limits of the microgrid components.

The cost function of the grid ($J_{grid}(t)$) minimizes the final cost of the energy purchased to the main grid while maximizes the revenue, deciding in which sample instants these processes have to be carried out.

$$\begin{aligned} J_{grid}(t+k|t) = \Big(-\hat{\Gamma}_{sale}^{DM}(t+k|t) \cdot P_{sale}(t+k|t) \\ + \hat{\Gamma}_{pur}^{DM}(t+k|t) \cdot P_{pur}(t+k|t)\Big) \cdot T_s \end{aligned} \quad (5.25)$$

where $\hat{\Gamma}_{sale}^{DM}(t+k|t)$ and $\hat{\Gamma}_{pur}^{DM}(t+k|t)$ represent the forecast values for the energy prices, $P_{sale}(t+k|t)$ and $P_{pur}(t+k|t)$ represent the sale and purchase of energy with the grid. Notice that the signal in the term of the cost function related to the selling of energy is negative. This is done with the objective to maximize the benefits of the revenue of selling energy in the day-ahead market. In order to differentiate the positive and negative values of P_{grid}, two logical variables are introduced in the controller δ_{sale} and δ_{pur}, which are active "1" or inactive "0", depending on the sign of the exchange of power with the main grid (P_{grid}). The next transformation is done:

$$P_{grid}(t) \leq 0 \Leftrightarrow \delta_{sale}(t) = 1 \tag{5.26}$$

$$P_{sale}(t) = -P_{grid}(t) \cdot \delta_{sale}(t) \tag{5.27}$$

$$P_{grid}(t) = P_{pur}(t) - P_{sale}(t) \tag{5.28}$$

$$\delta_{pur}(t) + \delta_{sale}(t) = 1 \tag{5.29}$$

where P_{pur} can be defined as a mixed product of the form:

$$P_{pur}(t) = P_{grid}(t) \cdot \delta_{pur}(t) \tag{5.30}$$

Using the proposition P7, defined in [4], the mixed product $P_{pur}(t)$ can be integrated into the controller with the following constraints:

$$-\infty \leq P_{pur}(t) - P_{grid}^{max} \delta_{pur}(t) \leq 0 \tag{5.31}$$

$$0 \leq P_{pur}(t) - P_{grid}^{min} \delta_{pur}(t) \leq \infty \tag{5.32}$$

$$-\infty \leq P_{pur}(t) - P_{grid}(t) - P_{grid}^{min}(1 - \delta_{pur}(t)) \leq 0 \tag{5.33}$$

$$0 \leq P_{pur}(t) - P_{grid}(t) + P_{grid}^{max}(1 - \delta_{pur}(t)) \leq \infty \tag{5.34}$$

The cost function of the batteries is given by expression (5.35). It pursues to minimize the economical cost related to the use of the batteries. The lifetime of batteries is given by manufactures as a number of charging and discharging cycles. Unfortunately, this limited number of cycles can be even reduced due to the degradation mechanisms detailed in Chap. 3 and summarized in Table 5.1. The main mechanism which has to be avoided is related with exposing the batteries to high-stress current ratio in the charging and discharging process. For this reason, a second term in the batteries' cost function is included, which penalizes high values of P_{bat}^2.

$$\begin{aligned} J_{bat}(t+k|t) = & \frac{\text{CC}_{bat}}{2 \cdot \text{Cycles}_{bat}} (P_{bat,ch}(t+k|t) + P_{bat,dis}(t+k|t)) \cdot T_s \\ & + \text{Cost}_{degr,ch} \cdot P_{bat,ch}^2(t+k|t) + \text{Cost}_{degr,dis} \cdot P_{bat,dis}^2(t+k|t) \end{aligned} \tag{5.35}$$

Parameter CC_{bat} expresses the capital cost of the battery, Cycles_{bat} are the number of life cycles of the battery, and $\text{Cost}_{degr,ch}$ and $\text{Cost}_{degr,dis}$ are the cost associated to the degradation mechanisms of the batteries.

$$\min J(t) = \sum_{k=1}^{SH} \left(J_{grid}(t+k \mid t) + J_{bat}(t+k \mid t) + J_{H_2}(t+k \mid t) \right) \tag{5.36}$$

The cost function of hydrogen storage (5.36) is the sum of the cost functions of its components: electrolyzer, fuel cell, and hydrogen tank. In order to simplify the cost, the compression cost of hydrogen (in case it exists) is not considered. As occurs with the batteries, electrolyzers and fuel cells have a limited lifetime. This lifetime is expressed as a number of working hours. This lifetime can also be reduced if the degradation aspects related to this technology are not minimized. For this reason, not only the working hours for electrolyzer and fuel cells are minimized but the startup/shutdown cycles and the fluctuations in the operation conditions are also included. Batteries have nearly no Operation and Maintenance (O&M) costs. But electrolyzer and fuel cell require of maintenance aspects which are included with an hourly cost in the cost function.

$$\begin{aligned} J_{elz}(t+k|t) &= \left(\frac{\mathrm{CC}_{elz}}{\mathrm{Hours}_{elz}} + \mathrm{Cost}_{o\&m,elz}\right)\delta_{elz}(t+k|t) \\ &+ \mathrm{Cost}_{start,elz}\cdot\sigma_{elz}^{on}(t+k|t) + \mathrm{Cost}_{degr,elz}\cdot\vartheta_{elz}^{2}(t+k|t) \end{aligned} \tag{5.37}$$

$$\begin{aligned} J_{fc}(t+k|t) &= \left(\frac{\mathrm{CC}_{fc}}{\mathrm{Hours}_{fc}} + \mathrm{Cost}_{o\&m,fc}\right)\delta_{fc}(t+k|t) \\ &+ \mathrm{Cost}_{start,fc}\cdot\sigma_{fc}^{on}(t+k|t) + \mathrm{Cost}_{degr,fc}\cdot\vartheta_{fc}^{2}(t+k|t) \end{aligned} \tag{5.38}$$

where CC_{elz} and CC_{fc} refer to the capital cost of the electrolyzer and the fuel cell, Hours_{elz} and Hours_{fc} are the lifetime hours of the electrolyzer and the fuel cell given by the manufacturers, $\mathrm{Cost}_{o\&m,elz}$ and $\mathrm{Cost}_{o\&m,fc}$ are the terms related to cost of operation and maintenance of the electrolyzer and fuel cell, $\mathrm{Cost}_{start,elz}$ and $\mathrm{Cost}_{start,fc}$ are the cost relatives to the degradation processes linked to the startup and shutdown of the devices, and finally, $\mathrm{Cost}_{degr,elz}$ and $\mathrm{Cost}_{degr,fc}$ express the costs related to the degradation processes associated to the power fluctuations in the electrolyzer and the fuel cell.

The startup ($\sigma_{\alpha}^{on}(t)$) state for a device can be defined with the following expressions:

$$\sigma_{\alpha}^{on}(t) = \delta_{\alpha}(t) \wedge (\sim \delta_{\alpha}(t-1))|_{\alpha=elz,fc} \tag{5.39}$$

Using P1 described in Table 5.2, this relationship can be introduced as constraints in the controller using the following expressions:

$$-\infty \le -\delta_{\alpha}(t) + \sigma_{\alpha}^{on}(t) \le 0 \tag{5.40}$$

$$-\infty \le -(1-\delta_{\alpha}(t-1)) + \sigma_{\alpha}^{on}(t) \le 0 \tag{5.41}$$

$$-\infty \le \delta_{\alpha}(t) + (1-\delta_{\alpha}(t-1) - \sigma_{\alpha}^{on}(t) \le 1 \tag{5.42}$$

In order to minimize the power fluctuations of electrolyzers and fuel cells, the third terms in the cost functions (5.37) and (5.38) are included. The variables $\vartheta_{\alpha}(t)$ are defined as the power variation in all the instants except those when the device moves from the startup state to the energized state or from the energized state toward switch off. This casuistic is defined by Eq. (5.43), giving as a result the MLD constraints (5.48)–(5.51). A new auxiliary variable called logic state of degradation by power variation ($\chi_\alpha(t_k)$), defined as the state of degradation by power variation is defined giving as results the constraints expressed in inequalities (5.45)–(5.47).

$$\vartheta_\alpha(t) = \Delta P_\alpha(t) \cdot (\delta_\alpha(t) \wedge \delta_\alpha(t-1)) \tag{5.43}$$

$$\chi_\alpha(t) = (\delta_\alpha(t) \wedge \delta_\alpha(t-1)) \tag{5.44}$$

Using the transformations given by P1 in Table 5.2, the relation between $\chi_\alpha(t)$ and $\delta_\alpha(t)$ can be introduced in the controller with the following constraints:

$$-\infty \le -\delta_\alpha(t) + \chi_\alpha(t) \le 0 \tag{5.45}$$

$$-\infty \le -\delta_\alpha(t-1) + \chi_\alpha(t) \le 0 \tag{5.46}$$

$$-\infty \le \delta_\alpha(t) + \delta_\alpha(t-1) - \chi_\alpha(t) \le 1 \tag{5.47}$$

Finally, $\vartheta_\alpha(t)$ is defined as a mixed product. Using P7 of Table 5.2, $\vartheta_\alpha(t)$ can be integrated into the controller with the following constraints:

$$-\infty \le \vartheta_\alpha(t) - \Delta P_\alpha^{max} \chi_\alpha(t) \le 0 \tag{5.48}$$

$$0 \le \vartheta_\alpha(t) - \Delta P_\alpha^{min} \chi_\alpha(t) \le \infty \tag{5.49}$$

$$-\infty \le \vartheta_\alpha(t) - \Delta P_\alpha(t) + \Delta P_\alpha^{min}(1 - \chi_\alpha(t)) \le 0 \tag{5.50}$$

$$0 \le \vartheta_\alpha(t) - \Delta P_\alpha(t) + \Delta P_\alpha^{max}(1 - \chi_\alpha(t)) \le \infty \tag{5.51}$$

Application Example

The tertiary controller can be applied to the Hylab microgrid used in the previous chapter. An ultracapacitor with $C_{uc} = 63$ F, $\eta_{uc}^{ch} = 0.97$, $\eta_{uc}^{dis} = 0.99$, and $P_{uc} = 3000$ W is considered, which has been emulated and will only be used in the secondary control level. The different cost factors utilized in the controller can be seen in Table 5.3.

The forecast model has been applied and validated with the renewable energy data provided by the meteorological station, from the company Geonica, model MTD 3008, which has been collecting data every ten minutes since 2009. The load profile corresponds to the collected data of a domestic home located in Puertollano (Ciudad Real, Spain) whose maximum contracted power with the electrical company is 5 kW. The sources and load have been emulated for the experimental results with the equipment. The maximum load emulation power is just 2.5 kW versus 5 kW of the real measurements, so the obtained load profile is divided by 2. The energy price data have been given by the Iberian market operator (OMIE)

The results of the forecast model and the Tertiary MPC are shown in Fig. 5.3. When a surplus of energy exists, the controller tries to sell energy to the grid in those instants when the energy prices are higher, storing energy in the batteries or producing

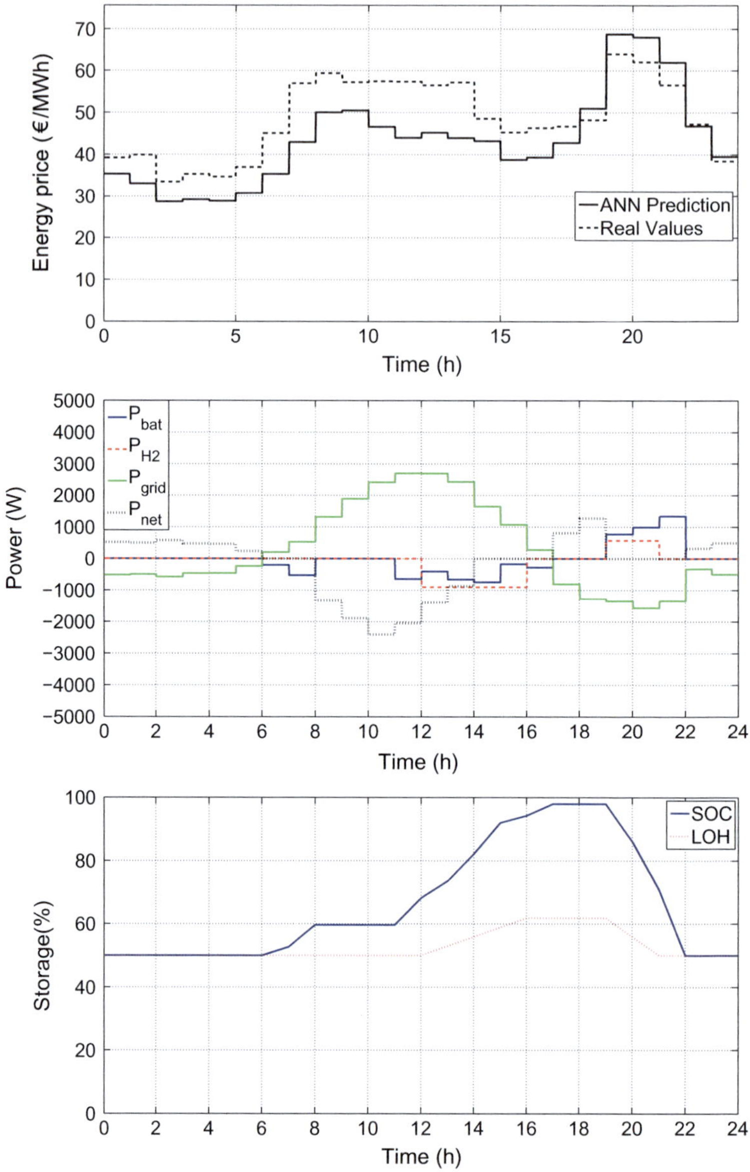

Fig. 5.3 Tertiary MPC controller schedule results in grid-connected mode

Table 5.3 Cost factor values utilized by the MPC scheduler. Data based on [8, 20, 40]

PEM Electrolyzer
$\eta = 0.23$ Nm^3/kWh, CC = 8.22 €/kW, $\text{Cost}_{o\&m,elz} = 2$ m€/h
$\text{Cost}_{start,elz} = 0.123$ €
$\text{Cost}_{degr,elz} = 0.05$ €/W, Lifetime = 10000 h
PEM Fuel Cell
$\eta = 1.320$ kWh/Nm^3, CC=30 €/kW, $\text{Cost}_{o\&m,fc} = 1$ m€/h
$\text{Cost}_{start,fc} = 0.01$ €,
$\text{Cost}_{degr,fc} = 0.01$ €/W, Lifetime = 10000 h
Batteries
$\eta_{ch} = 0.90$, $\eta_{dis} = 0.95$, CC = 125 €/kWh, Life cycles = 3000,
$\text{Cost}_{degr,dis} = 10^{-9}$ €/W^{2}h, $\text{Cost}_{degr,ch} = 10^{-9}$ €/W^{2}h
Ultracapacitor
$\eta_{ch} = 0.97$, $\eta_{dis} = 0.99$

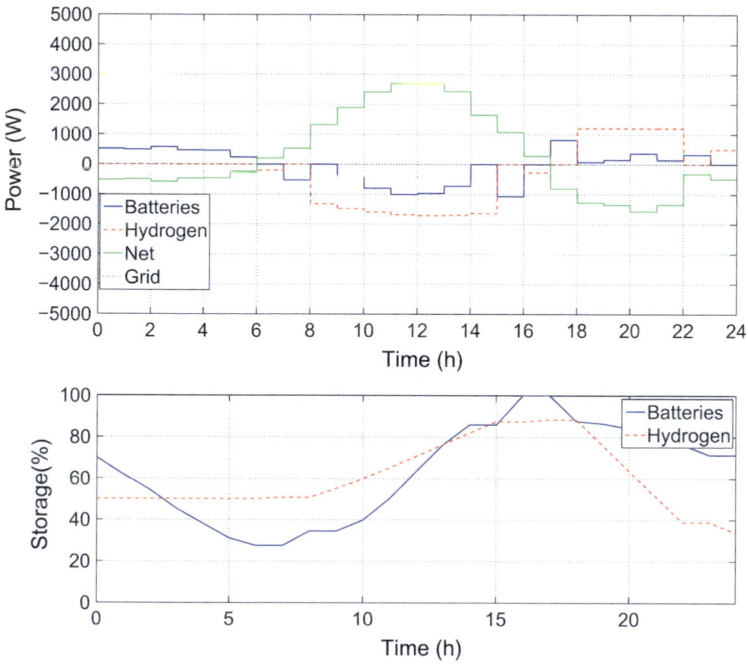

Fig. 5.4 Tertiary MPC controller schedule results in islanded mode

hydrogen in the rest. In case that there is a deficit of energy in the microgrid, the EMS tries to use the hybrid energy storage system when the energy prices are higher. As can be seen in the figure, the number of hours of use and the switching states of the electrolyzer and the fuel cell are minimized, as well as the peak power in the charge and discharge of the battery. The SOC of the batteries is also controlled imposing constraints and therefore protecting them from high states of charge or discharge. As can be seen in Fig. 5.3, the purchase of energy to the grid is done when the prices are lower, in the same way that the energy sale to the grid is done at maximum price periods. The power reference given to the fuel cell and the electrolyzer is maintained nearly constant, giving minimum variation to these devices and thus minimizing degradation due to the operation profile.

The controller has also been applied to the microgrid working in islanded mode. Similar results are found for this case in Fig. 5.4.

5.2.2 Secondary Control: Power Sharing

Real-time-operational scenario differs from that scheduled in the long-term optimization problem carried out by the tertiary control level. This control level has an intermediate position between the primary controllers of the electronic power converters and the long-term schedule carried out in the tertiary control level. It adapts the reference according to the real-time situation. The timescale of this controller is in the order of one second. In this timescale, the dynamics of the generators and the loads for all the sample instants of the control horizon can be assumed constant and equal to the sampled value. The functional cost of each ESS in this control level is based on the deviation from the power set points and the stored energy level from that set by the economical dispatch of the microgrid. Degradation or anomalous working conditions are avoided, introducing these terms in the objective function of the controller as explained below. The horizon N_p is selected as the maximum time delay to start or to stop of a device.

This control layer manages the short-term schedule and control of the microgrid [15]. As can be seen in Fig. 5.2, this control layer is divided into two blocks: (i) the plant model and (ii) the secondary MPC controller. There is not an energy forecast model since the controller uses directly the real-time measurements of P_{pv}^{meas}, P_{wt}^{meas}, and P_{load}^{meas} to calculate the net power in the microgrid. The controller assumes that these values are going to be constant during the prediction horizon. The plant model is similar to the one in the tertiary controller, but it incorporates the state of charge of the ultracapacitors. The measurements of the SOC of the batteries and ultracapacitors (SOC_{bat}^{meas}, SOC_{uc}^{meas}) and the LOH in the hydrogen tanks (LOH^{meas}) are inputs of the controller.

The secondary MPC controller receives as inputs the schedule given for the energy exchange with the main grid ($P_{grid}^{sch}(t+k)$), the schedule of the batteries power ($P_{bat}^{sch}(t+k)$), the schedule of the electrolyzer and fuel cell powers ($P_{elz}^{sch}(t+k)$ and

$P_{fc}^{sch}(t+k)$), for all the instants belonging to the prediction horizon ($k = 1, ..., CH$). Notice that in the case of the electrolyzer and the fuel cell, the on/off (logical) signals ($\delta_{elz}^{sch}(t+k)$) and ($\delta_{fc}^{sch}(t+k)$) are also used. As can be seen in Fig. 5.2, the secondary MPC controller outputs are sent via LAN to the primary controllers of the microgrid components P_{bat}^{ref}, P_{elz}^{ref}, P_{fc}^{ref}, and P_{grid}^{ref}.

The grid cost function is introduced in the controller giving a high penalty to the deviation of the exchange of power with the grid in real time (P_{grid}) with respect to the power scheduled (P_{grid}^{sch}) by the tertiary control level.

$$J_{grid}(t) = \sum_{k=1}^{N_p} w_{grid} \left(P_{grid}(t+k|t) - P_{grid}^{sch}(t+k)\right)^2 \tag{5.52}$$

where w_{grid} corresponds to the weighting factor given to this term of the global cost function.

As commented in Chap. 3, ultracapacitors storage possess a longer life expectancy in comparison to batteries/hydrogen in terms of charge/discharge cycles. For this reason, no costs have to be considered related with the number of charge and discharge cycles in the optimization process of the microgrid. They have a high specific power and an accurate transient response in comparison with batteries/hydrogen without nearly no degradation mechanisms associated to abrupt changes in the current demand. For these reasons, ultracapacitors can be used in this control level to compensate the effects of transient response in the rest of ESS technologies. Their use is limited by their low energy density. Due to this issue, ultracapacitors are desired to be maintained in an intermediate SOC_{uc}^{ref} to absorb or provide peak currents when abrupt changes in the balance of energy of the rest of components of the microgrids.

The cost function of the ultracapacitor is shown in Eq. (5.53). In order to be always available if required to compensate the rest of components of the microgrid, the ultracapacitors are kept in an intermediate state of charge, which also allows to protect them from undercharge or overcharge. If the second term is not included, suboptimal problem solutions can be found when the power calculated by the solver is near zero.

$$\begin{aligned} J_{uc}(t) = \sum_{k=1}^{N_p} \Big(& w_{uc}^{E} \left(SOC_{uc}(t+k|t) - SOC_{uc}^{ref}\right)^2 \\ & + w_{uc}^{P} \cdot (P_{uc}(t+k|t) - 0)^2 \Big) \end{aligned} \tag{5.53}$$

The terms w_{uc}^{E} and w_{uo}^{P} correspond to the weighting factors of the cost function of the ultracapacitor in the deviation in the energy reference and power reference, respectively.

The batteries' cost function is expressed by Eq. (5.54). The batteries have a double reference in power and energy. The power tracking is used to solve the current scenario with the forecast given for the following instants. The energy tracking is a way to correct the accumulation of errors in the tracking, not due to a bad behavior of the controller but due to mismatch caused by forecasting errors which result in an excess or deficit of energy stored in the batteries, which can be compensated in the current instant.

In comparison to hydrogen, batteries are more flexible due to the fact that startup and shutdown cycles do not affect them. The power tracking has a lower weight than in the case of the hydrogen, while the energy tracking has a higher weight than in the case of hydrogen. With the goal of protecting the batteries from high charging/discharging current ratio, high charging/discharging current values are penalized in the third term of the cost function. The last term of the cost function penalizes the AC current in the batteries.

$$\begin{aligned} J_{bat}(t) = \sum_{k=1}^{N_p} \Bigg(& w_{bat}^{P} \left(P_{bat}(t+k|t) - P_{bat}^{sch}(t+k)\right)^2 \\ & + w_{bat}^{E} \left(SOC_{bat}(t+k|t) - SOC_{bat}^{sch}(t+k|t)\right)^2 \\ & + w_{bat}^{degr} \cdot \left(\frac{P_{ch,bat}(t+k|t) + P_{dis,bat}(t+k|t)}{v_{dc,bat}(t)}\right)^2 \\ & + w_{bat}^{ripple} \cdot \left(\frac{\Delta P_{ch,bat}(t+k|t) + \Delta P_{dis,bat}(t+k|t)}{v_{dc,bat}(t)}\right)^2 \Bigg) \end{aligned} \tag{5.54}$$

where the terms w_{bat}^{E} and w_{bat}^{P} correspond to the weighting factors of the cost function of the batteries in the deviation in the energy reference and power reference respectively. The term w_{bat}^{degr} is the weighting factor to avoid high current values in the charging and discharging processes. Finally, as in this control level the ultracapacitor is integrated and due to the fact that AC current can damage the batteries (see Chap. 3), a term to penalize the variation of power in the batteries is also included: w_{bat}^{ripple}. The definition and formulation of $P_{dis,bat}$ and $P_{ch,bat}$ can be found in the previous section related to the tertiary controller.

The hydrogen cost function is defined by Eq. (5.55). As well as in the case of the batteries, the economical dispatch of the microgrid gives both references the schedule in energy and in power at each instant. In order to protect this ESS from the main causes of degradation, the startup and shutdown states are penalized in the controller. As commented in previous section, the power fluctuation is penalized in all the states except in those when the fuel cell or the electrolyzer change their state from startup to energized.

$$\begin{aligned} J_{H_2}(t) = \sum_{k=1}^{N_p} \Bigg(& w_{tank}^{E} \left(LOH(t+k|t) - LOH^{sch}(t+k)\right)^2 \\ & + w_{elz}^{P} \left(z_{elz}(t+k|t) - z_{elz}^{sch}(t+k)\right)^2 + w_{fc}^{P} \left(z_{fc}(t+k|t) - z_{fc}^{sch}(t+k)\right)^2 \\ & + w_{elz}^{ripple} \left(\vartheta_{elz}(t+k|t)\right)^2 + w_{fc}^{ripple} \left(\vartheta_{fc}(t+k|t)\right)^2 \\ & + w_{elz}^{startup} \cdot \sigma_{elz}^{on}(t+k|t) + w_{fc}^{startup} \cdot \sigma_{fc}^{on}(t+k|t) \Bigg) \end{aligned} \tag{5.55}$$

where w^E_{tank} is the weighting factor for the deviation in the energy stored in the hydrogen tank with respect to the schedule given in the tertiary controller, w^P_{elz} and w^P_{fc} are the terms to penalize the deviation of power from the schedule in the electrolyzer and the fuel cell. The terms w^{ripple}_{elz} and w^{ripple}_{fc} correct the fluctuations of power in the electrolyzer and fuel cell ϑ_{elz} and ϑ_{fc} (the definition of these variables are detailed in the previous section). Finally, $w^{startup}_{elz}$ and $w^{startup}_{fc}$ are the weighting factors used to penalyze the startup/shutdown cycles (the definition of σ^{on}_{elz} and σ^{on}_{fc} can be found in previous section related with the tertiary controller).

As can be seen in Fig. 5.2, the logical signals used in the tertiary control level are δ_{elz} and δ_{fc} corresponding to the energized state of the electrolyzer and the fuel cell. But in the secondary controller the logical signals Λ_{elz} and Λ_{fc} are used, which correspond to the logical signals to switch on/off these devices. This is due to the fact of a time delay existing since these devices are switched on until they can generate or absorb energy from the microgrid.

The internal controller of the electrolyzer is charged to maintain the correct water level in the separators in order to avoid drying conditions in the membrane, feeding the stack with enough water to produce the electrolysis reaction. Before the electrolyzer is in the state of hydrogen production, it has to pass by the stages of water filling of the separators and nitrogen purge in all the gas circuit. This procedure, known as *startup* sequence, takes a starting time of around 10 s. This delay is also an issue to be incorporated in the EMS of the micogrid. In a similar way, fuel cells also have a startup sequence. During the *stand-by* state, the air fan is switched off and the hydrogen inlet valve is closed. The period required for nominal conditions to be reached is known as the startup sequence. During startup, the fan switches on and the hydrogen inlet valve opens. During this time, no current is delivered by the fuel stack. The starting time of the fuel cell is typically around 2 s.

This startup sequence can be expressed as a delay between the logical control signal $\Lambda(t)$ to switch on/off the electrolyzer until the electrolyzer is ready or in an energized state, defining $\Lambda_{elz} = 1$ when the logical control signal is "ON" and $\Lambda_{elz} = 0$ if the logical control signal is "OFF". The energized state is defined by the logical variable δ_{elz} whose value is set to "1" in this state and "0" in the rest of the states. Due to the startup sequence, $\delta_{elz}(t)$ can be expressed as function of $\Lambda_{elz}(t)$. The energized state is reached when Λ_{elz} is active in all the instants of the required period ψ_{elz} for the starting sequence. The relation between $\delta_{elz}(t)$ and $\Lambda_{elz}(t)$ is defined by

$$\delta_{elz}(t) = 1 \Leftrightarrow \psi_{elz} - \sum_{k=0}^{\psi_{elz}} (\Lambda_{elz}(t-k)) \leq 0 \tag{5.56}$$

Using the conversions defined in P6 of Table 5.2, this expression can be transformed into the constraints expressed in inequalities below:

$$\epsilon \leq \psi_{elz} - \sum_{k=0}^{\psi_{elz}} (\Lambda_{elz}(t-k)) - (m-\epsilon) \cdot \delta_{elz}(t) < \infty \tag{5.57}$$

$$-\infty \leq \psi_{elz} - \sum_{k=0}^{\psi_{elz}} (\Lambda_{elz}(t-k)) + M \cdot \delta_{elz}(t) \leq M \tag{5.58}$$

In previous equations M and m refer to the maximum and minimum value of $\psi_{elz} - \sum_{k=0}^{\psi_{elz}} (\Lambda_{elz}(t-k))$, being $M = \psi_{elz}$ and $m = 0$.

Similar procedure has to be followed to integrate the delay in the fuel cell. In this controller, the signals corresponding to σ_{elz}^{on} and σ_{fc}^{on} have to be introduced with Λ_{elz} and Λ_{fc}, respectively.

$$\sigma_{\alpha}^{on}(t) = \Lambda_{\alpha}(t) \wedge (\sim \Lambda_{\alpha}(t-1))|_{\alpha=elz,fc} \tag{5.59}$$

Notice that the power fluctuations ϑ_{elz} and ϑ_{fc} only affect the energized state of the electrolyzer and fuel cell. So they have to be introduced as done in the tertiary controller.

Application Example

Real operational scenario in renewable energy microgrids differs from the forecast carried out in the long-term managed by the tertiary control. The secondary control has to follow the contracted schedule established with the grid. This MPC controller is responsible for tracking the schedule carried out protecting the hybrid ESS.

The secondary controller is tested on Hylab, using a sampling time of $T_s = 1$ s (more details can be found in [15]). The weighting factors and constraint limits of the different components of the microgrid managed by the controller are exposed in Table 5.4. The weighting factor assignment criterion is that the maximum weight is given to the schedule tracking with the main grid. The second level in power tracking importance is given to the hydrogen ESS in order to minimize the number of working hours, although the energy tracking has lower priority than the battery. In the degradation costs, the biggest importance is given to the hydrogen facilities, followed by the battery and finally the ultracapacitor.

In Fig. 5.5, three graphs are exposed. The first one corresponds to the schedule carried out by the tertiary controller imposed to the secondary controller. The second one corresponds to the experimental results obtained for the secondary controller applied to the Hylab microgrid. The results of a 24-h experimental test applied to this controller can be observed for both power results and stored energy results. A day with a very fluctuating weather profile which has periods of wind early in the morning, cloudy in the noon, and sunny in the evening is selected in order to show the potentiality of the controller. As can be seen in Fig. 5.5, due to the stochastic behavior of the renewable energy and load consumption, only the power schedule applied to the grid is followed exactly at all the instants of the day. The day begun with smooth power variations due to the wind, as can be seen from 0–5 h, a smooth power profile is demanded to the batteries, while the power fluctuations are absorbed by the ultracapacitor, whose SOC is maintained over SOC_{uc}^{ref} given by the controller. In the period from 5 h to 6 h, high power fluctuations in the P_{net} of the microgrid were found while the electrolyzer was scheduled to be activated. As can be observed in this period,

Table 5.4 Constraints limits and weighting factors imposed to the controller

Microgrid component	Controller's values
Grid	$P_{grid}^{max} = 6000$ W, $P_{grid}^{min} = -2500$ W, $w_{grid}^{P} = 10^{10}/(P_{grid}^{max})^2$
Ultracapacitor	$P_{uc}^{max} = 3000$ W, $P_{uc}^{min} = -3000$ W, $SOC_{uc}^{max} = 1 p.u.$ $SOC_{uc}^{min} = 0.1 p.u.$ $w_{uc}^{P} = 1/(P_{uc}^{max})^2$, $w_{uc}^{E} = 10^8$ $SOC_{uc}^{ref} = 0.5$
Electrolyzer	$P_{elz}^{max} = 900$ W, $P_{elz}^{min} = 300$ W, $w_{elz}^{P} = 10^7/(P_{elz}^{max})^2$ $\Delta P_{elz}^{max} = 20$ Ws^{-1}, $\Delta P_{elz}^{min} = -20$ Ws^{-1}, $w_{elz}^{ripple} = 10^9/(P_{elz}^{max})^2$ $w_{elz}^{startup} = 10^6$, $w_{elz}^{shutdown} = 10^{10}$
Fuel Cell	$P_{fc}^{max} = 750$ W, $P_{fc}^{min} = 300$ W, $\Delta P_{fc}^{max} = 10$ Ws^{-1}, $\Delta P_{fc}^{min} = -10$ Ws^{-1}, $w_{fc}^{P} = 10^7/(P_{fc}^{max})^2$ $w_{fc}^{ripple} = 10^9/(P_{fc}^{max})^2$ $w_{fc}^{startup} = 10^6$, $w_{fc}^{shutdown} = 10^{10}$
Metal Hydride	$LOH^{max} = 7$ Nm^3, $LOH^{min} = 0$ Nm^3 $w_{tank}^{E} = 10^6/(LOH^{max})^2$
Pb-Acid Battery	$P_{bat}^{max} = 2500$ W, $P_{bat}^{min} = -2500$ W $SOC_{bat}^{max} = 1 p.u.$ $SOC_{bat}^{min} - 0.2 p.u.$ $w_{bat}^{P} = 10^2/(P_{bat}^{max})^2$, $w_{bat}^{E} = 10^8$ $w_{bat}^{degr} = 10^5/(P_{bat}^{max}/U_{bat}^{nom})^2$, $w_{bat}^{ripple} = 10^4/(P_{bat}^{max})^2$

the ultracapacitor absorbs the highest requirement in power fluctuations. A smoother power profile is demanded to the battery, while the startup and shutdown cycles of the electrolyzer are minimized. This can be observed at 5.45 when, although the net power of the microgrid is decreasing, the electrolyzer shutdown is delayed until 6.00. In a similar way, the control system is able to start up the electrolyzer although it was not scheduled at 6.30 in order to minimize P_{bat}^{ch} at this moment. Although high changes in the remaining power profile appear at 7.00, the electrolyzer is smoothly shutdown and the battery power is smoothly changed while all the power oscillation is absorbed by the ultracapacitor. In all the cases, although high power oscillations exist, the schedule with the grid is followed satisfactorily.

Similar results can be observed for the cases when the fuel cell is started up, such as the interval of 18.00 when the fuel cell startup is delayed until 19.00. At 20.30, a high load step is demanded, the MPC changes the power of the ultracapacitor and smoothly the fuel cell power and the battery. Finally, the ultracapacitor is again charged to be available for the next high power step fluctuation. The third graph in Fig. 5.5 corresponds to the evolution of the state variables SOC_{uc}, SOC_{bat}, and LOH.

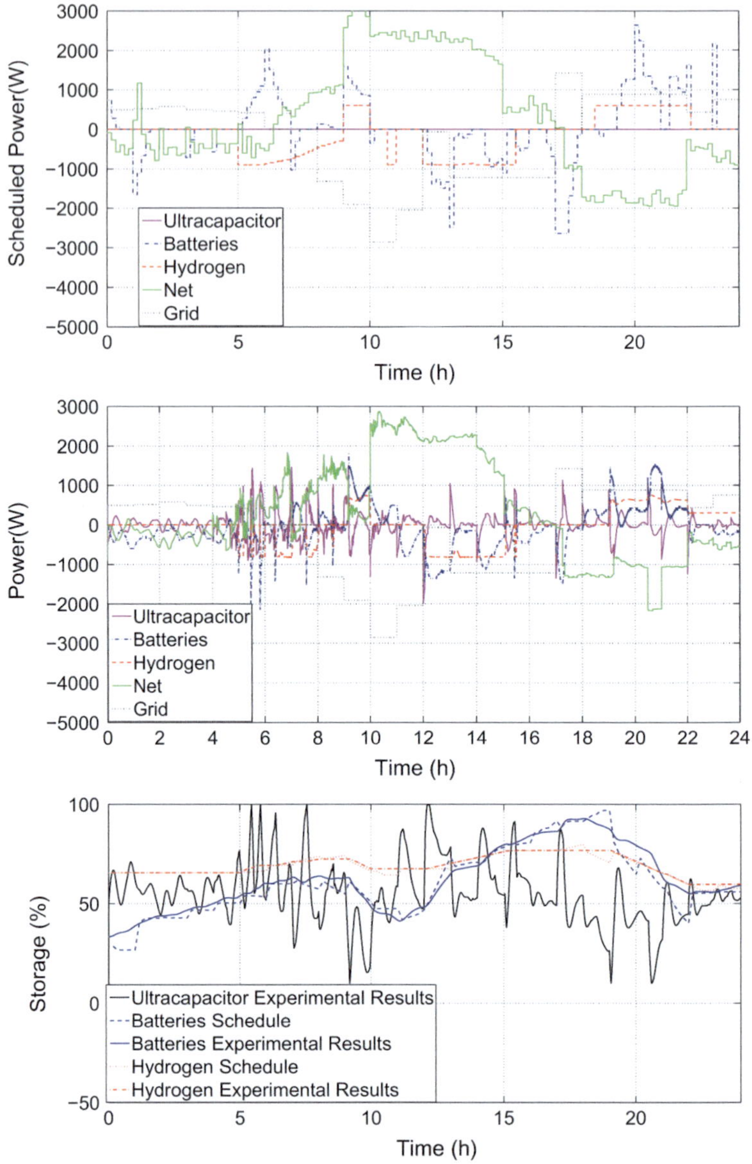

Fig. 5.5 Schedule and experimental results of the secondary controller

While a correct tracking in the SOC_{bat} and LOH is achieved, high oscillations are found over the reference of SOC_{uc}. The maximum and minimum levels of the state variables imposed to the controller are respected.

The complementarity among the three ways of storing energy is also demonstrated in this chapter. When the hydrogen storage is active, the overcharge and the undercharge of the battery and the ultracapacitor can be controlled. The high-stress current ratio in the battery is also neglected with the use of the hydrogen ESS. The fluctuating operation conditions in hydrogen are avoided with the use of batteries and ultracapacitor. The results obtained from cost function minimization conclude that the hydrogen ESS is appropriate for great quantities of energy storage requirements in short-functional periods, while batteries are appropriate for low quantities in the energy storage requirements with long-functional periods. Ultracapacitors are the most appropriate technology to correct rapid fluctuations in the disposal of energy in the microgrid. When the hydrogen storage is active, the overcharge and the undercharge of the battery and the ultracapacitor can be controlled.

The proposed MPC algorithm can contribute to improve the lifetime of the ESS. The method can be expanded to other technologies used to store energy.

5.3 Integration in Electrical Market of Microgrids Using MPC

This section presents the different stages of the electrical market and how MPC can be formulated in order to improve the economic competitiveness of microgrids through their integration in the electrical market. First, an overview of the operation of advanced electrical markets is exposed. Later on, the different aspects of the formulation of a tertiary MPC controller to integrate the different stages of the electrical market are detailed and validated.

5.3.1 Electrical Market Operation

The electricity market is one of the most complex processes due to size of the system to be operated. The large amount of consumers and suppliers make it necessary accurate long-term and real-time markets to carry out the schedule which matches generation and consumption in order to avoid unbalances in the power grid. But besides these mentioned long-term and real-time matching processes, the end-user energy price has to be optimized at each instant. For these issues, electricity markets have several steps. The matching process between suppliers and consumers begins with the day-ahead (spot) market, where participants[1] (sellers and buyers) propose, before market closure, their bids concerning a quantity, and a price of energy over the following delivery horizon. Participants have to be compromised under contract of being responsible for any deviation with respect to the schedule proposed. The forecasts

[1] Market participants are undertakings that are authorized to act directly in the electric power market as buyers and sellers of electricity.

of energy production and consumption are more accurate when the prediction horizon is shorter. For this issue, advanced electrical markets include intraday markets, where it is possible to correct the schedule carried out in the day-ahead market. The regulation market is developed to manage real-time deviations with respect to the day-ahead and intraday schedules. This market, managed by the System Operator,[2] ensures real-time balance between generation and load. For the fast load variations, and unforeseen problems with production capacity, there are reserves at the system operators disposal [19, 35].

The mentioned markets are referred to active power, but it should be noticed that this is only a part of the problem. Safety and reliable operation is required at each instant. The economic schedules closed for each hour of the next day have also to be validated with relevant bilateral contracts with the system operator. When congestion occurs, these schedules are modified according to technical constraints. In order to ensure the balance of the power system, the system operator has at his disposal several services called the ancillary services.

It would be quite difficult to establish an international electricity market model description in order to apply advanced control strategies. The high penetration of renewable energy in the Iberian (Spain and Portugal) electricity market makes it a good reference model on which to develop the control algorithm which tries to solve the economic dispatch of a smart grid highly composed of renewable energy. There are several advanced electricity markets with similar characteristics, such as the Australian market, the USA market, the German market, or the Norwegian market. The Iberian market sequence is shown in Fig. 5.6. As can be observed, there are two main actors in the market: the Market Operator and the System Operator. Although it is out of the scope of this book, there exist two kinds of Systems Operators, the Distribution System Operator who manages the distribution power grids and the Transmission System Operator who manages the transmission power grid.

In several markets, it exists a special regime which includes renewable generation and cogeneration. Electricity generation under the special regime is already a key part of markets with high penetration of renewable energy as the Iberian. It should be considered as essential actions to integrate aspects concerning environmental protection, energy efficiency, and independence in the process of electricity energy liberalization. This special regime should not only contribute to the final energy production, but also it should be considered for the available capacity. The energy transition requires a solid contribution to the development of the market and to the safe operation of the system considering high penetration of renewable energies. Currently, this special regime takes a particular consideration regarding renewable energy. This kind of producers can participate in the market with a regulated tariff or in the free market. If the regulated tariff is selected, the average or reference tariff is established by law. Its value is mainly based on the revenues needed to satisfy

[2]The role of the system operator in a wholesale electricity market is to manage the security of the power system in real time and coordinate the supply of and demand for electricity, in a manner that avoids fluctuations in frequency or interruptions of supply. On the other hand, the market operator is in charge of the wholesale transactions (bids and offers) in the day-ahead and intraday markets.

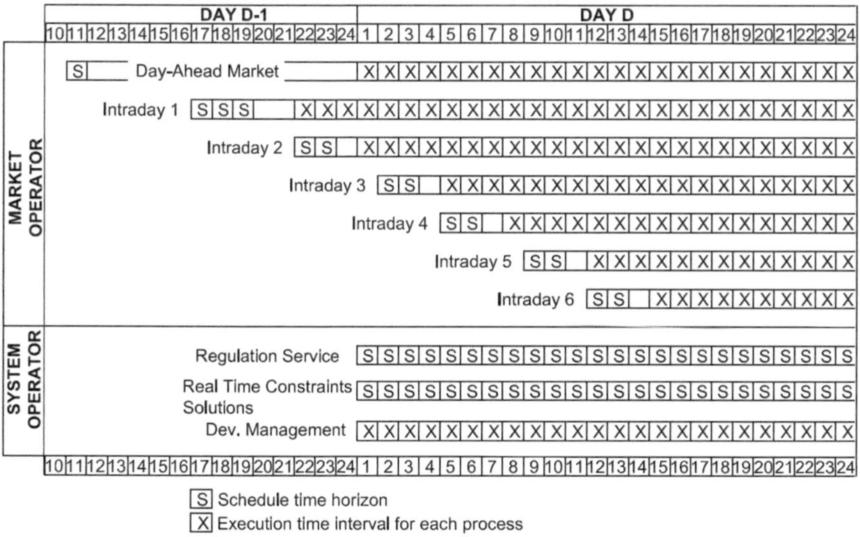

Fig. 5.6 Market sequence in the Iberian electricity market

the economic remuneration of each electric activity and the estimation given for the consumption. If the second case is chosen, these producers will receive the spot market price plus a regulated premium which is used in order to compensate the marginal costs supported by these producers and the risk of electricity generation with a less mature technology. In the current transition, renewable producers will have to compare their technologies with classic producers. So these technologies will have to participate only in the free market (assuming more difficult rules). Microgrids and ESSs, as well as the use of advanced controller in the EMS emerge as technological solution. In the next sections, the way to integrate microgrids in the market using advanced MPC controllers is described, as well as the operating procedure for the participation in the mentioned markets.

Day-Ahead Market

The purpose of the day-ahead market is to establish, in an hourly interval, a Locational Marginal Pricing (LMP) which is established for the next operating day based on generation offers, demand bids, and scheduled bilateral transactions (see Fig. 5.7). In the day-ahead market, the market operator handles the electrical transactions which will take place in the next day by means of the presentation of electricity bids corresponding to the processes of selling and purchasing carried out by market participants. This process is based on bids presented to the market operators, which are included in a matching procedure which will have as result the daily programming schedule corresponding to the day after the deadline date for receiving bids for the day ahead session. This session is comprised of 24 consecutive programming hours corresponding to the whole following day. The role of the market operator is to

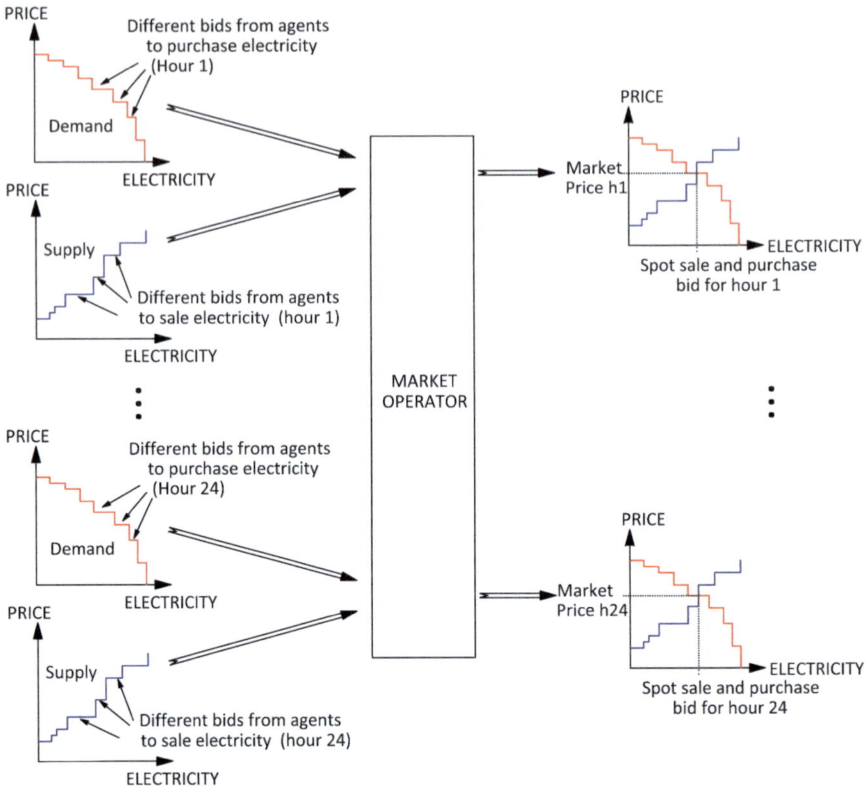

Fig. 5.7 Locational marginal price formation in the day-ahead spot market

match electricity power purchase and sale bids. The closing hour for this process is 10.00 a.m. of the previous day when these electricity transactions are going to be carried out. The calculation of the energy price for each hour corresponds to the price of the last block of the sale bid of the last production unit which has been accepted to match the demand. The market operator publishes the matching result from this process, representing the schedule for the hourly production and demand given. It should be noticed that there exist several concepts which increments the price of energy purchase with respect to the price of energy sale. This increment in the prices is mainly due to aspects as payments by capacity, payment for energy losses, net marketing, traps, VAT, distribution tolls, test equipment rental, or the electricity excise duty [32].

Intraday Market

Electricity markets with high penetration of renewable energies incorporate intraday market sessions to readjust the final viable daily schedule. The purpose of these sessions is to respond to changes in the energy forecast scenario through the presentation of new power sale and purchase bids. Taking as example the Iberian market, the

Table 5.5 Intraday market sessions (Source [32])

Session	First (h)	Second (h)	Third (h)	Fourth (h)	Fifth (h)	Sixth (h)
Opening	16.00	21.00	01.00	04.00	08.00	12.00
Horizon	21–24	1–24	5–24	8–24	12–24	16–24
Schedule	28	24	20	17	13	9

intraday market is structured into six sessions whose hourly distribution and schedule horizon are shown in Table 5.5. Agents may only participate in those hourly periods corresponding to which they have participated in the day-ahead market [32].

After each intraday market session, the final schedule is the result of the complete acceptance of the bid plus the previous schedule (day-ahead or previous intraday market session) of the selling or purchasing unit. The limitations imposed by the system operator for the schedule horizon have always to be respected. As done in the day-ahead market, the market operator matches electricity power purchase and sale bids. As proceeded in the day-ahead market, the final price for each hourly schedule will be the price of the last block of the sale bid of the last production unit whose acceptance has been carried out in order to meet, partially or totally, the purchase bids at a price equal to, or greater than, the marginal price [32].

Ancillary Services

The system operator is responsible for these required activities which ensure not only the continuity but also the security of supply. The system operator is also in charge of coordinating producers with the electrical transmission system. It has to be ensured that the energy produced by the generators is transported through the transmission and distribution networks accomplishing the quality conditions defined by the current regulations [32]. All the steps and coordination in the different actions between system and market operators are summarized in Fig. 5.6.

The system operator has several tools at his disposal to carry out the balance of the power system. In first instance, the momentary imbalances are regulated by primary regulation. This regulation reserves are obligatory but not remunerated. The primary regulation automatically corrects instantaneous imbalances between generation and demand, adjusting the speed/power of the generators.

If the imbalance is prolonged during more than 30 s, the secondary regulation reserves are used. The secondary regulation manages the imbalances with a time horizon of between 30 s and 15 min. The dynamic response has to be corresponding to a time constant requirement of 100 s. Secondary regulation is marginally paid based on two concepts: availability (power band) and use (energy), both to increase and to decrease power generation. When secondary reserves are activated, the primary reserves are again free for the regulation of new imbalances. If the imbalance problem continues, the tertiary regulation is activated, making secondary regulating resources available again [32]. The secondary and tertiary regulations are external reserves to the system operator, which are remunerated at the same price. These reserves are not obligatory. These complementary services are coordinated at the end of the day-ahead market by the system operator [32].

5.3.2 Design of a Tertiary MPC Controller for Electrical Market Integration

In this section, an MPC controller to be integrated into the different electrical market stages (day-ahead, intraday sessions, and regulation service) is designed. The different timescales of the electricity markets are integrated into a whole tertiary controller. The mentioned controller is up to manage the long-term horizon schedule required by the day-ahead market (38 h ahead) but also managing the different sessions of the intraday market (28-9 h). Finally, the MPC is also valid to act in the regulation service market with a schedule horizon of 3 h, using a sampling time of 10 min. Finally, it sends the schedule references required for the power sharing carried out by the secondary controller (see previous section). The schedule carried out by the tertiary controller will determine the most appropriate use of ESSs in the microgrid, considering operational and degradation costs and constraints of each ESS.

The block diagram of the different levels of the MPC controller is detailed in Fig. 5.8. In order to clarify the algorithm explanation, the tertiary MPC controller has been divided into three modules: day-ahead, intraday, and regulation service. The

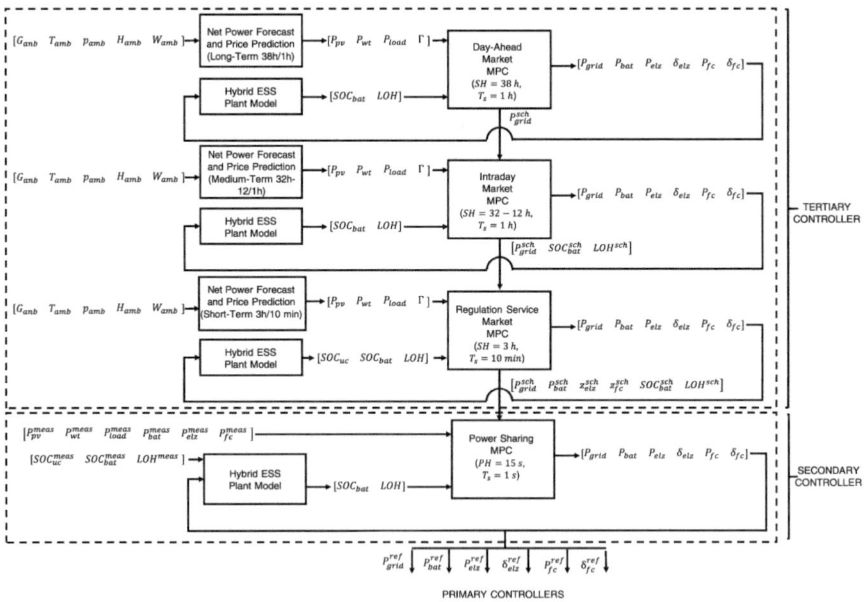

Fig. 5.8 Block diagram for the tertiary and secondary MPC controller

secondary controller corresponds to power sharing. The whole block diagram of the tertiary controller with all the market steps and the secondary controller is exposed in the figure.

Day-Ahead Market MPC

The day-ahead MPC controller has a similar formulation as the one developed for the tertiary controller given in the previous section. The only difference is based on the fact that the schedule is done at 10.00 a.m of the previous day, as can be seen in Fig. 5.6.

$$\begin{aligned} J_{grid}(t) = \sum_{k=D}^{k=D+SH} & \big(-\Gamma_{sale}^{DM}(t+k|t)\cdot P_{sale}(t+D+k|t) \\ & +\Gamma_{pur}^{DM}(t+k|t)\cdot P_{pur}(t+D+k|t)\big)\cdot T_s \end{aligned} \tag{5.60}$$

In Eq. (5.60), D represents the number of remaining sampling instants to begin the mentioned schedule at the moment it is developed; $\Gamma_{sale}^{DM}(t+D+k|t)$ and $\Gamma_{pur}^{DM}(t+D+k|t)$ represent the forecast value for the energy price in the day-ahead market. $P_{sale}(t+D+k|t)$ and $P_{pur}(t+D+k|t)$ represent the sale and purchase of energy with the grid. As done in the tertiary controller in previous section, the signal related to the selling of energy is negative. This is done with the objective to maximize the benefits of the revenue of selling energy in the day-ahead market. The cost function for the batteries is similar to previous section, but considering the delay D until the schedule is carried out.

$$\begin{aligned} J_{bat}(t) = \sum_{k=D}^{k=D+SH} & \left(\frac{\text{CC}_{bat}}{2\cdot\text{Cycles}_{bat}}(P_{bat,ch}(t+k|t)+P_{bat,dis}(t+k|t))\cdot T_s\right. \\ & \left.+\text{Cost}_{degr,ch}\cdot P_{bat,ch}^2(t+k|t)+\text{Cost}_{degr,dis}\cdot P_{bat,dis}^2(t+k|t)\right) \end{aligned} \tag{5.61}$$

The same aspects have to be considered to reformulate the cost function developed in Eq. (5.37) and (5.38), for the electrolyzer and the fuel cell:

$$\begin{aligned} J_{elz}(t) = \sum_{t=k+D}^{k+D+SH} & \left(\left(\frac{\text{CC}_{elz}}{\text{Hours}_{elz}}+\text{Cost}_{o\&m,elz}\right)\delta_{elz}(t+k|t)\right. \\ & \left.+\text{Cost}_{start,elz}\cdot\sigma_{elz}^{on}(t+k|t)+\text{Cost}_{degr,elz}\cdot\vartheta_{elz}^2(t+k|t)\right) \end{aligned} \tag{5.62}$$

$$\begin{aligned} J_{fc}(t) = \sum_{t=k+D}^{k+D+SH} & \left(\left(\frac{\text{CC}_{fc}}{\text{Hours}_{fc}}+\text{Cost}_{o\&m,fc}\right)\delta_{fc}(t+k|t)\right. \\ & \left.+\text{Cost}_{start,fc}\cdot\sigma_{fc}^{on}(t+k|t)+\text{Cost}_{degr,fc}\cdot\vartheta_{fc}^2(t+k|t)\right) \end{aligned} \tag{5.63}$$

Intraday Market MPC

As commented before, the intraday market sessions correct the final schedule of energy exchange with the main grid. The first session of the intraday market is used to reschedule the energy plan established by the day-ahead market session. The second intraday session corrects the schedule of the first intraday session. The rest of intraday sessions adjust the energy schedule of the corresponding previous session. The main rule of intraday market sessions is that market participants may only participate for the hourly periods corresponding to those included in the day-ahead market. The final exchange of energy with the main grid is given by Eq. (5.64). The sampling time used for this control level is similar to the day-ahead, $T_s = 1$ h. The schedule horizon of the controller corresponds to the established in Table 5.5.

$$P_{grid}(t) = P_{grid}^{DM}(t) + \sum_{session=1}^{N_s} P_{grid}^{IM,session}(t) \tag{5.64}$$

where N_s is the number of sessions of the intraday markets which have occurred in the current sample instant. Taking into account the considerations made for the intraday market control level of the microgrid, the next global cost function is introduced, whose Schedule Horizon (SH) is given by the prediction horizon of each intraday session. The same components which acted in the day-ahead market work in the intraday market.

$$\min J(t) = \sum J_i(t)|_{i=grid,bat,H_2} \tag{5.65}$$

All the terms of the cost functions are analogous to the case of the day-ahead market (changing the N_p accordingly), except the terms related to the grid. The intraday market schedule algorithm can only participate in those instants when energy exchange with the grid in the day-ahead market exists. For this reason, active/inactive binary coefficients are included in the cost function, being actives $\alpha_{grid}^{DM}(t) = 1$ in those instants when an exchange of energy is carried out in the day-ahead market.

$$\begin{aligned} J_{grid}(t) = & \sum_{k=D}^{k=D+SH} \big(-\Gamma_{sale}^{IM}(t+k|t) \cdot z_{sale}(t+k|t) \\ & +\Gamma_{pur}^{IM}(t+k|t) \cdot z_{pur}(t+k|t)\big) \cdot \alpha_{grid}^{DM}(t+k) \cdot T_s \end{aligned} \tag{5.66}$$

Similar cost functions at the intraday market are used for the electrolyzer, the fuel cell, and the batteries as given for the day-ahead MPC.

The results of the application of the MPC controller for the intraday market are exposed in Fig. 5.9. As can be seen, the forecast scenario for the net power in the microgrid differs between the day-ahead market and the intraday market. This difference can be done by rescheduling the ESS but also modifying the scheduled energy exchange with the main grid. This action can be done only for those instants when the microgrid has acted exchanging energy in the day-ahead market. The rest

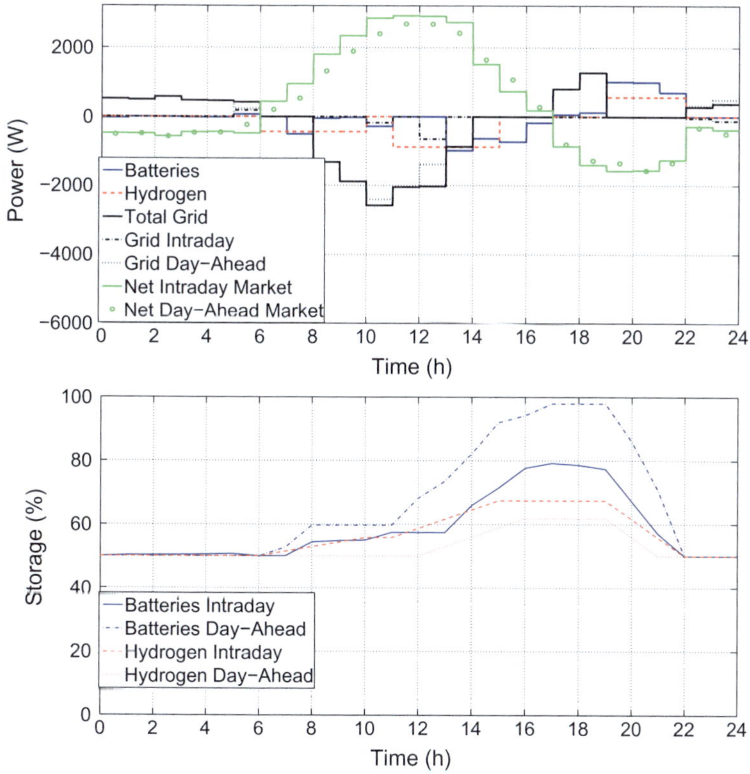

Fig. 5.9 Intraday MPC controller schedule results

of reschedule has to be done with the ESS since it is not possible to buy or sell energy in the intraday market in those hourly periods in which the microgrid has not participated in the day-ahead market. The total energy exchange with the main grid is given, adding the results of the day-ahead market and the intraday-ahead market.

Regulation Service Market MPC

As commented in previous section, the regulation service market is coordinated by the system operator. It helps to avoid imbalances between generation and consumption in order to maintain the nominal frequency of the system. Market participants are penalized for these deviations with respect to the schedule carried out in the day-ahead and intraday markets. This penalty cost is used as a way to encourage market participants to maintain their power balance. The sample time of this controller is 10 min, while its schedule horizon is $SH = 18$. In order to avoid further deviations in these instants after this controller schedule, the final references for the level of stored energy have to be followed. The sample time $T_s = 10$ min is chosen according to the sampling time of data supplied by the meteorological station. At this schedule level, all the components of the microgrid are working, including the ultracapacitor. The cost function of this controller is given in Eq. (5.67).

$$\min J(t) = J_{grid}(t) + J_{uc}(t) + J_{bat}(t) + J_{H_2}(t) \tag{5.67}$$

The penalties of the regulation market can be for excess or deficit of energy exchange with the main grid. Both aspects will affect the balance between generation and consumption in the power system. The penalties are different in the two cases: positive (downregulation required) or negative (upregulation required). In order to formulate this, two auxiliary Boolean variables are introduced $\delta^{down}(t)$ and $\delta^{up}(t)$:

$$(P_{grid}(t) - P_{grid}^{sch}(t)) \leq 0 \Leftrightarrow \delta_{down}(t) = 1 \tag{5.68}$$

$$1 \leq \delta_{up}(t) + \delta_{down}(t) \leq 1 \tag{5.69}$$

Using the proposition P6 defined in Table 5.2, these expressions result in the introduction of the corresponding constraints: Finally, two variables are defined as mixed products to compute the power exchange given in the real-time market:

$$z_{down}(t) = -(P_{grid}(t) - P_{grid}^{sch}(t)) \cdot \delta_{down}(t) \tag{5.70}$$

$$z_{up}(t) = (P_{grid}(t) - P_{grid}^{sch}(t)) \cdot \delta_{up}(t) \tag{5.71}$$

The cost function of the grid (5.72) is defined by the economic costs given for the upregulation z_{up} and downregulation z_{down}, in order to follow the schedule given by the intraday market $P_{grid}^{sch}(t)$. Both processes have different prices which can be fixed or predicted: $\Gamma_{up}(t+k|t)$ and $\Gamma_{down}(t+k|t)$.

$$\begin{aligned} J_{grid}(t) = \sum_{k=0}^{SH} & \left(\Gamma_{up}(t+k|t) z_{up}(t+k|t)\right. \\ & \left. + \Gamma_{down}(t+k|t) z_{down}(t+k|t)\right) \end{aligned} \tag{5.72}$$

The ultracapacitor main mission is to absorb abrupt power changes in the microgrid, since the other ESSs will incur in degradation processes. Its associated cost function tries that the ultracapacitor will always be in an intermediate state of charge.

$$J_{uc}(t) = \sum_{k=0}^{SH} w_{uc} \left(SOC_{uc}(t+k|t) - SOC_{uc}^{ref}\right)^2 \tag{5.73}$$

The weighting factor w_{uc} has to be selected depending on the desired flexibility of the ultracapacitor.

The cost function for the battery manages the deviation in the state of charge of the batteries at the end of the schedule horizon. All the aspects concerning the operation and degradation costs related to the batteries are also considered by the controller. At this control step, the power fluctuation of the batteries is penalized with the last term of the cost function due to the presence of the ultracapacitor.

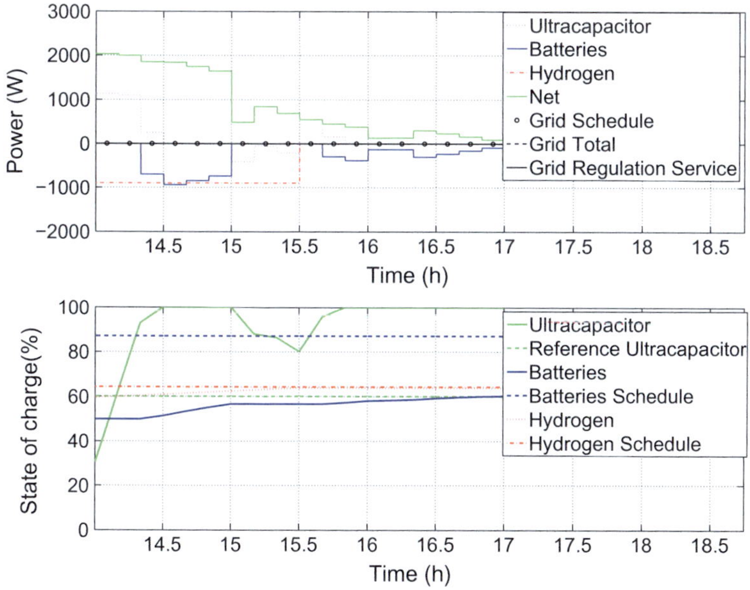

Fig. 5.10 Regulation service MPC controller schedule results under normal scenario

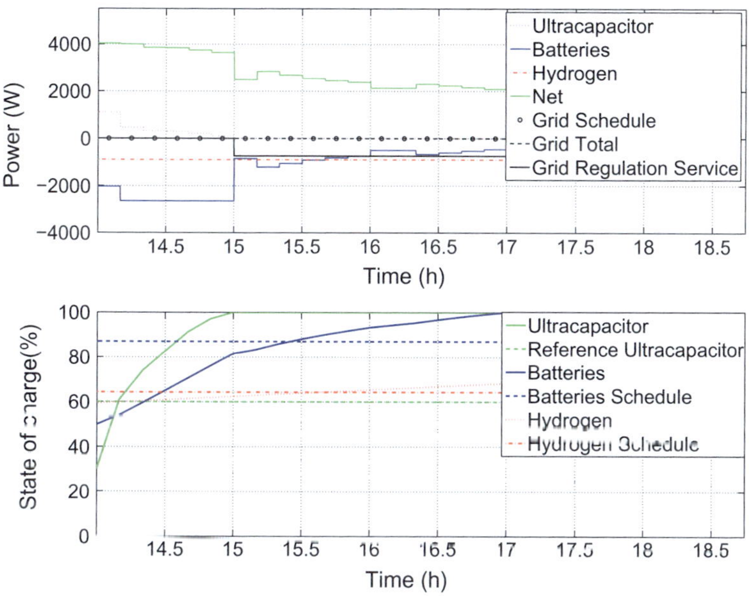

Fig. 5.11 Regulation service MPC controller schedule results under surplus energy scenario

$$\begin{aligned} J_{bat}(t) &= w_{bat}\left(SOC_{bat}(t+SH|t) - SOC_{bat}^{sch}(t+SH)\right)^2 \\ &+ \frac{1}{6}\sum_{k=0}^{SH}\left(\frac{\mathrm{CC}_{bat}}{2\cdot \mathrm{Cycles}_{bat}}(P_{bat,dis}(t+k|t) + P_{bat,ch}(t+k|t))\right. \\ &+ \mathrm{Cost}_{degr,ch}\cdot P_{bat,ch}^2(t+k|t) + \mathrm{Cost}_{degr,dis}\cdot P_{bat,dis}^2(t+k|t) \\ &\left.+ w_{bat}^{ripple}(\Delta P_{bat}(t+k|t))^2\right) \end{aligned} \tag{5.74}$$

The regulation service cost function for the hydrogen ESS has similar procedure of the one given for the batteries. It penalizes the deviation of the LOH at the last sample instant of the schedule horizon. As done with the batteries, the operation and degradation costs for the electrolyzer and the fuel cell are also considered.

$$\begin{aligned} J_{H_2}(t) &= w_{H_2}\left(LOH(t+SH|t) - LOH^{sch}(t+SH|t)\right)^2 \\ &+ \sum_{k=0}^{SH}\left(\frac{1}{6}\left(\frac{\mathrm{CC}_{elz}}{\mathrm{Hours}_{elz}} + \mathrm{Cost}_{o\&m,elz}\right)\delta_{elz}(t+k|t)\right. \\ &+ \mathrm{Cost}_{startup,elz}\cdot \sigma_{elz}^{on}(t+k|t) + \mathrm{Cost}_{degr,elz}\cdot \vartheta_{elz}^2(t+k|t) \\ &\frac{1}{6}\left(\frac{\mathrm{CC}_{fc}}{\mathrm{Hours}_{fc}} + \mathrm{Cost}_{o\&m,fc}\right)\delta_{fc}(t_{k+j}) \\ &\left.+\mathrm{Cost}_{startup,fc}\cdot \sigma_{fc}^{on}(t+k|t) + \mathrm{Cost}_{degr,fc}\cdot \vartheta_{fc}^2(t+k|t)\right) \end{aligned} \tag{5.75}$$

The regulation service MPC is the last schedule control level before the real scenario carried out in the power sharing given by the secondary controller. This controller is executed every ten minutes sending the energy and power references to the secondary controller. The schedule of this controller under a normal scenario can be observed in Fig. 5.10.

The results of this controller in an energy surplus scenario can be observed in Fig. 5.11, when the exchange of power with the main grid has to be rescheduled acting in the regulation service market due to the fact that ESSs reach their maximum capacity. As can be seen, in both figures all the degradation issues related to batteries, electrolyzer, and fuel cell are minimized. The ultracapacitor is always maintained in an intermediate state of charge when normal conditions are found in Fig. 5.10.

References

1. Adika CO, Wang L (2014) Autonomous appliance scheduling for household energy management. Smart Grid IEEE Trans 5(2):673–682
2. Arce A, del Real AJ, Bordons C (2009) Mpc for battery/fuel cell hybrid vehicles including fuel cell dynamics and battery performance improvement. J Process Control 19(8):1289–1304
3. Awerbuch JJ, Sullivan CR (2008) Control of ultracapacitor-battery hybrid power source for vehicular applications. In: Energy 2030 conference, 2008. ENERGY 2008. IEEE, pp 1–7. IEEE

4. Bemporad A, Morari M (1999) Control of systems integrating logic, dynamics, and constraints. Automatica 35(3):407–427
5. Bordons C, Ridao MA, Pérez A, Arce A, Marcos D (2010) Model predictive control for power management in hybrid fuel cell vehicles. In: Vehicle power and propulsion conference (VPPC), 2010 IEEE, pp 1–6. IEEE
6. De Angelis F, Boaro M, Fuselli D, Squartini S, Piazza F, Wei Q (2013) Optimal home energy management under dynamic electrical and thermal constraints. Ind Inform IEEE Trans 9(3):1518–1527
7. del Real AJ, Arce A, Bordons C (2007) Hybrid model predictive control of a two-generator power plant integrating photovoltaic panels and fuel cell. In: IEEE conference on decision and control, pp 5447–5452
8. US DOE (2014) Multi-year research, development and demonstration plan (hydrogen production). Department of Energy (DOE)
9. Dufo-López R, Bernal-Agustn JL, Contreras J (2007) Optimization of control strategies for stand-alone renewable energy systems with hydrogen storage. Renew Energy 32(7):1102–1126
10. Fathima AH, Palanisamy K (2015) Optimization in microgrids with hybrid energy systems-a review. Renew Sustain Energy Rev 45:431–446
11. Ferrari-Trecate G, Gallestey E, Letizia P, Spedicato M, Morari M, Antoine M (2004) Modeling and control of co-generation power plants: a hybrid system approach. Control Syst Technol IEEE Trans 12(5):694–705
12. Fu C, Lin J, Song Y, Zhou Y, Mu S (2017) Model predictive control of an integrated energy microgrid combining power to heat and hydrogen. In: Energy internet and energy system integration (EI2), 2017 IEEE conference on, pp 1–6. IEEE
13. Garcia FS, Ferreira AA, Pomilio JA (2009) Control strategy for battery-ultracapacitor hybrid energy storage system. In: Applied power electronics conference and exposition, 2009. APEC 2009. Twenty-Fourth Annual IEEE, pp 826–832. IEEE
14. Garcia-Torres F, Bordons C (2015) Optimal economical schedule of hydrogen-based microgrids with hybrid storage using model predictive control. Ind Electron IEEE Trans 62(8):5195–5207
15. Garcia-Torres F, Valverde L, Bordons C (2016) Optimal load sharing of hydrogen-based microgrids with hybrid storage using model-predictive control. IEEE Trans Ind Electron 63(8):4919–4928
16. Geer T, Manwell JF, McGowan JG (2005) A feasibility study of a wind/hydrogen system for marthas vineyard, massachusetts. In: American wind energy association windpower 2005 conference
17. Greenwell W, Vahidi A (2010) Predictive control of voltage and current in a fuel cell-ultracapacitor hybrid. Ind Electron IEEE Trans 57(6):1954–1963
18. Gu W, Wu Z, Yuan X (2010) Microgrid economic optimal operation of the combined heat and power system with renewable energy. In: Power and energy society general meeting, 2010 IEEE, pp 1–6. IEEE
19. Holttinen H (2005) Optimal electricity market for wind power. Energy Policy 33(16):2052–2063
20. Howel D (2012) Battery status and cost reduction prospects. In EV everywhere grand challenge battery workshop, Department of Energy (DOE)
21. Jiang B, Fei Y. Smart home in smart microgrid: a cost-effective energy ecosystem with intelligent hierarchical agents
22. Jiang Q, Xue M, Geng G (2013) Energy management of microgrid in grid connected and stand-alone modes. Power Syst IEEE Trans 28(3):3380–3389
23. Kanchev H, Lu D, Colas F, Lazarov V, Francois B (2011) Energy management and operational planning of a microgrid with a pv-based active generator for smart grid applications. Ind Electron IEEE Trans 58(10):4583–4592
24. Khodaei A (2014) Microgrid optimal scheduling with multi-period islanding constraints. Power Syst IEEE Trans 29(3):1383–1392
25. Korpås M (2004) Distributed energy systems with wind power and energy storage. PhD thesis, Norwegian University of Science and Technology

26. Lin W, Zheng C (2011) Energy management of a fuel cell/ultracapacitor hybrid power system using an adaptive optimal-control method. J Power Sour 196(6):3280–3289
27. Malysz P, Sirouspour S, Emadi A (2014) An optimal energy storage control strategy for grid-connected microgrids. Smart Grid IEEE Trans 5(4):1785–1796
28. MIBEL (2009) Description of the operation of the mibel. Technical report, Regulatory Council, Madrid
29. Tung Nguyen D, Le LB (2014) Risk-constrained profit maximization for microgrid aggregators with demand response
30. Olama A, Mendes PRC, Camacho EF (2018) Lyapunov-based hybrid model predictive control for energy management of microgrids. IET Gen Transm Distrib 12(21):5770–5780
31. Olivares DE, Mehrizi-Sani A, Etemadi AH, Cañizares CA, Iravani R, Kazerani M, Hajimiragha AH, Gomis-Bellmunt O, Saeedifard A, Palma-Behnke R, Jimenez-Estevez GA, Hatziargyriou ND (2014) Trends in microgrid control. IEEE Trans Smart Grid 5(4):1905–1919
32. OMI-Polo Español S.A. (OMIE)
33. Palma-Behnke R, Benavides C, Lanas F, Severino B, Reyes L, Llanos J, Sáez D (2013) A microgrid energy management system based on the rolling horizon strategy. IEEE Trans Smart Grid 4(2):996–1006
34. Pineda I, Wilczek P, Moccia J, Bourgeois S, Wilkes J (2012) Creating the internal energy market in Europe. European Wind Energy Association
35. Pinson P, Chevallier C, Kariniotakis GN (2007) Trading wind generation from short-term probabilistic forecasts of wind power. Power Syst IEEE Trans 22(3):1148–1156
36. Pukrushpan JT, Stefanopoulou AG, Peng H (2004) Control of fuel cell power systems: principles, modeling, analysis and feedback design. Springer
37. Sachs J, Sawodny O (2016) A two-stage model predictive control strategy for economic diesel-pv-battery island microgrid operation in rural areas. IEEE Trans Sustain Energy 7(3):903–913
38. Samson GT, Undeland TM, Ulleberg O, Vie PJS (2009) Optimal load sharing strategy in a hybrid power system based on pv/fuel cell/battery/supercapacitor. In: Clean electrical Power, 2009 international conference on, pp 141–146. IEEE
39. Schaltz E, Rasmussen PO (2008) Design and comparison of power systems for a fuel cell hybrid electric vehicle. In: Industry applications society annual meeting, 2008. IAS'08. IEEE, pp 1–8. IEEE
40. Spendelow J, Marcinkoski J, Satyapal S (2014) Doe hydrogen and fuel cells program record 14012. Department of Energy (DOE)
41. Strbac G, Mancarella P, Pudjianto D, Schwaegerl C, Tao L, Vasiljevska J, Bessa R, Matos M, Anastasiadis A, Hatziargyriou N, et al (2009) More microgrids dh3: Business cases for microgrids. Available at More Microgrids website http://www.microgrids.eu/documents/682.pdf
42. Suh KW (2006) Modeling, analysis and control of fuel cell hybrid power systems. PhD thesis, University of Michigan
43. Thounthong P, Chunkag V, Sethakul P, Sikkabut S, Pierfederici S, Davat B (2011) Energy management of fuel cell/solar cell/supercapacitor hybrid power source. J Power Sour 196(1):313–324
44. Thounthong P, Raël S, Davat B (2009) Energy management of fuel cell/battery/supercapacitor hybrid power source for vehicle applications. J Power Sour 193(1):376–385
45. Vaccaro A, Loia V, Formato G, Wall P, Terzija V (2015) A self organizing architecture for decentralized smart microgrids synchronization, control and monitoring. IEEE Trans Ind Inform 11(1):289–298
46. Vahidi A, Greenwell W (2007) A decentralized model predictive control approach to power management of a fuel cell-ultracapacitor hybrid. In: American control conference, 2007. ACC'07, pp 5431–5437. IEEE
47. Valverde L, Bordons C, Rosa F (2012) Power management using model predictive control in a hydrogen-based microgrid. In: Annual conference on IEEE industrial electronics society, pp 5669–5676

48. Yang P, Nehorai A (2014) Joint optimization of hybrid energy storage and generation capacity with renewable energy. Smart Grid IEEE Trans 5(4):1566–1574
49. Zhang Y, Jiang Z, Yu X (2008) Control strategies for battery/supercapacitor hybrid energy storage systems. In: Energy 2030 conference, 2008. ENERGY 2008. IEEE, pp 1–6. IEEE
50. Zhou T, Francois B (2009) Modeling and control design of hydrogen production process for an active hydrogen/wind hybrid power system. Int J Hydrog Energy 34(1):21–30
51. Zhou T, François B (2011) Energy management and power control of a hybrid active wind generator for distributed power generation and grid integration. Ind Electron IEEE Trans 58(1):95–104

Chapter 6
Demand-Side Management and Electric Vehicle Integration

Abstract This chapter extends the energy management systems developed in previous chapters to the case of controllable loads and electric vehicles. EVs are loads for the microgrid but, due to their storage capability, they can also supply energy to the microgrid when needed and thus they can be considered as prosumers. An appropriate management of loads and EV charging can help improve the operation of the microgrid. The concept of Demand-Side Management (DSM) is introduced, and the main Demand Response (DR) techniques are described and illustrated. The integration of EVs in the microgrid is approached, customizing the MPC techniques to this situation and contemplating the notion of Vehicle-to-Grid (V2G). The chapter presents some simulations to illustrate load shifting and curtailment and several experiments performed in a pilot-scale microgrid to demonstrate V2G capabilities.

6.1 Demand-Side Management

Demand-Side Management (DSM) is a relevant function in electrical networks that allows customers to make decisions regarding their energy consumption and helps operators to reduce the peak load demand and to reshape the load profile. DSM includes everything regarding the demand side of an energy system. It refers to a variety of activities that are related to energy consumption, not only the modification of energy use but also the behaviors that are involved in these processes (such as regulation, promotion, or education) [22]. It comprises programs implemented by utility companies to control energy consumption at the customer side, which are employed to use the available energy more efficiently without installing new infrastructure. The use of DSM provides several benefits, such as improvement in the efficiency of the system, security of supply and reduction in overall operational costs, and environmental impact. On the other hand, the problem becomes more complex since new degrees of freedom appear. While DSM was utility-driven in the past, it might move toward a customer-driven activity in the near future [30].

C. Bordons et al., *Model Predictive Control of Microgrids*,
Advances in Industrial Control, https://doi.org/10.1007/978-3-030-24570-2_6

In DSM framework, Demand Response (DR) refers to the actions taken by customers that use information (mainly prices) to adjust their loads. The concept of DR encompasses the discretionary changes by consumers of their usual consumption patterns in response to external conditions (usually price signals) and has attracted the attention of many researchers [18]. Besides the savings regarding electricity bills, this kind of schemes can be used to avoid undesirable peaks in the demand curve that takes place in some time periods along the day, resulting in a more beneficial rearrangement [24]. Although DSM is a general concept and DR is related to the control actions taken on the load side, in many occasions the terms are used indistinctly.

This section focuses on DR techniques applicable to microgrids. In these systems, the volatility introduced by intermittent resources on the generation side can be compensated for by a more responsive and controllable consumption behavior [1]. With DR, some loads can be adjusted (both in amplitude and in connection times) to contribute to fulfill microgrid operation objectives. The flexibility added by DR entails the introduction of new manipulated variables (both continuous and binary) in the problem formulation, which makes the optimization problem more complex.

6.1.1 Demand Response Techniques

In a microgrid, loads can be manipulated up to a certain point. There are some critical loads whose demand must always be met. These uncontrollable loads are to be operated at a certain power and a certain time that cannot be deferred. But, on the other hand, there exist controllable loads whose total consumption or occupied time duration can be modified, such as Electric Vehicles (EV) or Heating, Ventilating and Air Conditioning (HVAC). Then, some loads can be reduced, shed, or postponed during supply constraints or emergency situations or just to optimize the operation of the microgrid. Controllable loads can be classified as follows [22]:

- Deferrable or shiftable: Their activation can be stopped, restarted, or shifted to other time slots. On account of electricity tariffs or operational needs, they can be shifted from peak to off-peak hours. Shiftable loads are flexible within the time window but their demands cannot be adjusted, and they cannot work before the earliest start time and the latest finish time. In addition, once their work is started they cannot be stopped before completed. Washing machines or electric vehicles belong to this category while lighting systems do not.
- Adjustable or curtailable: Their consumption can be adjusted to a lower level if necessary. Although these loads have a nominal level, their magnitude is flexible so that the demand level can be lowered when needed (e.g., at peak hours or in islanded mode). Heating systems and in general thermal loads are examples of adjustable loads.

However, reducing consumption or shifting the load to some other point in time can affect customer's comfort, which can be measured by the Quality of Experience (QoE) [2]. Therefore, a certain cost must be associated to load curtailment/shifting when implementing the DR procedure.

The primary objective of the DR techniques found in literature is the reduction of system peak load demand and operational costs. The first related techniques used in electrical grids (called Direct Load Control) were developed using dynamic programming [13] and linear programming [20]. Nowadays, load shedding/shifting can also be used in microgrids to prevent system instability under emergency conditions [11] or for frequency regulation [12], and different control strategies are used. The architectural models, technology infrastructure, and communication and control protocols that are currently in use in microgrids are detailed in [35], where some projects for commercial buildings and microgrids are described. This section focuses on DR techniques for EMS during normal operation.

Two main methods can be used to manage loads in a microgrid:

- Curtailment/Shedding: This strategy consists of adjusting the magnitude of the power that can be demanded by loads when necessary. Therefore, it is considered that loads have a certain degree of manipulation, so the demanded power can be lowered during certain times in order to improve the operation of the microgrid or during contingencies. A maximum allowable level of curtailment must be specified for each load and some benefit for the load must be established.
- Shifting: This is a strategy that considers the shifting of certain amounts of energy demand from some time periods to others with lower expected demand, typically in response to price signals. The EMS has flexibility to defer some energy packets, but the total amount of power required by the load must be satisfied for the desired period.

These mechanisms can be used both in grid-connected and in islanded mode. In both cases, they can be used to improve the economic benefit, but in the case of islanded mode they can be crucial, since the grid is not available to supply the loads when power deficit exists. In this case, the required amount of curtailed load must be chosen according to the estimated deficit. Although they can be used at different timescales and control layers, in general load shifting is more oriented to scheduling and load curtailment to power sharing.

There are several ways to address DR. The paper [24] presents a DSM strategy based on load shifting technique for smart grids with a large number of devices of several types. The day-ahead load shifting technique proposed is mathematically formulated as a minimization problem and solved with a heuristic-based evolutionary algorithm. Genetic algorithms are used in [17] for load shifting: the inconvenience caused to the customer is modeled as a polynomial function of the shifting time depending on the type of load. The objective is to minimize the combination of generation cost and the inconvenience caused to the customer. A scenario-based stochastic optimization approach is developed in [4] for real-time price-based DR management of residential appliances, which can be embedded into smart meters, considering time-varying electricity price uncertainties. A multi-objective optimization method and a Linear Matrix Inequality (LMI) approach are used in a case study of three microgrids connected to the grid in [22].

Minimizing the operation cost while minimizing the inconvenience caused due to shifting or curtailment of loads is a multi-objective optimization problem that may include binary variables associated to the connection or disconnection of loads. Therefore, a mixed-integer optimization problem must be solved. This can be done in an open-loop fashion or using feedback as done in MPC.

There are several works using MPC for DR, as the one presented in [19] for a Combined Heat and Power (CHP) residential microgrid. In that work, unit commitment and economic dispatch are performed, considering energy and power requirement for home appliances and using thermal inertia to buffer electricity consumption of the refrigerator and of the living space. Simulation results for 1 year are presented, showing that the use of MPC can reduce the annual operation costs by 7.2%. The MILP solver is embedded in MPC, as done in [31, 32]. In these papers, load curtailment is integrated into the mixed logical dynamical framework. The methodology is extended to the stochastic case in [33], considering flexible loads with an associated variable that represents the percentage of preferred power level to be curtailed at each time.

6.1.2 Formulation of MPC for DR

A general formulation of MPC for DR is formulated in this section for the deterministic case. The strategies described in previous chapters can be extended to include load curtailment and shifting.

Load Curtailment

The basic EMS presented in Chap. 4 can be modified in order to consider load curtailment in a simple way. If the load (or at least part of it) is adjustable, its associated power (P_{load}) can be manipulated by the EMS. Therefore, P_{load} is part of the vector of manipulated variables instead of a disturbance.

Some additional constraints must be added to the optimization problem, since the load can only be adjusted up to a certain point, so its limits must be set:

$$P_{load}^{\min}(t) \leq P_{load}(t) \leq P_{load}^{\max}(t) \qquad \forall t \tag{6.1}$$

where the minimum and maximum values may change at each instant (and can be set to avoid any curtailment if needed). The rest of constraints are those shown in Chap. 4: energy balance and amplitude and rate constraints. Also, since curtailment can lead to inconvenience to users, a set point P_{load}^{ref} can be used in order to avoid great deviations from a desired value and its associated weight ($\gamma(t)$) can be set to a high value to prevent curtailment at a certain time instant or interval. Then, the optimization problem can be solved using QP as all the variables are continuous.

Load Shifting

Although load curtailment can be solved using continuous-valued variables, the consideration of load shifting requires the introduction of logical (binary) variables, as done in Chap. 5. The MILP algorithm presented there can be modified to include DR functionalities. Changes in cost function and constraints are needed to consider DR.

In order to contemplate the possibility of deferring loads, the following variables and constants must be included in the formulation (one for each deferrable load):

- Initial and final times of connection of the deferrable load: t_{ini} and t_{end}.
- Constant power demanded by the load when connected: $P_{load,i}$.
- Total number of active instants: N_t. The load may be activated or not, but if so, it must be connected during these required instants (it cannot stop until its work is completed).
- State of connection of the load: $\delta_{on}(t)$, which takes value 1 when the load is connected and 0 otherwise. This variable takes the value 0 at instants outside the interval $[t_{ini},\ t_{end}]$.
- State of transition of the load: $\sigma_{on}(t)$, which can be defined as[1]

$$\sigma_{on}(t) \Leftrightarrow \delta_{on}(t) \wedge \sim \delta_{on}(t-1)$$

and the following constraints must be defined:

- Energy balance:

$$\sum_{i=1}^{n_g} P_{gen,i}(t) + \sum_{i=1}^{n_e} P_{ext,i}(t) + \sum_{i=1}^{n_s} P_{sto,i}(t) - \sum_{i=1}^{n_l} P_{load,i}\delta_{on,i}(t) = 0$$

- Total active instants:

$$0 \leq \sum_{k=1}^{k=SH} \delta_{on}(k) \leq N_t$$

  where SH is the schedule horizon (usually 24 h).
- Only one transition:

$$0 \leq \sum_{k=1}^{k=SH} \sigma_{on}(k) \leq 1$$

More constraints can be established for different home appliances, as detailed in [4]. Then, the optimization problem must consider continuous and binary variables, and therefore mixed-integer programming must be used.

An example of load curtailment is presented below, whereas load shifting will be demonstrated in the next sections on EV charging applications.

6.1.3 Example: Load Curtailment

In this section, an illustrative example of load curtailment is presented. It shows how DR can help manage the microgrid when the external grid cannot supply energy, that

[1] $\wedge$ stands for the AND logical operator and $\sim$ for NOT.

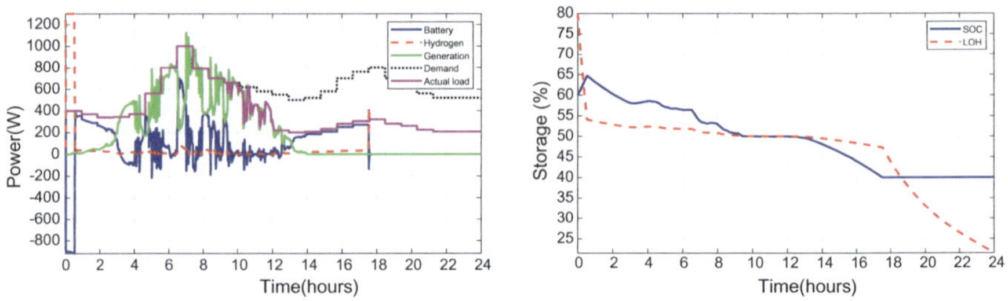

Fig. 6.1 Powers and storage levels with load curtailment

is, the microgrid is working in islanded mode. If the loads are home appliances or HVAC, this adjustment can be easily assumed.

The Hylab microgrid used in the case study 1 of Chap. 4 (see Sect. 4.4.1) is used as an example, considering the possibility of working in islanded mode. Under certain circumstances, the load can be curtailed in order to keep the microgrid in operation even if the load demand is not completely met. Then, the control-oriented model of the microgrid used in Chap. 4 is modified by including the power demanded by the load, P_{load}, as a manipulated variable, with $T_s = 30$ s, and is given by

$$\begin{bmatrix} SOC(t+1) \\ LOH(t+1) \end{bmatrix} = \begin{bmatrix} SOC(t) \\ LOH(t) \end{bmatrix} + \begin{bmatrix} 0.046 & 0.046 & -0.046 \\ -0.169 & 0 & 0 \end{bmatrix} \begin{bmatrix} P_{H_2}(t) \\ P_{grid}(t) \\ P_{load}(t) \end{bmatrix} + $$
$$+ \begin{bmatrix} 0.046 \\ 0 \end{bmatrix} P_{gen}(t)$$

This model contemplates the case of grid-connected mode ($P_{grid} \neq 0$) or islanded mode ($P_{grid} = 0$). In the following simulation, the microgrid is working in islanded mode during all the experiments, so $P_{grid} = 0$ and this variable could be eliminated in the model.

A simulation has been done for a cloudy day (the same as shown in Fig. 4.7). Now, the microgrid is islanded and, in order to keep it operating in the absence of generation, the load can be adjusted. The load profile can then be modified accordingly. A maximum curtailment of 60% is allowed.

Figure 6.1 displays the evolution of the powers and the stored energy. Notice that at the beginning of the day, the battery is used to supply the load with some contribution of the fuel cell, but when the storage units are about to reach their lower limits, the load is curtailed in order to fulfill the operational objectives. The degree of load curtailment will depend on the weights of the cost function and on the constraints. If no curtailment is done, it can be checked by simulations that, although the load is fulfilled at 100% during the first hours, later the storage devices are depleted and the load cannot be fed at all. Consequently, the microgrid has to be shut down.

6.2 Integration of Vehicles in Microgrids: V2G

The connection of electric vehicles to the grid is a tendency for the near future. Consequently, the development of EMS for managing the use of vehicle batteries is a key research field. EVs charging can be included in DSM strategies (since EVs are loads for the microgrid) but, due to their storage capability, EVs can also provide energy to the grid when needed and therefore can be considered as prosumers.

Vehicle-to-Grid (V2G) systems consist of the use of EVs batteries, during periods when they are not being used, as energy storage for an electrical network. It is estimated that a vehicle is in motion only 4% of the time [9], so the rest of the time it could be available as an electrical energy storage unit. Moreover, in normal use, the batteries are recharged overnight (which is the period of low electricity demand) and are parked in the workplace during periods of high electrical demand, so the stored energy could be used to meet peak demand. V2G systems will enable new business models providing services to vehicles, buying or selling energy and building new connections with the network operator. In recent years, control algorithms for charging electric vehicles in intelligent networks have appeared in literature, on one hand, to offer better charging service to attend drivers' demand preferences, and, on the other hand, to ensure a given power profile on the grid, also considering various constraints in vehicles, the charging station, and the grid. The integration of V2G systems can be a key factor in microgrid stability to guarantee against load and generation fluctuations.

Recent studies have focused their research on the optimization of the interaction of EVs and the grid. DSM for EVs is addressed in [28] formulating the problem as a convex optimization, proposing a solution by means of a decentralized algorithm. A moving horizon approach is used to handle the random arrival of EVs and the inaccuracy of the forecast of non-EV load through the use of a distribution grid capacity market scheme. A stochastic optimization strategy that is capable of handling uncertain outputs of EVs and renewable generation is formulated in [36], while other works such as [27] propose a closed-form solution with which to schedule optimally time-shiftable loads with uncertain deadlines, with a focus on charging EVs with uncertain departure times. A multiple MPC strategy applied to the bidirectional charging/discharging of plug-in hybrid electric vehicles by controlling the SOC of the batteries in order to control the microgrid frequency stabilization is developed in [29]. The coordination problem among the EV owner, the aggregator, and the System Operator (SO) is dealt with in [14] by describing a theoretical market framework in which the congestion problem is solved by coordinating SO and aggregators through the use of a distribution grid capacity market scheme. A game theoretical analysis [21] is used to study the competition as regards price among EVs charging stations. In [5, 34], the problem is solved by real-time optimization algorithms, whereas in [10] an MPC-based algorithm is presented. Also, solutions based on hierarchical distributed algorithms have been presented [3, 6, 7]. There are several works investigating how to embed the EVs in Home EMS (HEMS). A HEMS which integrates a prediction of future vehicle usage and home load, optimization of the

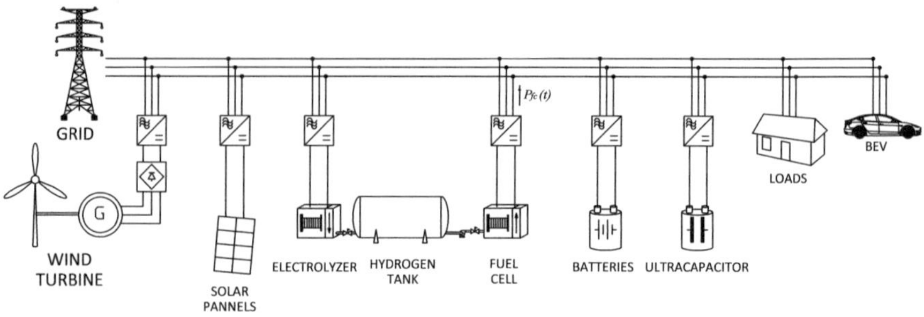

Fig. 6.2 Microgrid with electric vehicles

charge/discharge profile of the in-vehicle batteries, and real-time execution by using MPC framework is presented in [16].

In charging stations, the uncertainty associated to the arrival of EVs to be charged is a key factor; although MPC can deal with it thanks to its intrinsic feedback mechanism, its performance can be improved using a stochastic formulation, which will be described in next chapter.

EVs charging can be done using MPC methodologies described along the book, depending on the particular problem to be solved. Two situations are addressed below: (i) charge management during a known interval of parking time, which is solved by load shifting, and (ii) V2G, where vehicles battery can collaborate with the ESS of the microgrid to maximize benefits.

6.2.1 *Example: Microgrid with an EVs Charging Station*

The charge of EVs can be done using the load shifting mechanism. If cars are parked during a period of time, the charging process can be optimized contemplating energy prices and microgrid operational costs. Considering that charge is done at a constant power, the optimization can be achieved by calculating the best charging interval (inside the period that the car is parked). This example illustrates the management of EVs charging, without V2G capabilities, that is, the EVs do not supply energy to the microgrid. That issue is addressed in the next section.

The microgrid used in Chap. 5 is adopted here to illustrate the procedure. Figure 6.2 shows the microgrid, where the additional capability of charging EVs has been added.

Then, given a parking time interval, the optimization supplies the values of δ_{ev}, indicating the best connection interval. Notice that δ_{ev} takes value 1 if the vehicle is connected at instant t and "0" if not and $\sigma(t)_{ev}$ is used to indicate the transition from disconnected to connected (as indicated in Sect. 6.1.2). The optimization problem is formulated considering the following costs (as done in Chap. 5):

1. The cost of energy exchanged with the main grid is defined by the price of energy in the day-ahead market.
2. The cost of using the batteries as ESS will depend on the number of charge and discharge cycles. A penalty factor is also used to smooth the charging and discharging process of the battery.
3. The cost of using the electrolyzer and the fuel cell will depend on the number of working hours of these devices. Since fluctuation operations in the electrolyzer and the fuel cell, as well as the startup and shutdown states degrade these components, they are penalized.

The following additional constraints are added to accomplish the charging process:

- Fulfill the necessary energy E_{ev} for the desired charge at a constant power $P_{ev,ch}$:

$$\sum_{k=1}^{N_p} P_{ev,ch} T_s \delta_{ev}(t) = E_{ev}$$

- Charge during a total number of instants:

$$0 \leq \sum_{k=1}^{N_p} \delta_{on}(k) \leq N_{ev}$$

- Charge without interruptions (only one transition):

$$0 \leq \sum_{k=1}^{N_p} \sigma_{ev}(k) \leq 1$$

- Energy balance at each instant t:

$$\sum_{i=1}^{n_g} P_{gen,i}(t) + \sum_{i=1}^{n_e} P_{ext,i}(t) + \sum_{i=1}^{n_s} P_{sto,i}(t) - P_{ev,ch}\delta_{ev}(t) = 0$$

The formulation can be extended to any number of vehicles, just adding as many δ (for the connection states) and σ (for the transitions) as the number of EVs and the corresponding constraints, but it is not done here for the sake of simplicity.

The solver finds an optimal solution for the microgrid providing a set of the control variables, which are continuous and binary, so MIQP is used. The output signals generated by the solver are the values of exchange power with the main grid (P_{grid}), the power of batteries, electrolyzer and fuel cell (P_{bat}, P_{elz} and P_{fc}), the activation signals for the electrolyzer and the fuel cell (δ_{elz} and δ_{fc}) and the activation and transition of the EV (δ_{ev} and σ_{ev}). The schedule horizon is 38 h and the sampling time is 1 h, as occurred in the day-ahead market.

Fig. 6.3 Power flows without EV charge

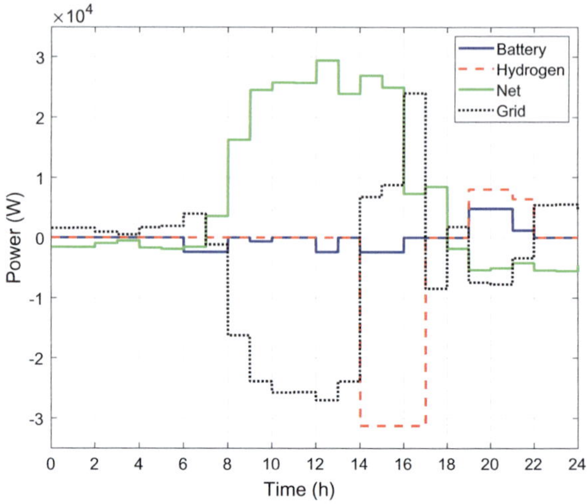

Fig. 6.4 Power flows when midnight charge

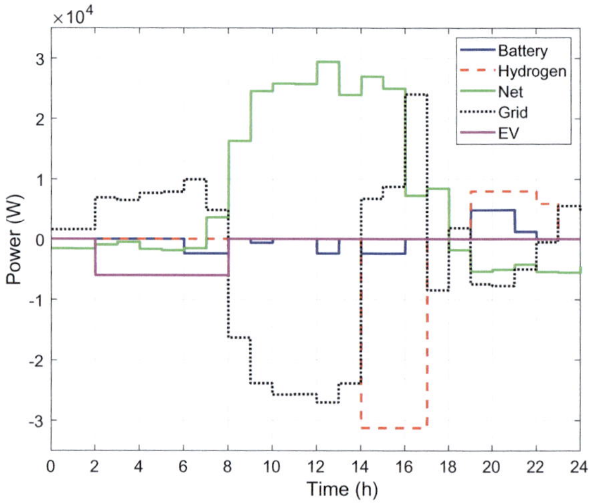

Three scenarios for a sunny day are analyzed: (i) microgrid operation without EVs, (ii) the EV is parked from midnight to 8 a.m. (and it has to be fully charged at 8 a.m.), and (iii) the EV is parked all day and can be charged at any interval along the whole day. The net power is calculated as the difference between solar generation and the loads connected to the microgrid.

Simulation results show that the operation of the microgrid is slightly different in the three cases, although the percentage of power need to charge the EV is not too big with respect to the rest of the microgrid. In the first case (Fig. 6.3), the surplus of energy during the morning is mostly sold to the grid, since price of electricity is high. In the second case (Fig. 6.4), the EV must be charged during the night, in order to be charged at 8 a.m., which implies that most of the energy must be purchased from the

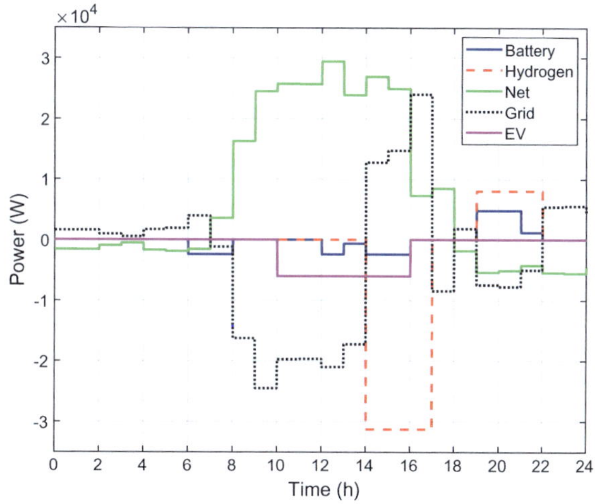

Fig. 6.5 Power flows with flexible EV charge

grid and the recharge cost is 3.52 €. In the third case (Fig. 6.5), since the charging interval can be chosen at any time of the day, the optimizer shifts the load around midday, where a surplus of energy exists and therefore the cost is lower (1.91 €). The conclusion is that load shifting can be used to choose the best charging interval for EVs considering time constraints and optimizing operational costs.

An extension of this example to Fuel Cell Electric Vehicles (FCEVs) and analysis of different market sessions can be found in [8]. The paper also outlines the importance of the schedule for the EV/FCEV in order to consider the best time period in which to recharge/refuel the vehicle, finding lower prices for the recharge of the EV or the refueling of the FCEV if they are planned before the day-ahead market session. This can be extended to other external agents such as the reserves of Transmission/Distribution System Operators, aggregators/prosumers or other microgrids, which can benefit from exchanging energy with microgrids using local markets. The interconnection of several microgrids will be addressed in Chap. 8.

6.2.2 *Case Study: EMS of a Microgrid Coupled to a V2G System*

The objective of this section is to present an MPC strategy for optimizing a microgrid coupled to a V2G system consisting of four charge points for electric vehicles. The proposed algorithm performs the management of renewable energy sources, energy storage units, vehicles charge, and the purchase and sale of electric power with the grid. In this application, EVs can act as loads and also as generators, so they can be considered as prosumers. Up to four vehicles can be at the charging station, and they can exchange energy with the microgrid, which can in turn buy or sell energy

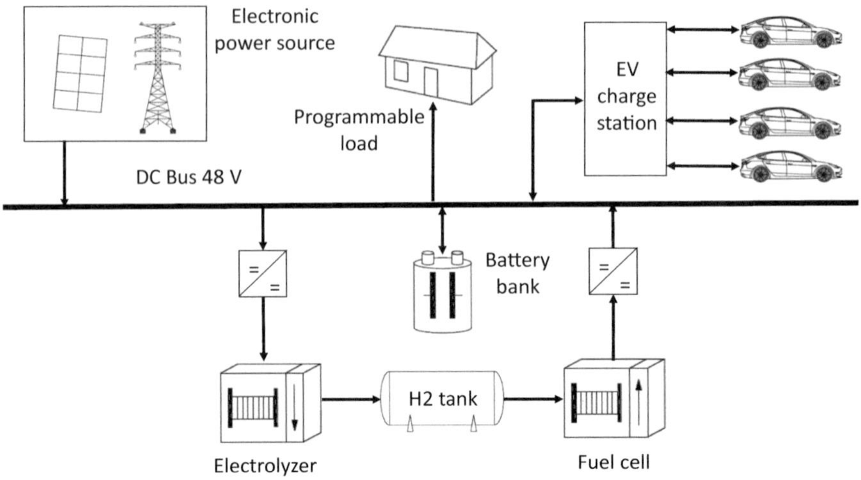

Fig. 6.6 Experimental microgrid coupled to a V2G system

from/to the grid depending on the tariffs. In this sense, when the cars are parked, their batteries can be used by the microgrid to expand the buffer capacity during fast transients. This application is an extension of the previous example: now power exchange with the EV batteries is bidirectional and besides, the charging process can be interrupted when necessary, provided the car is fully charged at the scheduled pickup time.

This will be demonstrated on an extension of the laboratory-scale microgrid presented in Chap. 4. In this case, a charging station for four EVs is added, as depicted in Fig. 6.6 (see [26] for additional details). For the experiments, the EVs batteries are emulated by means of programmable electronic source and load. The objective of the EMS is then to compute the different powers P_{grid}, P_{bat} (the power of the battery bank), P_{H_2} (the power of the hydrogen storage), and P_{ev1}, P_{ev2}, P_{ev3}, P_{ev4} (the powers of vehicle batteries) in such a way that the performance of the overall system is optimized.

The proposed solution has two control layers: the upper layer comprises a scheduler that aims at the economical benefit of the charging station and the Charging Station Management Unit (CSMU), which manages the EV charging depending on the parking time and the type of charge (slow or fast). The lower level layer is a fast power sharing strategy, which runs every second. The upper layer takes into account electrical tariffs and load shifting and is solved by MIQP, while the lower layer is responsible for tracking the power targets computed by the upper layer and is solved using a fast QP algorithm.

Control-Oriented Model

In this section, the procedure used in Chap. 4 is applied to the microgrid and EV charging station modeling. For the sake of simplicity, the battery bank is considered to have the same charging/discharging efficiency, so it is not necessary to define binary variables for it. The batteries of EVs are modeled in the same way as the microgrid battery bank, but with the addition of a binary variable ϵ, which indicates the physical connection between the vehicle and the charging station. This variable provides a change in the prediction model without using hybrid modeling, as the value of ϵ is informed by the vehicle connection and is not a decision variable. If the vehicle is connected $\epsilon = 1$ and a state related to EV battery SOC is enabled in the prediction model; if not, $\epsilon = 0$, and the state is disabled.

To model the hydrogen storage dynamics, it is necessary to define the variable $z_{H_2}(t) = P_{H_2}(t)\delta_{H_2}(t)$, which is related to charging/discharging the hydrogen storage. P_{H_2} is positive when the fuel cell is injecting power into the bus and δ_{H_2} takes the value 1 when the fuel is operating. To manage the purchase and sale of energy to the grid for the economical optimization, different weights for sale and purchase were used. In order to make this possible, a new variable $z_{grid}(t) = P_{grid}(t)\delta_{grid}(t)$ is defined and the corresponding MLD constraints were introduced (see Chaps. 2 and 5). Taking into account that the microgrid battery has to balance the power at the bus, it must fulfill that

$$P_{bat}(t) = P_{load}(t) + P_{elz}(t) - P_{fc}(t) - P_{grid}(t) - P_{gen}(t) + \sum_{i=1}^{4} P_{evi}(t) \quad (6.2)$$

where $d(t) = P_{gen}(t) - P_{load}(t)$ is the measurable disturbance. Therefore, P_{bat} is not a manipulated variable but a combination of the others. Thus, the complete decision vector (manipulated variables) is

$$u = \left[\, P_{grid}\ P_{H_2}\ P_{ev1}\ P_{ev2}\ P_{ev3}\ P_{ev4}\ \delta_{H_2}\ \delta_{grid}\ z_{H_2}\ z_{grid} \,\right]^T \quad (6.3)$$

where P_{H_2} is the power supplied by the hydrogen storage system and P_{evi} is the power that is charged into electric vehicle i. Then the model can be rewritten in a condensed form:

$$\begin{aligned} x(t+1) &= Ax(t) + Bu(t) + B_d d(t) \\ y(t) &= x(t) \end{aligned} \quad (6.4)$$

where the state vector is composed of the SOC of the batteries (the one for microgrid storage and those of the EVs) and the LOH of the hydrogen storage:

$$x = \left[\, SOC\ LOH\ SOC_{ev1}\ SOC_{ev2}\ SOC_{ev3}\ SOC_{ev4} \,\right]^T \quad (6.5)$$

Then, the systems matrices are given by

$$A = \begin{bmatrix} 1 & 0 & 0 & 0 & 0 & 0 \\ 0 & 1 & 0 & 0 & 0 & 0 \\ 0 & 0 & \epsilon_1 & 0 & 0 & 0 \\ 0 & 0 & 0 & \epsilon_2 & 0 & 0 \\ 0 & 0 & 0 & 0 & \epsilon_3 & 0 \\ 0 & 0 & 0 & 0 & 0 & \epsilon_4 \end{bmatrix} \quad B = \begin{bmatrix} \theta_1 & 0 & -\theta_1 & -\theta_1 & -\theta_1 & -\theta_1 & 0 & 0 & 0 & 0 \\ 0 & \mu_1 & 0 & 0 & 0 & 0 & 0 & 0 & \mu_2 & 0 \\ 0 & 0 & \tau_1 & 0 & 0 & 0 & 0 & 0 & 0 & 0 \\ 0 & 0 & 0 & \tau_2 & 0 & 0 & 0 & 0 & 0 & 0 \\ 0 & 0 & 0 & 0 & \tau_3 & 0 & 0 & 0 & 0 & 0 \\ 0 & 0 & 0 & 0 & 0 & \tau_4 & 0 & 0 & 0 & 0 \end{bmatrix} \quad B_d = \begin{bmatrix} \theta_1 \\ 0 \\ 0 \\ 0 \\ 0 \\ 0 \end{bmatrix}$$

with

$$\theta_1 = \frac{\eta_{bat} T_s}{C_{\max}}, \ \mu_1 = \frac{\eta_{elz} T_s}{V_{\max}}, \ \mu_2 = \frac{(-\frac{1}{\eta_{fc}} - \eta_{elz}) T_s}{V_{\max}}, \ \tau_i = \frac{\eta_{Bevi} T_s}{C_{\max i}} \tag{6.6}$$

where η is the charging/discharging efficiency of the storage units, $C_{\max}$ and $V_{\max}$ are the maximum storage capacities, and T_s is the sampling time (see Chap. 4 for details).

Control Strategy

A two-level hierarchical control structure based on MPC, as shown in Fig. 6.7, is proposed. The controllers act in different timescales. The secondary controller is executed with a sampling period of one second ($T_s = 1$ s), and it is responsible for ensuring the fulfillment of constraints, operation conditions, and power flows exchange among the microgrid elements and the charging station, tracking the targets supplied by the tertiary controller. This runs at the upper level control layer and manages the use of EVs batteries as storage while guaranteeing that the charging constraints (charging type and time) are satisfied. The CSMU is designed to achieve the microgrid economic optimization managing energy selling and purchasing. This controller uses a sampling time of five minutes ($T_s = 5$ min).

Secondary MPC

The low-level MPC is in charge of power sharing minimizing an objective function similar to that used in the previous chapters, with the addition of a term that penalizes deviation of the manipulated variables from their targets (computed by the high-level control layer):

$$J = \sum_{j=1}^{N_p} \|\hat{x}(t+j \mid t) - x^{ref}(t+j)\|_R^2 + \sum_{j=1}^{N_c} \|u(t+j-1)\|_{Q_1}^2 + \\ + \sum_{j=1}^{N_c} \| \Delta u(t+j-1)\|_{Q_2}^2 + \sum_{j=1}^{N_p} \|u(t+j-1) - u^{ref}(t+j-1)\|_{Q_3}^2 \tag{6.7}$$

subject to local dynamics and amplitude and rate constraints, where x^{ref} is the reference value for the storage units (set to 50% in the experiments) and u^{ref} is the target value for the power that is computed by the upper layer. Matrix R is used

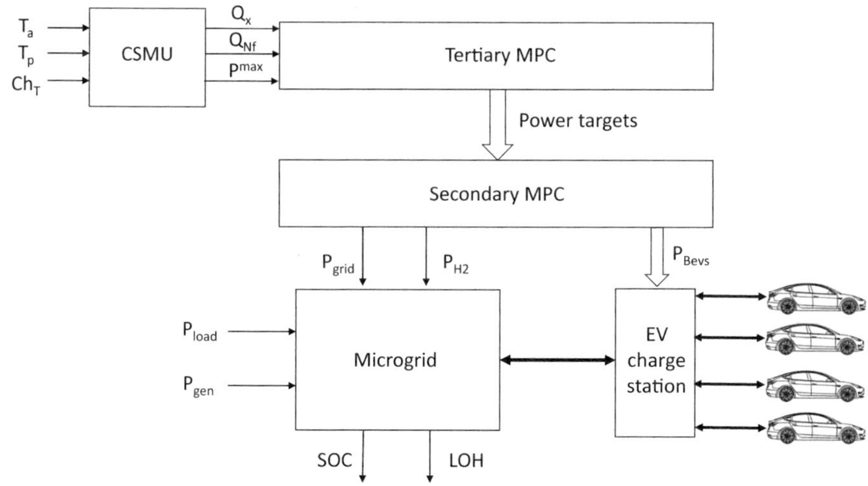

Fig. 6.7 Control Structure

to maintain the value of SOC and LOH next to a reference value, matrix Q_1 is tuned to minimize the use of the grid and the storage units, matrix Q_2 is adjusted to minimize the control increments in the electrolyzer and fuel cell (reducing degradation), and matrix Q_3 is set to guarantee that the controller will track the targets computed by the upper layer.

Tertiary MPC

This control level includes the CSMU and a second MPC that maximizes the economic benefit of sales/purchase with the grid and guarantees that the EVs batteries are fully charged when needed. Therefore, the objective function must include these two terms:

- The part of the objective function related to the energy exchanged with the grid uses different weights for sale and purchase (subscripts *sale* and *purchase* in the following equation) in order to manage the purchase and sale of energy to the grid:

$$
\begin{aligned}
J_{grid} = & \sum_{l-1}^{N_p} \hat{P}_{grid}(t+l)^T Q_{sale} \hat{P}_{grid}(t+l) + \\
& \hat{z}_{grid}(t+l)^T \left(Q_{purchase} - Q_{sale}\right) \hat{z}_{grid}(t+l) + \\
& f_{sale} \hat{P}_{grid}(t+l) + \left(f_{purchase} - f_{sale}\right) \hat{z}_{grid}(t+l)
\end{aligned}
\tag{6.8}
$$

Note that when power $\hat{P}_{grid} > 0$ then $\delta_{grid} = 1$ and $\hat{z}_{grid} = \hat{P}_{grid}$, which means that energy is purchased from the grid and therefore the current purchase weight is used. Otherwise, $\hat{P}_{grid} < 0$ implies $\delta_{grid} = 0$ and $\hat{z}_{grid} = 0$ and the sale weight is used. This makes it possible to use different weights for the same variable. The values of the weights are adjusted according to the price of energy.

- A third term relative to the final state weights is introduced to ensure that the vehicle batteries will be fully charged at the end of the charging time $(t + N_f)$. The following term must be considered:

$$\left(\hat{x}(t + N_f) - \hat{x}_{ref}(t + N_f)\right)^T Q_{N_f} \left(\hat{x}(t + N_f) - \hat{x}_{ref}(t + N_f)\right) \tag{6.9}$$

Charging Station Management Unit

A charging station management unit is designed to manage the use of the electric vehicle batteries over the microgrid. At the time of vehicle connection to the charging station, the user must inform of the charging type Ch_T (slow or fast), the arrival time T_a, and the parking time T_p. If slow charge is chosen, the battery is charged during the parking time, using low-power charge. If fast charge is chosen, the battery is available for use as a storage for the microgrid, and it is charged with maximum power only half an hour before the preset pickup time. During the charging period (either slow or fast), the weights Q_x and Q_{N_p} are set to a positive value in order to ensure that the load is charged on time. When the battery is used as storage, these weights take null values. The CSMU determines the operation mode of each EV battery, the weights, and the limits, and communicates them to the upper level MPC. This module is implemented by Algorithm 6.1.

Algorithm 6.1 Charging Station Management

Input: $T_a, T_p, Ch_T, \epsilon, Time$
Output: $Q_x, Q_{N_f}, P^{max}_{Bev}$
for $i = 1$ to N_{ev} **do**
  if $\epsilon(i) = 1$ {Test if there is a parked vehicle} **then**
    if $Ch_T(i) = 1$ {Fast Charge} **then**
      if $Time \geq T_a + T_p - 30$ minutes **then**
        Set $Q_x(i, i)$ to the fast charge value
        Set $Q_{N_f}(i, i)$ to the fast charge value
        Set $P^{máx}_{Bevi}$ to the fast charge value
      else {Use the battery as a grid storage}
        Set $Q_x(i, i)$ to zero
        Set $Q_{N_f}(i, i)$ to zero
      end if
    else {Slow Charge}
      Set $Q_x(i, i)$ to the slow charge value
      Set $Q_{N_f}(i, i)$ to the slow charge value
      Set $P^{máx}_{Bevi}$ to the slow charge value
    end if
  end if
end for

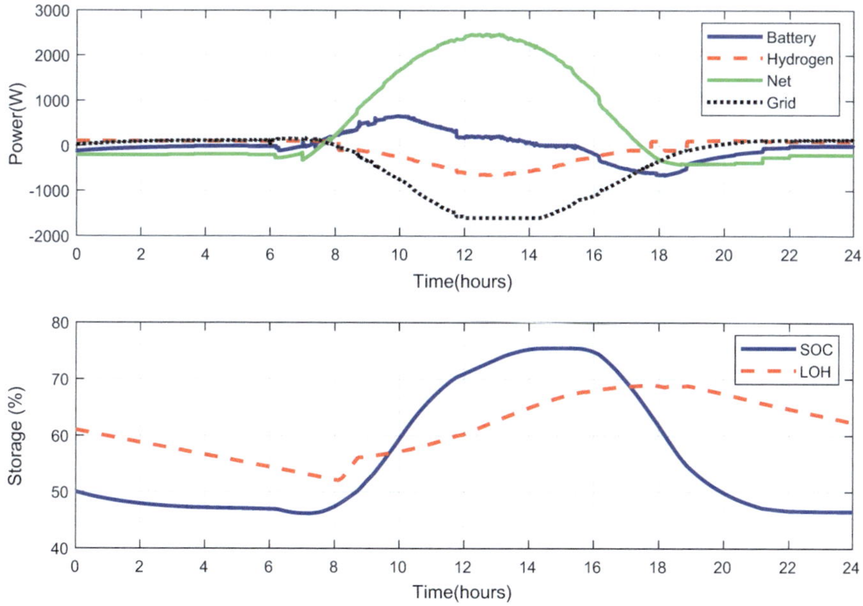

Fig. 6.8 Energy sources and storage units—Sunny day

Experimental Results

The EMS of a microgrid coupled to a V2G system is demonstrated on the laboratory-scale microgrid Hylab that includes a charging station where the vehicles are emulated by a Hardware-In-the-Loop (HIL) methodology. The dynamics of the electric vehicle batteries are simulated and interfaced with the microgrid through the programmable electronic load (for charging) and power supply (for discharging).

The control goals are to maximize the use of RES, make the purchase and sale management of electricity to the grid, coordinate the use of the battery bank and the hydrogen storage to minimize the unbalance between generation and demand, and perform the charging of electric vehicles while fulfilling the microgrid load demand at all periods of time.

In this section, the experimental results of the proposed control strategy applied to the physical microgrid are presented. The control strategy was implemented using the software MATLAB [25] with Yalmip toolbox [23] and solver CPLEX [15], running on a computer connected to a Programmable Logic Controller (PLC) through OLE for Process Control (OPC).

The experiments were performed for a 24-h period. Two of the vehicles (1 and 3) draw fast charge, while 2 and 4 receive slow charge.

Three different experiments were carried out for the controller validation in different scenarios. In particular, for the first experiment, the programmable electronic source was used to emulate a solar photovoltaic generator on a sunny day as shown in Figs. 6.8 and 6.9. This is the same generation profile as used in Case Study 1 of Chap. 4.

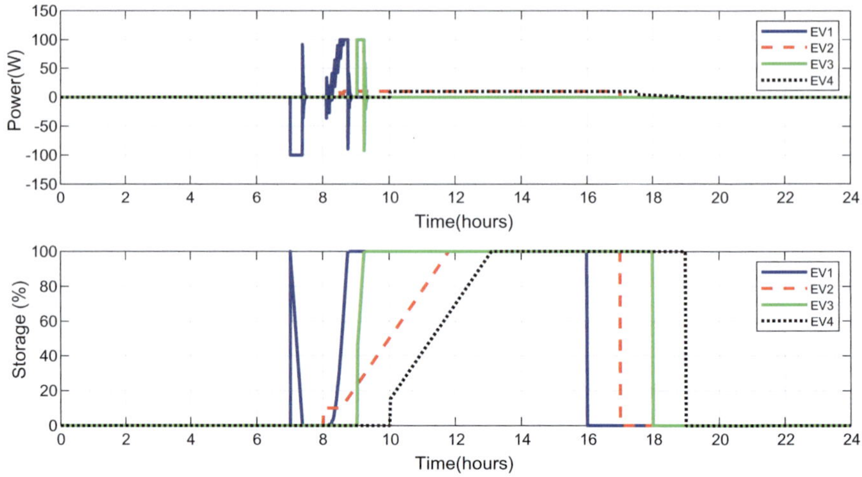

Fig. 6.9 Electric vehicles charge management—Sunny day

The behavior of the ESSs of the microgrid (battery bank and hydrogen) changes along the day. In the periods of low irradiance (from 0 h to 8 h and from 19 h to 24 h), they worked uninterruptedly to provide most part of the energy required to meet the demand and to reduce the amount of energy purchased to the grid. During the high irradiance period (8 h to 18 h), part of the energy surplus is sold to the grid and the rest is used to charge the battery and the hydrogen storage through the electrolyzer operation. At the end of the day (18 h to 19 h), when there is less irradiance, a switching between the fuel cell and the electrolyzer happened, respecting the minimum operation time of each equipment.

Additionally, EVs batteries are used to avoid fluctuations, as shown in Fig. 6.9. When EV1 (which accepts fast charge) is parked at 7 h, its battery is used as microgrid storage, and after 8 h EV1 it is fully charged. There exist some oscillations in the power of EV1 and EV3 during the charge procedure that are produced by an aggressive tune of the controller parameters. EV2 and EV4 are charged in slow-charge mode in the most convenient way in order to be ready at pickup time.

The behavior in a cloudy day scenario is slightly different, as shown in Figs. 6.10 and 6.11. A cloudy day represents a great challenge for the microgrids, since the control has to cope with high power fluctuations. In this sense, the storage system must have the ability to absorb such fluctuations. As the microgrid is composed of a hybrid storage system, it is expected that the battery bank absorbs high-frequency oscillations while the hydrogen storage provides energy for a long time when the irradiance is not sufficient to meet the demand. Another challenge is to minimize the switching in the hydrogen storage which may be caused by oscillatory conditions.

In cloudy conditions, between 12 h and 17:30 h, the battery bank and the EVs batteries absorb most of the power fluctuation. A short switching between the electrolyzer and the fuel cell, caused by the irradiance oscillation, is observed from 12

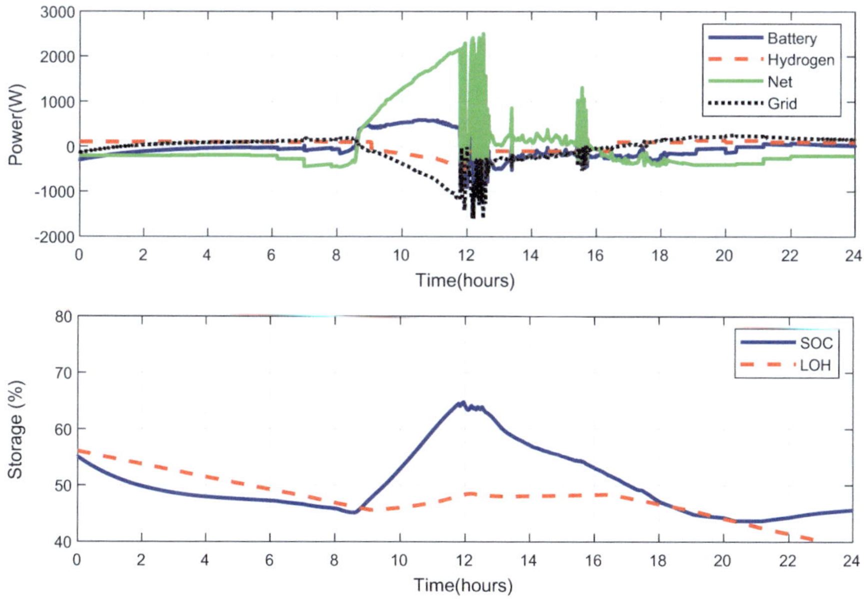

Fig. 6.10 Energy sources and storage units—Cloudy day

h to 13 h. In the rest of the experiment, the hydrogen storage provided energy to the system at night and stored energy during the day. The difference between the initial and the final LOH is greater than compared to the sunny day experiment, which is expected once the irradiance during the day is not sufficient to replace the energy expended at night. The irradiance oscillation, caused by the clouds, affects directly the profile of energy sold to the grid and the battery life as the number of charge/discharge cycles increases with fluctuation conditions.

When EV1 was parked at 7 h, its battery is used to provide energy to the microgrid load, and after 9 h EV1 is already charged, as displayed in Fig. 6.11. During the cloudy period, the batteries of EV1 and EV3 (which allow fast charge) are used to mitigate the power fluctuations. A slow charge is applied in EV2 and EV4.

The third experiment is performed using a wind turbine as RES, and the results are shown in Figs. 6.12 and 6.13. In this experiment, the fluctuations are not so abrupt as in the cloudy day. However, it still presents a high stochastic behavior in the wind turbine production. There are some switching in the hydrogen storage, at 10 h when a wind fluctuation happens and after 20 h when it is necessary to provide energy to the load. In this case, the final value of LOH at the end of the day is greater than at the beginning, as a consequence of the electrolyzer operation in most part of the day. The EV batteries, although available for balancing, are scarcely used by the microgrid as displayed in Fig. 6.13.

In all experiments, the controller has been able to manage the energy in the microgrid, supplying the internal load as well as charging the EVs as requested.

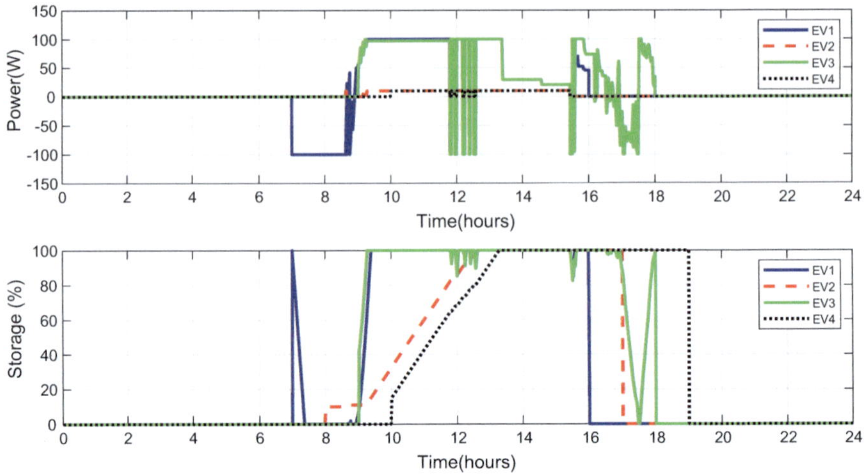

Fig. 6.11 Electric vehicles charge management—Cloudy day

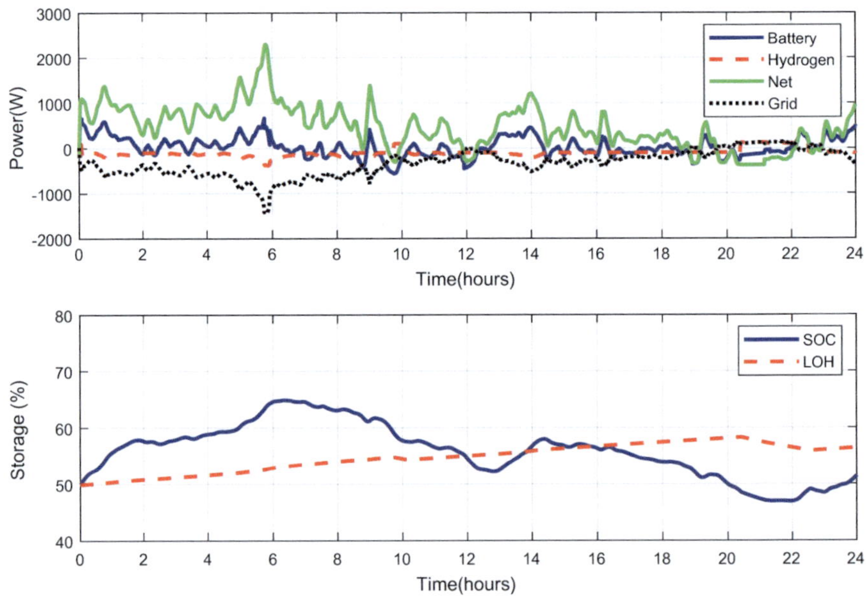

Fig. 6.12 Energy sources and storage units—Windy day

Besides, this has been done minimizing energy costs and avoiding degradation of the ESSs. The use of EVs batteries with V2G mechanism has contributed to the solution.

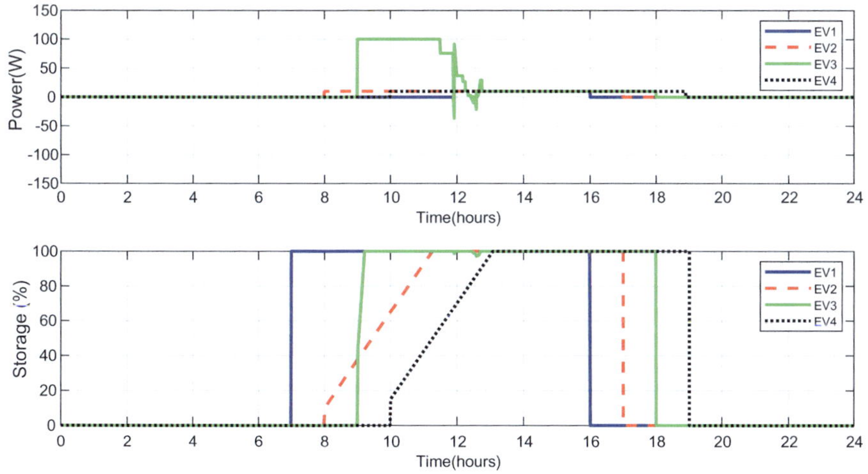

Fig. 6.13 Electric vehicles charge management—Windy day

References

1. Alizadeh M, Li X, Wang Z, Scaglione A, Melton R (2012) Demand-side management in the smart grid: Information processing for the power switch. IEEE Signal Process Mag 29(5):55–67
2. Ballesteros LGM, Álvarez O, Markendahl J (2015) Quality of experience (QOE) in the smart cities context: an initial analysis. In: 2015 IEEE first international smart cities conference (ISC2), pp 1–7
3. Bashash S, Fathy H (2011) Robust demand-side plug-in electric vehicle load control for renewable energy management. In: American control conference (ACC), pp 929–934
4. Chen Z, Wu L, Fu Y (2012) Real-time price-based demand response management for residential appliances via stochastic optimization and robust optimization. IEEE Trans Smart Grid 3(4):1822–1831
5. Deilami S, Masoum AS, Moses PS, Masoum M (2011) Real-time coordination of plug-in electric vehicle charging in smart grids to minimize power losses and improve voltage profile. IEEE Trans Smart Grid 2(3):456–467
6. Fan Z (2012) A distributed demand response algorithm and its application to phev charging in smart grids. IEEE Trans Smart Grid 3(3):1280–1290
7. Galus MD, Andersson G, Art S (2012) A hierarchical, distributed pev charging control in low voltage distribution grids to ensure network security. In: Power and energy society general meeting, 2012 IEEE, pp 1–8
8. Garcia-Torres F, Vilaplana DG, Bordons C, Roncero-Sanchez P, Ridao MA (2018) Optimal management of microgrids with external agents including battery/fuel cell electric vehicles. IEEE Trans Smart Grid, 1–1
9. Gautschi M, Scheuss O, Schluchter C (2009) Simulation of an agent based vehicle-to grid (v2g) implementation. Electr Power Syst Res 120:177–183
10. Giorgio AD, Liberati F, Canale S (2014) Electric vehicle charging control in smartgrids: a model predictive control approach. Control Eng Pract 22:147–162
11. Gouveia C, Moreira J, Moreira CL, Peças Lopes JA (2013) Coordinating storage and demand response for microgrid emergency operation. IEEE Trans Smart Grid 4(4):1898–1908
12. Harley R, Haboter T (2013) Utilizing building-level demand response in frequency regulation of actual microgrids. In: IECON 2013—39th annual conference of the IEEE industrial electronics society, pp 2205–2210

13. Hsu Y, Su C (1991) Dispatch of direct load control using dynamic programming. IEEE Trans Power Syst 6(3):1056–1061
14. Hu J, You S, Lind M, Østergaard J (2014) Coordinated charging of electric vehicles for congestion prevention in the distribution grid. IEEE Trans Smart Grid 5(2):703–711
15. IBM ILOG. Cplex. (2007)
16. Ito A, Kawashima A, Suzuki T, Inagaki S, Yamaguchi T, Zhou Z (2018) Model predictive charging control of in-vehicle batteries for home energy management based on vehicle state prediction. IEEE Trans Control Syst Technol 26(1):51–64
17. Jayadev V, Swarup KS (2013) Optimization of microgrid with demand side management using genetic algorithm. In: Proceedings of the IET conference on power in unity: a whole system approach
18. Kostkova K, Omelina L, Kycina P, Jamrich P (2013) An introduction to load management. Electr Power Syst Res 95:184–191
19. Kriett PO, Salani M (2010) Optimal control of a residential microgrid. Energy 42(1):321 – 330, 2012. 8th World Energy System Conference, WESC (2010)
20. Kurucz CN, Brandt D, Sim S (1996) A linear programming model for reducing system peak through customer load control programs. IEEE Trans Power Syst 11(4):1817–1824
21. Lee W, Xiang L, Schober R, Wong VWS (2015) Electric vehicle charging stations with renewable power generators: a game theoretical analysis. IEEE Trans Smart Grid 6(2):608–617
22. Li D, Chiu WY, Sun H (2017) Microgrid. Advanced control methods and renewable energy system integration, chapter Demand Side Management in Microgrid Control Systems. Elsevier
23. Lofberg J (2004) Yalmip: a toolbox for modeling and optimization in matlab. In: IEEE international symposium on computer aided control systems design, pp 284–289
24. Logenthiran T, Srinivasan D, Shun TZ (2012) Demand side management in smart grid using heuristic optimization. IEEE Trans Smart Grid 3(3):1244–1252
25. Mathworks. Matlab. (2009)
26. Mendes PRC, Valverde L, Bordons C, Normey-Rico JE (2016) Energy management of an experimental microgrid coupled to a v2g system. J Power Sour 327:702–713
27. Mohsenian-Rad H et al (2015) Optimal charging of electric vehicles with uncertain departure times: a closed-form solution. IEEE Trans Smart Grid 6(2):940–942
28. Mou Y, Xing H, Lin Z, Fu M (2015) Decentralized optimal demand-side management for phev charging in a smart grid. IEEE Trans Smart Grid 6(2):726–736
29. Pahasa J, Ngamroo I (2015) Phevs bidirectional charging/discharging and soc control for microgrid frequency stabilization using multiple mpc. IEEE Trans Smart Grid 6(2):526–533
30. Palensky P, Dietrich D (2011) Demand side management: demand response, intelligent energy systems, and smart loads. IEEE Trans Ind Inform 7(3):381–388
31. Parisio A, Glielmo L (2011) A mixed integer linear formulation for microgrid economic scheduling. In: Smart grid communications (SmartGridComm), 2011 IEEE international conference on, pp 505–510. IEEE
32. Parisio A, Rikos E, Glielmo L (2014) A model predictive control approach to microgrid operation optimization. IEEE Trans Control Syst Technol 22(5):1813–1827
33. Parisio A, Rikos E, Glielmo L (2016) Stochastic model predictive control for economic/environmental operation management of microgrids: an experimental case study. J Process Control 43:24–37
34. Richardson P, Flynn D, Keane A (2012) Optimal charging of electric vehicles in low-voltage distribution systems. IEEE Trans Power Syst 27(1):268–279
35. Samad T, Koch E, Stluka P (2016) Automated demand response for smart buildings and microgrids: the state of the practice and research challenges. Proc IEEE 104(4):726–744
36. Wang G, Zhao J, Wen F, Xue Y, Ledwich G (2015) Dispatch strategy of phevs to mitigate selected patterns of seasonally varying outputs from renewable generation. IEEE Trans Smart Grid 6(2):627–639

Chapter 7
Uncertainties in Microgrids

Abstract Uncertainties in the supply or load is an important issue that must be tackled in Energy Management Systems (EMS) of a microgrid. Renewable generation (solar or wind) and consumer loads typically are not controllable but a forecast of their time evolution is of great interest, especially if control techniques as MPC are applied, where the prediction in a future time horizon plays a crucial role. Prediction of renewable production is an active field of research, based on weather forecast and historical data, analyzed by a range of statistical methods or alternatives as neural networks, machine learning, etc. Nevertheless, uncertainty in these values is unavoidable, and the approach in this chapter is the explicit characterization and introduction in the control problem of those uncertainties, that is, the deterministic decision-making of conventional controllers is replaced by a stochastic process. MPC is essentially a deterministic approach, and can be troublesome in systems where uncertainty is an important topic. This chapter is devoted to the application of Stochastic MPC (SMPC) to the EMS problem. SMPC is based on an explicit statistical representation of the uncertainties, i.e., probabilistic distribution, and including it in the optimization problem formulation. Also, constraints can be defined stochastically and some violations are allowed with a determined probability criteria. Next sections describe some of these stochastic MPC algorithms and its application to a laboratory-scale microgrid.

7.1 Stochastic MPC Concept and Mathematical Formulation

The feedback mechanism of MPC provides some robustness to the control of systems with uncertainties, but when they are significant, other alternatives have to be considered, although some MPC schemes guarantee feasibility and stability when disturbances are considered [3]. With this objective, robust MPC algorithms have received significant interest in last years. In robust MPC, uncertainties are considered deterministic and bounded.

C. Bordons et al., *Model Predictive Control of Microgrids*,
Advances in Industrial Control, https://doi.org/10.1007/978-3-030-24570-2_7

The design of robust deterministic MPC approaches has many solutions [2, 13, 22]; however, in these algorithms feasibility and stability are usually oriented to a worst-case analysis. Consequently, their solutions are conservative or even unfeasible in the case of unbounded uncertainties and they may require the solution of min-max optimization problems that are computationally very demanding. Also, these approaches do not take into account any knowledge of the statistical properties of the disturbances, as probability distribution functions, which can be estimated in many problems, i.e., wind and irradiance forecast in renewable-energy-related problems.

This section presents the formulation of the stochastic MPC problem. Two main elements are first considered: the used model and the stochastic characterization of constraints. Next, the complete optimization problem will be formulated for several different approaches.

7.1.1 *Models and Constraints*

Most of the algorithms in MPC literature are based on linear models, and stochastic MPC is not an exception. The optimization problem is more complex than in deterministic MPC, and that explains the use of simpler models. Nevertheless, a few works on nonlinear stochastic MPC approaches can be found, as [28, 41].

Let us consider a state-space discrete-time linear model. Models must include the effect of uncertainty in the dynamic process. Basically, two approaches are used: additive, where the uncertainty is added to the deterministic part of the dynamics and multiplicative, where the uncertainty multiplies state or inputs variables. In the case of additive uncertainty:

$$x(t+1) = Ax(t) + Bu(t) + B_d d(t) \tag{7.1}$$

with $x \in \mathbb{R}^{n_x}$, $u \in \mathbb{R}^{n_u}$, and $d \in \mathbb{R}^{n_d}$. For linear systems with multiplicative uncertainties, the state-space matrices are time-varying with known probability distribution:

$$x(t+1) = Ax(t) + Bu(t) + \sum_{j=1}^{q} (\bar{A}_j x(t) + \bar{B}_j u(t)) d_{t,j} \tag{7.2}$$

where $\{d_{t,j}\}_{j=1}^{q}$ is a sequence of q zero mean independent and identically distributed (i.i.d.) random variables with known variance [28]. Multiplicative uncertainty in MPC framework increases the complexity of the problem, because of the product of future state or inputs and the uncertain variable, all of them random variables.

In this chapter, only discrete linear systems with additive uncertainty will be considered. As it will be seen in examples, an additive term models adequately the uncertainty typically found in microgrid systems.

Regarding constraints, taking into account that uncertainties of the state and input disturbances are characterized as stochastic processes, they should be reformulated in a probabilistic way. Notice that a deterministic constraint $Gx \leq g$, taking into account uncertainty, can lead to infeasibility, for example, when using unbounded Gaussian distributions, and obviously it is impossible to ensure deterministic constraints on the state variable.

The most common stochastic constraints are [12] as follows:

- Expectation constraints: $\mathbb{E}[Gx] \leq 0$,
- Chance constraints: $\mathbb{P}[Gx \leq 0] > 1 - \delta_x$,

where operator $\mathbb{E}$ is the expected value and $\mathbb{P}$ denotes probability. Notice that in both cases, constraints can be violated, but they are satisfied on averages (expectation constraints) or with a given probability (chance constraints).

7.1.2 Stochastic MPC Formulation

As mentioned above, in a stochastic framework, not only the MPC optimization problem has to be reconsidered but also how the uncertainty is modeled, generated, and managed. In this respect, mainly two alternatives can be considered as given below:

- Scenario-based approaches: The stochastic system dynamics can be characterized by a finite set of random realization of uncertainties, utilized to solve the MPC optimization problem. The number of scenarios is a key issue, because a high number of them increases the computational burden of the MPC optimization problem. Scenario reduction techniques and more efficient scenarios implementation (i.e., tree-based structures) acquire great importance in these approaches. Many of the solutions in literature are based on stochastic programming techniques.
- Analytic approaches: These methods are based on the knowledge of mathematical models of uncertainty (i.e., probabilistic functions). Cost function and chance constraints are reformulated in a deterministic way by the use of statistical elements: mean, variance, cumulative distribution functions, etc.).

The MPC cost function in a stochastic framework is a random variable since the states are also random processes, so it must be treated from a statistical point of view. Some of the alternatives are as follows:

- Cost Function Expectation:

$$\Phi = \mathbb{E}\left[\sum_{i=1}^{N_p} J(\hat{x}(t+i \mid t), u(t+i \mid t)\right] \tag{7.3}$$

- If N_s scenarios are considered, then an approximate cost function expectation can be computed as follows:

$$\Phi = \frac{1}{N_s} \sum_{j=1}^{N_s} \left[\sum_{i=1}^{N_p} J(\hat{x}_j(t+i \mid t), u(t+i \mid t) \right] \tag{7.4}$$

- Worst-case selection in scenario-based approach:

$$\Phi = \max_{j=1,\ldots N_s} \left[\sum_{i=1}^{N_p} J(\hat{x}_j(t+i \mid t), u(t+i \mid t) \right] \tag{7.5}$$

Next section describes different stochastic MPC approaches. First, the stochastic programming method as a key concept in SMPC is described. Next, two scenario-based MPC approaches will be studied: the first one is a simple Multiple-Scenario Algorithm (MS-MPC) also called multiple MPC in [42], and the second one is a scenario approach organized in a tree-based structure MPC (TB-MPC). Finally, an MPC based on an analytical approach using chance constraints (CC-MPC) will be considered. These last three approaches will be tested, validated, and compared on a real plant, specifically the laboratory-scale microgrid described in previous chapters.

7.2 Stochastic MPC Approaches

This section presents three SMPC methods that will be applied to a laboratory-scale microgrid in next sections. The described methods are not an exhaustive list of SMPC algorithms that can be found in literature, but some of them with relevance in microgrid applications. The section begins with a brief description of stochastic programming, a general method for optimization of stochastic problems used by some of the SMPC techniques. A more complete list of SMPC methods can be found in [27, 37].

7.2.1 Stochastic Programming

A widespread method to solving stochastic MPC is the application of stochastic programming techniques. The classic two-stage stochastic program with recourse assumes that some problem data are uncertain and a probabilistic description of the random variable is available. The are two types of decision variables [5]:

- First-stage decisions (x variables): The decision is taken before the realization of the random events d. The period where these decisions are taken is called *first stage*.

- Second-stage decisions ($y(d)$) variables: The decision has to be taken after the realization of the random event and the first state decision, that is, $y(d)$ where d is the disturbance, describes the consequences of x decision.

In order to clarify these concepts, let us consider the following example of a micro-grid [23]. The optimal management of a microgrid needs to consider uncertainties in generation (wind and solar) and load consumption. If the microgrid participates in the electricity market, the first decision consists of the cost of the electricity sold or purchased in advance. Obviously, real values of random variables (realization of random events) are unknown at this moment, so only probabilistic values of random variables can be used in this stage.

The inevitable unbalance between generation and load due to random variables will be solved using storage units (or spinning reserves) with operational costs. The cost of using storage units or the use of spinning reserves are the second-stage variables, whose values depend on the electricity purchase in the first stage and on the realization of random variables. The two-stage stochastic program can be formulated as the following optimization problem [37]:

$$\min J = f_1(x) + \mathbb{E}[f_2(y(d), d)] \tag{7.6}$$

subject to

$$g_m(x) \leq 0 \tag{7.7a}$$

$$h_n(x, y(d)) \leq 0 \tag{7.7b}$$

where x and $y(d)$ are the set of first-stage and second-stage decision variables, respectively. $g_m(x) \leq 0$ is the set of first-stage constraints (i.e., limits of electricity purchase in the previous example). On the other side, constraints $h_n(x, y(d)) \leq 0$ link first- and second-stage variables (i.e., balance equation in the example).

The previous problem is quite complex to solve, in particular, the computation of the expected value of the second stage. In practical applications, the problem can be simplified by choosing a set of samples of the random variables and forming scenarios. Then, the problem is decomposed in a set of deterministic optimization problems.

7.2.2 *Scenario-Based MPC Approaches*

The main idea of the scenario-based techniques is to draw samples of the uncertainty or to use historical data to formulate the MPC control problem. In this way, the stochastic optimization is transformed into a deterministic one with a larger number of deterministic constraints corresponding to the original constraints evaluated for every scenario.

A suitable modeling of scenarios is necessary to manage the stochastic optimization problem properly. The selected scenarios must correspond to real situations and can be associated to an occurrence probability. Otherwise, the number of scenarios is a critical issue, because statistically, a high number of scenarios are needed to reduce uncertainty but, on the other hand, it creates a more complex optimization problem. Scenario reduction techniques and the use of more efficient methods to manage a high number of scenarios, as fan- or tree-based approaches, are active fields of research.

In this section, two scenario-based approaches are described. The first one is a simple solution where a collection of different scenarios produces a single control sequence. The second one uses a tree-based approach to handle disturbances throughout the prediction horizon, with also the control actions structured in the tree.

Multi-scenario MPC

Multi-Scenario MPC (MS-MPC) approach is an intuitive and simple to formulate technique that generates a unique control sequence considering simultaneously all the selected scenarios, as a result of a deterministic convex optimization, even when the original problem is not convex [39]. This technique does not need to know statistical information of the random variables in advance (although such knowledge can be exploited). This approach has been applied in [42] for water systems. In this work, three scenarios have been considered. Two of them are maximum and minimum scenarios, both with small probabilities of occurrence, while the third one is the most likely average scenario. A unique set of control actions is generated for all the scenarios. An application on microgrid system can be found in [30], where a finite set of scenarios is used to solve the first stage of a two-stage stochastic-programming-based MPC.

In this approach, a finite set of scenarios is considered, and the same dynamical system is used for different disturbance realizations, resulting in the following optimization problem:

$$\min \sum_{j=1}^{N_s} \left(\sum_{i=1}^{N_p} J(\hat{x}_j(t+i \mid t), u(t+i \mid t) \right) \tag{7.8}$$

subject to

$$x_j(t+1) = Ax_j(t) + Bu(t) + B_w d_j(t) \tag{7.9a}$$

$$x_j(t+1) \in \mathcal{X} \tag{7.9b}$$

$$u(t) \in \mathcal{U} \tag{7.9c}$$

where N_s is the finite number of scenarios, $d_j(t)$ is the random disturbance forecast for scenario $j = 1...N_s$, and $\mathcal{X}, \mathcal{U}$ are the sets of state and input constraints, that is [14],

$$\mathcal{X} \equiv \{x(t) \mid Gx(t) \leq g\} \quad \forall t \tag{7.10a}$$

$$\mathcal{U} \equiv \{u(t) \mid Fu(t) \leq f\} \quad \forall t \tag{7.10b}$$

being c_x and c_u the number of state and input constraints, respectively. N_s different state trajectories over the prediction horizon are implicit in the solution, each for a different scenario but all of them obtained with the same input sequence. The cost function measures an average over all scenarios. As mentioned above, each scenario can be associated to an occurrence probability.

Unlike the simple scenario generation of [42], a more systematic scenario generation, with a careful selection of the number of scenarios N_s, can be used to find limits of the probability of violation of the constraints as a function of this number [6]. Nevertheless, due to its nature, this method cannot guarantee feasibility.

The scenario-based approach converts a stochastic control problem into multiple deterministic problems by substituting each scenario by a set of constraints and considering a new cost function as an average of the cost functions of individual scenarios. Nevertheless, as mentioned in [39], the solution can produce a *randomization* of the control law, and occasionally an erratic behavior could be shown.

A formal selection of N_s is based on the early works on sampled convex programs in [6, 7] and later works in [39]. Notice that a higher value of N_s makes the solution more robust but the computational complexity increases. The aim of these works is to find a lower bound of N_s which fulfills a set of statistical conditions about the satisfaction of state constraints, with assumptions about the distribution function of the uncertainties. The demonstration of these results is out of the scope of this book, but an interested reader will find it in previous references.

In order to determine a lower bound of N_s, let us consider $\delta_x \in [0, 1]$ as the acceptable risk level of violation of the constraints for the system states in the next sampling time:

$$\mathbb{P}[x_j(t+1) \in \mathcal{X}] > 1 - \delta_x \tag{7.11}$$

and let us define a quite small confidence level $\beta \in [0, 1]$. According to [7], the lower bound on N_s depends on the number of optimization variables (z), δ_x and β, and has to satisfy the following:

$$\sum_{i=0}^{z-1} \binom{N_s}{i} {\delta_x}^i (1 - \delta_x)^{N_s - i} \leq \beta \tag{7.12}$$

based on a binomial distribution function. The limit value of N_s can be obtained from this expression as indicated in [12]

$$N_s \geq \frac{z + 1 + \ln(1/\beta) + \sqrt{2(z+1)\ln(1/\beta)}}{\delta_x} \tag{7.13}$$

As mentioned above, this approach will result in a single control action that satisfies all the constraints of the different scenarios with a specified probability.

Tree-Based MPC

A more efficient way to manage the uncertainty of stochastic systems is the use of trees to model uncertainty and control actions. The inherent concept of this technique is that uncertainty increases with time, that is, an accurate prediction of energy demand and energy generation by a renewable source is easier to obtain in a short-term horizon. In this approach, the different feasible evolutions of the disturbances are represented by a tree, where a branching point is established when possible disturbances lead to more than one different trajectories. Therefore, distinct control actions can be decided in each branch, and the result is a tree of control actions. This method has been applied to a semi-batch reactor in [24] to open water systems in [25], and finally to an energy management system of a microgrid with hydrogen storage in [35].

Therefore, this approach consists of confining the different considered evolutions of the disturbances in a tree, whose root is in the current sampling time. Using the terminology of [29], let us consider $d(t)$ as a stochastic disturbance with values restricted to a finite set $\{\bar{d}_1, \bar{d}_2, ...\bar{d}_s\}$ and a known $p(t)$ probability mass function of d, where

$$\forall t,\ p_j(t) = \mathbb{P}[d(t) = \bar{d}_j], \quad p_j(t) \geq 0, \quad \sum_{j=1}^{s} p_j(t) = 1 \tag{7.14}$$

in such a way that it can be predicted in the future using only the information at time t.

At time t, in MPC problems with prediction horizon N_p, a tree of s alternatives in each bifurcation point and a depth of N_p steps produces s_p^N scenarios, each of one being a sequence of N_p consecutive disturbance values. This approach is impractical for the large number of scenarios, because all possible disturbance sequences are being considered. Consequently, scenario reduction techniques are needed, in such a way that only the more likely sequences are considered through pruning of the tree.

In the method proposed in [4, 29], a path from the root to a leaf node defines a scenario and consists of a sequence of disturbance realizations. In a first step, from the root node, all the s possible values of the disturbances are considered and added to the tree. After this step, only the leaf node with the highest probability of reaching that node from the root is expanded. The result is a tree expanded in the most promising direction, where the paths from the root to leaf nodes have different lengths, and consequently different prediction horizons. Typically, the algorithm ends when the tree reaches a predetermined value of maximum number of nodes. The algorithm is repeated at every sampled time (See Fig. 7.1). Other techniques for the reduction of the number of scenarios can be found in [39].

Nevertheless, as in the previous multi-scenario approach, the selected number of scenarios plays an important role in the robustness of the method and in the probability of constraint violations.

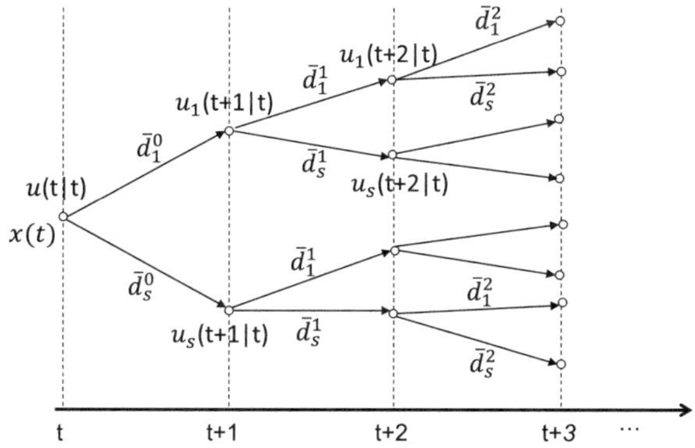

Fig. 7.1 Scenario tree over the prediction horizon

Unlike the MS-MPC approach, a different control sequence in the prediction horizon is computed for each scenario in the tree, and consequently, more decision variables are needed. Due to causality, if two scenarios share the same path in the tree from the root to level m, both scenarios necessarily will have the same input sequence until that level m, and as mentioned previously, the solution is also a tree of control inputs. Obviously, this MPC only applies the first input of this tree, the root element, common for all the scenarios. Let us consider N_s as the number of different scenarios in the disturbances tree. The tree-based MPC (TB-MPC) problem can be formulated as follows:

$$\min \sum_{j=1}^{N_s} \left(\sum_{i=1}^{N_p} J(\hat{x}_j(t+i \mid t), u_j(t+i \mid t) \right) \tag{7.15}$$

subject to

$$x_j(t+1) = Ax_j(t) + Bu(t) + B_d d_j(t) \tag{7.16a}$$
$$x_j(t+1) \in \mathcal{X} \tag{7.16b}$$
$$u_j(t) \in \mathcal{U} \tag{7.16c}$$

Another important issue in this approach is the need for the introduction of non-anticipative constraints to guarantee that the controller computes its input actions considering only the observed disturbances before the branching points [36]. These non-anticipative constraints can be represented by

$$u_i(t+k) = u_j(t+k) \quad if \quad d_i(t+k) = d_j(t+k); \quad \forall k = 0, .., m \quad \forall i \neq j \tag{7.17}$$

Equation (7.17) can be introduced in the optimization problem using equality constraints.

7.2.3 Analytical-Based SMPC

Analytical approaches are based on the explicit use of mathematical probabilistic models of the disturbances. This strategy implies the reformulation of the MPC elements, mainly cost functions and constraints, in a deterministic mode using statistical concepts, i.e., cumulative distribution or probability density functions.

Different applications of these techniques can be found in literature. The work in [15] uses an analytical-based SMPC to a drinking water network. In addition, [14] shows a comparison between analytical- and tree-based approaches applied to those systems. In the context of the control of microgrids, these techniques are having a remarkable interest among researchers, for example, the works presented in [19, 21, 26, 48]. Also, this approach has been used in [46] to a unit commitment problem with uncertainty in the wind power, with a solution based on two-stage stochastic programming. The work in [31] presents a prediction based on an ARMA model for load and generation uncertainties.

In this section, an analytical method named chance-constraint MPC (CC-MPC) is presented. Let us consider the linear dynamics with stochastic disturbances.

$$x(t+1) = Ax(t) + Bu(t) + B_d d(t) \tag{7.18}$$

Considering the set of state constraints $\mathcal{X}$ in (7.10), an individual constraint i, that is, the ith row of $\mathcal{X}$, can be notated as follows:

$$G_{(i)}x \leq g_{(i)} \tag{7.19}$$

Due to the stochastic nature of the problem, in CC-MPC, state constraints are relaxed with a predefined probability as follows:

$$\mathbb{P}[x(t) \in \mathcal{X}] \geq 1 - \delta_x \tag{7.20}$$

This probability of constraint violation has a considerable effect on the solution of the SMPC. A low value will have as a consequence a more conservative solution, with a reduced number of unsatisfied constraints, only in very unlikely situations, but normally with an increment in the value of the cost function. A selection of this probability is an important issue in SMPC and a compromise solution must be adopted. If the probability distribution is known, chance constraints can be transformed in deterministic constraints as follows:

$$F^{-1}_{G_{(i)}x}(1 - \delta_x) \leq g_{(i)} \tag{7.21}$$

where $F^{-1}_{Gx}(\cdot)$ is its inverse cumulative distribution. Probabilistic constraints can be defined in two different ways [15]:

- Individual chance constraints: Each row of the constraint set has to be validated with its respective probabilities

$$\mathbb{P}[G_{(i)}x(t) < g_{(i)}] > 1 - \delta_{x,i}, \quad \forall i = 1...c_x \tag{7.22}$$

where $G_{(i)}$ and $g_{(i)}$ are the ith row of G and g, respectively.

- Joint chance constraints: There is an unique probability of constraint violation δ_x that has to be fulfilled jointly by all constraints:

$$\mathbb{P}[G_{(i)}x < g_{(i)}, \forall i = 1...c_x] > 1 - \delta_x \tag{7.23}$$

Let us consider a known quasi-concave probability distribution and consequently a known cumulative distribution function [15, 43, 44]. Equation (7.23) can be transformed to a deterministic equivalent constraint when applied to the prediction horizon in the controller in the following way:

$$\begin{aligned} &\mathbb{P}[G_{(i)}x(t+1) < g_{(i)}] > 1 - \delta_x \\ \Leftrightarrow\ & F_{G_{(i)}B_d d(t)}(g_{(i)} - G_{(i)}(Ax(t) + Bu(t))) > 1 - \delta_x \\ \Leftrightarrow\ & G_{(i)}(Ax(t) + Bu(t)) < g_{(i)} - F^{-1}_{G_{(i)}B_d d(t)}(1 - \delta_x) \end{aligned} \tag{7.24}$$

Here, $F_{G_{(i)}B_d d(t)}(\cdot)$ is the cumulative distribution function of the random variable $G_{(i)}B_d d(t)$, and $F^{-1}_{G_{(i)}B_d d(t)}(\cdot)$ is its inverse cumulative distribution function.

Then, CC-MPC must solve an optimization problem defined as follows:

$$\min \sum_{j=1}^{N_p} \mathbb{E}[J(x_(t + j|t), u_j(t + j|t)] \tag{7.25}$$

subject to

$$x(t+1) = Ax(t) + Bu(t) + B_d d(t) \tag{7.26a}$$

$$G_{(i)}(Ax(t) + Bu(t)) < g_{(i)} - F^{-1}_{G_{(i)}B_d d(t)}(1 - \delta_x) \tag{7.26b}$$

$$u_j(t) \in \mathcal{U} \tag{7.26c}$$

7.3 Stochastic MPC Applied to Microgrids

This section describes different stochastic MPC approaches applied to energy management systems with renewable energy sources, and specifically to microgrids.

A previous problem when using multi-scenarios, tree-based MPC or approaches based on stochastic dynamic programming is the selection of scenarios. MPC with a large number of scenarios requires an unacceptable calculation effort in most practical applications, so scenario reduction without loss of critical information of the stochastic process is needed. Two interesting general papers about scenario creation and reduction are [10, 18]. Scenario reduction and scenario construction approaches with application to power management systems are presented in [16]. A set of scenarios for electric load and wind power generation is computed in [1].

In [20] an islanded microgrid composed of wind farm, a fuel cell, a battery storage system, and the customer load are controlled. The uncertainty in the wind profile is modeled using a Gaussian distribution and the disturbance term in the load is assumed to have a normal distribution. The authors present a method to solve an MPC problem using an approximate expected value of a quadratic cost function and dynamic programming.

A scenario-based stochastic MPC algorithm to solve in real time, the optimal power dispatch of a problem considering the market is presented in [33]. Loads, power generation, and energy prices are defined as stochastic processes without a priori assumption on its distribution. A grid-connected system consisting of conventional and renewable generators and energy storage units is used.

Another scenario-based stochastic MPC approach is shown in [49] for the economic dispatch problem. The authors decomposed the SMPC problem into subproblems which can be solved in parallel using the optimality condition decomposition [9]. The scenario enumeration approach of [3] is applied in [38] to a similar energy management problem: hybrid electric vehicle power management.

In [17], a tree-based approach for isolated microgrids is proposed. Fans of forecast scenarios are created, and then a scenario tree is produced to reduce the number of scenarios to keep the computing effort of the SMPC problem manageable, using the forward selection algorithm of [18].

The work in [30] proposes a two-stage algorithm for an EMS of isolated microgrids. First stage computes a unit commitment using a stochastic approach. It is formulated as a fixed-recourse, two-stage stochastic problem, resulting in a mixed-integer linear programming problem, considering different scenarios for the renewable energy generation. Second stage performs an optimal power flow using a nonlinear shrinking horizon control algorithm.

The paper [32] tackles an optimal operation management of a microgrid under generation and load uncertainties with economic and environmental objectives, considering thermal and electricity demand. The approach is based on MPC with mixed-integer linear programming and stochasticity managed with a two-stage stochastic programming approach. A chance-constrained MPC approach is proposed in [31], incorporating ARMA prediction models, and considering the correlation between wind generation and demand to obtain a probability distribution of forecast based on current and past observations. The obtained probability distribution is used to evaluate the probability of constraint violations. A similar approach is applied to a hybrid wind–solar–thermal plant in [47].

Also, a chance-constraint MPC is used in [8] focused on a microgrid composed of gas microturbine, a battery, a PV generator, customer load, and grid connection. The authors propose a two-layer control system, where the high-level controller works with the nominal operating conditions of the components, optimizing an economic performance index. Uncertainties are taken into account in the low-level controller, whose objective is to compensate disturbances on load and renewable generation and to guarantee operational constraints. A CC-MPC approach with uncertainties modeled with Gaussian distribution functions is proposed.

7.4 Case Study

7.4.1 Plant Description

The microgrid used in this section is HyLab, the laboratory-scale plant described in Chap. 4. [40, 45]. The system (see Fig. 7.2) consists of renewable generation (solar or wind), emulated by a programmable power source. Consumption profiles can also be emulated by an electronic load. The system includes two types of storage: batteries and hydrogen to manage the excess or deficit of energy between generation and demand. The hydrogen storage system includes a fuel cell to supplement the lack of generation and an electrolyzer to store hydrogen when generation is higher than load. Hydrogen is stored in a metal hydride tank. Additionally, the microgrid allows sale or purchase energy from the main grid. The technical characteristics of the equipment are summarized in Chap. 4, Table 4.1.

7.4.2 MPC Problem Statement

As described in previous chapters, MPC controllers require a control-oriented model to predict the evolution of the system throughout the prediction horizon. This model must be simple because it will be included in a mathematical optimization problem

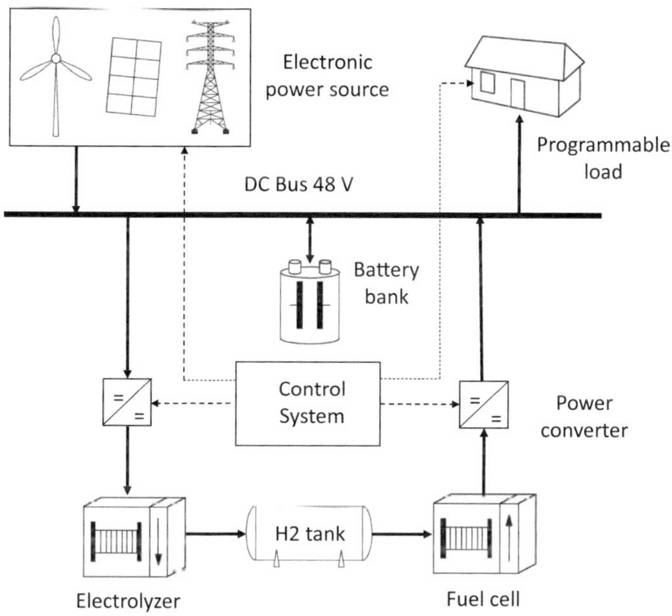

Fig. 7.2 Hylab scheme

at each sampling time. A linear model will be used as control-oriented model in these experiments:

$$x(t+1) = x(t) + Bu(t)B_d d(t) \tag{7.27}$$

In the proposed model, the state vector is $x(t) = [SOC(t)\ LOH(t)]^T$, where SOC is the state of charge of the battery and LOH is the level of hydrogen in the hydride tanks. A fixed value of the efficiency of the battery is used to avoid the use of binary variables.

$$SOC(t+1) = SOC(t) - \frac{\eta_{bat} T_s}{C_{\max}} P_{bat}(t) \tag{7.28}$$

$$LOH(t+1) = LOH(t) + \frac{\eta_{elz} T_s}{V_{\max}} P_{elz}(t) - \frac{T_s}{\eta_{fc} V_{\max}} P_{fc}(t) \tag{7.29}$$

P_{bat} is the power supplied by the battery and $V_{\max}$ is the capacity of H_2 (in normal cubic meters) in the metal hydride tanks. As described in Chap. 4, and additional simplification is performed: power of fuel cell and electrolyzer is condensed in an unique variable $P_{H_2} = P_{fc} - P_{elz}$.

The manipulated variables are the power that can be exchanged with the hydrogen system (P_{H_2}) and the grid (P_{grid}). The disturbances are defined as $d(t) = P_{net}(t) = P_{gen}(t) - P_{load}(t)$. Finally, the battery is directly connected to the DC bus and absorbs the unbalance, so P_{bat} can be obtained from the balance equation:

$$P_{bat}(t) = -P_{H_2}(t) - P_{grid}(t) - P_{net}(t) \tag{7.30}$$

In this case study, the control objective is similar but simpler than the one used in Chap. 4:

$$\begin{aligned} J(x(t), u(t)) &= \sum_{k=1}^{N_c} \alpha_1 P_{H_2}^2(t+k) + \alpha_2 P_{grid}^2(t+k) + \\ &+ \sum_{k=1}^{N_p} \gamma_1 (SOC(t+k) - SOC_{ref})^2 + \gamma_2 (LOH(t+k) - LOH_{ref})^2 \end{aligned} \tag{7.31}$$

that is, a multi-objective weighted cost function looking for an efficient use of the power of the microgrid components and tracking the hydrogen and battery reference levels.

Upper and lower values of levels of energy storage systems and power variables are the physical constraints in this application example. Also, constraints on power increments will be taken into account. The hydrogen-related powers (electrolyzer and fuel cell) are limited to 0.9 kW. This value is lower than their nominal operational points, with the objective of the protection of hydrogen equipment, looking for an increase in their lifespan. On the other hand, the minimum production of electrolyzer and fuel cell powers has been fixed to 0.1 kW. The power limits of the electronic units determine the constraints for $P_{grid}(t)$. Also, constraints on their incremental signals $\Delta P_{H_2}(t)$ and $\Delta P_{grid}(t)$ have been included.

$$-0.9\,kW \leq P_{H_2}(t) \leq 0.9\,kW \tag{7.32a}$$
$$-2.5\,kW \leq P_{grid}(t) \leq 2\,kW \tag{7.32b}$$
$$-20\,Ws^{-1} \leq \Delta P_{H_2}(t) \leq 20\,Ws^{-1} \tag{7.32c}$$
$$-2.5\,kWs^{-1} \leq \Delta P_{grid}(t) \leq 2\,Ws^{-1} \tag{7.32d}$$

With respect to state constraints, operational limits have been considered to battery and metal hydride levels. Batteries are responsible for maintaining the bus voltage level, and consequently undercharge must be avoided. Also, the metal hydride tank must maintain a minimum pressure, corresponding to a minimum value of LOH. Upper bound is imposed by safety reasons. The state constraints $x(t)$ in these experiments are given below:

$$40\% \leq SOC(t) \leq 90\% \tag{7.33a}$$
$$10\% \leq LOH(t) \leq 90\% \tag{7.33b}$$

7.4.3 Stochastic MPC Algorithms Setup

Setting the values of the parameters of the controller according to the control objectives is a key issue in MPC. The main objective is an equilibrate use of the devices, and consequently the weight associated to power exchange with the grid and the weight corresponding to hydrogen power are similar. The tracking of reference levels has been considered as a secondary objective, and consequently the weights corresponding to the power of hydrogen equipment and main grid are much higher than those of the storage levels. The specific values γ_i and α_i of the cost function have been tuned using the studies and results of [11, 34, 40].

A prediction horizon of $N_p = 5$ and a sampling time of 30 s has been considered. Constant references have been selected for battery SOC and metal hydride level, $SOC_{ref} = 65\%$ and $LOH_{ref} = 40\%$.

A probability of violation of constraint $\delta_x < 10\%$ has been selected for the tests in all the approaches. δ_x is a direct parameter in CC-MPC, but it is the result of the number of scenarios in the other two approaches. According to Eq. (7.13), the number of scenarios has been set at 316 for both scenario-based approaches. These scenarios have been determined using the electricity demand and the solar generation obtained from historical data, published by the TSO of the Spanish electricity system Red Eléctrica de España (REE), including sunny and cloudy days. Historical generation and demand from the same source have also been used to obtain the cumulative

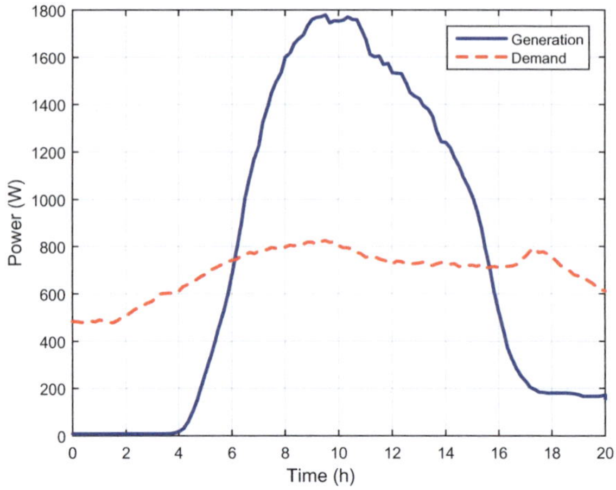

Fig. 7.3 Generation corresponding to a sunny day and demand considered in experiments

distribution function needed for the chance-constraint method. Next sections present experimental results on the plant carried out in [43, 44].

7.4.4 Experimental Results

Each experiment was carried out in the microgrid real plant during a period of 20 hours. The controllers were tested with solar generation profile corresponding to a sunny day, with high irradiance during the central hours of the day. Load profile used in the experiments are scaled data (adapted to microgrid power values) of the real demand on May 23, 2014, registered by the REE.[1] Generation and demand are shown in Fig. 7.3.

The initial values of the state variables were $SOC(0) = 70\%$ and $LOH(0) = 50\%$ for all the experiments. Figures 7.4, 7.5, and 7.6 show the behavior of the microgrid using MS-MPC, TB-MPC, and CC-MPC, respectively. Power evolution of the different components is presented on the left and the storage level evolution on the right.

The qualitative behavior of all controllers is similar. They have to manage the excess of energy during the central part of the day and the deficit during the first and last parts of the day. The deficit of energy during the first hours is supplied by the batteries, fuel cell, and to a lesser extent, the main grid. The energy surplus in the following hours is used mainly to charge the batteries, but when the SOC is reaching its upper value, a portion of the energy is used to produce hydrogen with the electrolyzer and part of the energy is sold to the main grid. The deficit during the

[1]REE demand historical data are available at: https://demanda.ree.es/movil/peninsula/demanda/total.

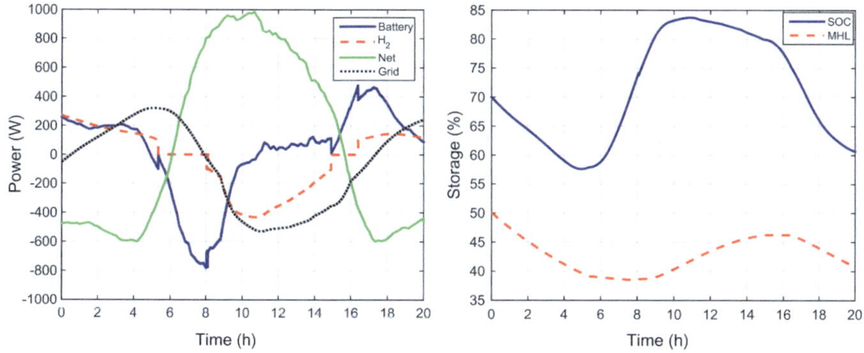

Fig. 7.4 Power flows and storage using MS-MPC controller

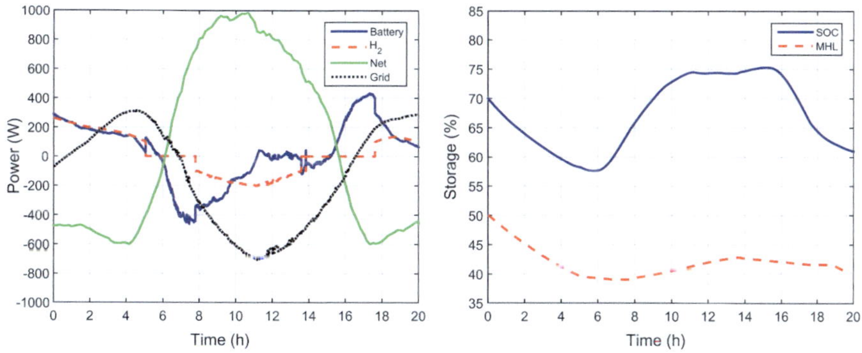

Fig. 7.5 Power flows and storage using TB-MPC controller

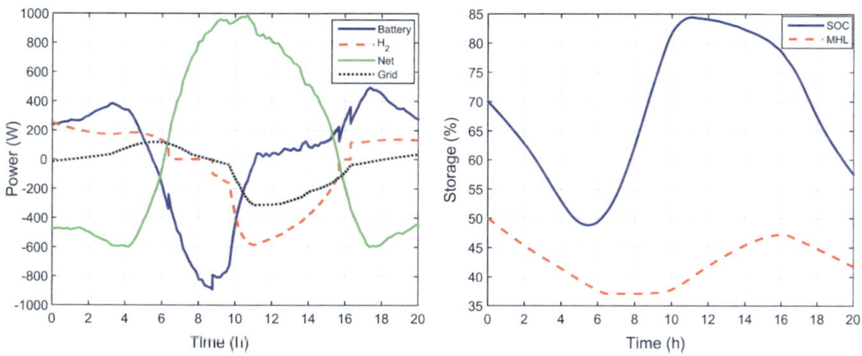

Fig. 7.6 Power flows and storage using CC-MPC controller

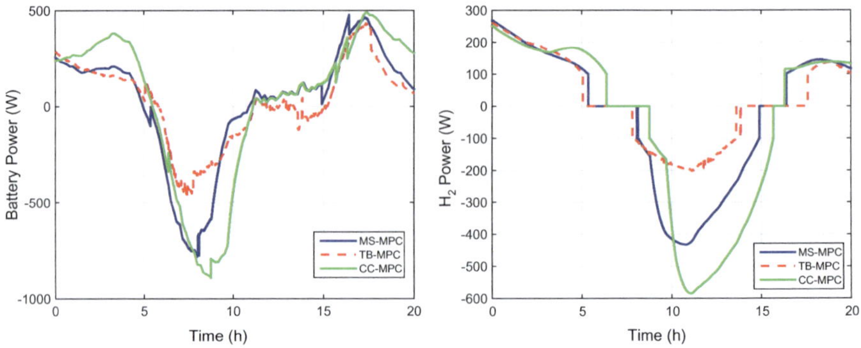

Fig. 7.7 Battery (left) and hydrogen system (right) evolutions when the stochastic MPC approaches are used

last hours of the day is supplied by the battery, fuel cell, and main grid in different ways depending on the controller. The evolution of the SOC and LOH is almost free between their operational limits since the weights utilized in the cost function for reference tracking are small.

As can be noticed, each approach shows particular quantitative differences. CC-MPC controller carries out a more intensive use of battery and electrolyzer during the central hours of the day with an important energy excess (see Fig. 7.7), with the consequence of an approach to the upper operational limit of the SOC. Nevertheless, the state variable constraints are never violated in any of the controllers. On the other side, TB-MPC shows a smoother use of the battery and electrolyzer. Finally, MS-MPC technique has a behavior between the two other methods.

As a consequence of the limited use of the self-consumption components (battery and hydrogen system), TB-MPC carries out a more intensive energy exchange with the grid (see Fig. 7.8-left). The positive face of this type of action is the smoothing of the use of the microgrid equipment, and consequently a possible limitation of the degradation effects.

Figure 7.8 (right) shows a comparison in the SOC evolution for the three approaches. A deeper cycle of batteries is performed in MS-MPC and specially in CC-MPC. In both of them, a value near the upper limit is reached at the end of the energy excess period at central hours of the day.

Table 7.1 summarizes the cumulative value of the cost function for the complete experiment period and the computation time for the three tested SMPC algorithms.

Regarding the cumulative cost of the objective function, the highest value corresponds to MS-MPC approach. This is due to the nature of the procedure, where a conservative optimal solution is obtained as a consequence of the fact that the computed sequence of control actions is valid for all considered scenarios, that is, those close to the present disturbance conditions but also to less favorable ones.

The constraint of the considered scenarios in a tree structure permits a more effective implementation of disturbance occurrence probabilities, resulting in a lesser value of the cost function in the TB-MPC approach. But as a drawback of the method, an increment in the number of control variables involved in the optimization problem,

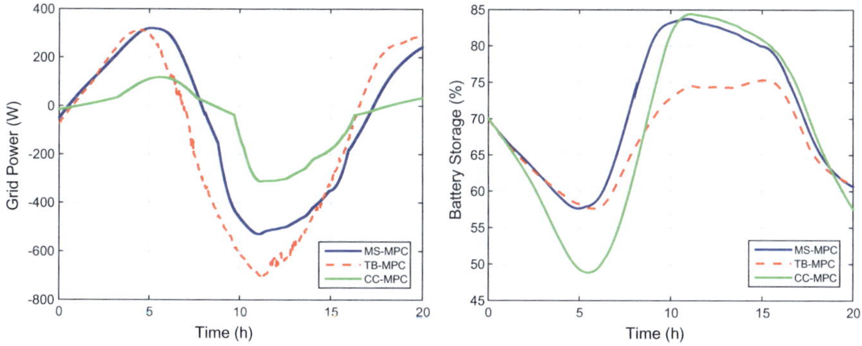

Fig. 7.8 Main grid energy (left) and SOC (right) evolutions when the stochastic MPC approaches are applied

Table 7.1 Performance indicators obtained with the three SMPC methods

	MS-MPC	TB-MPC	CC-MPC
Cumulative cost	3.89×10^{12}	2.75×10^{12}	2.44×10^{12}
Computation time (s)	7.76	18.15	1.04

Table 7.2 Comparison of the energy produced by the fuel cell, consumed by the electrolyzer, and exchanged by batteries and main grid during the experiments

	MS-MPC	TB-MPC	CC-MPC
$E_{fc}(Wh)$	302.0	261.0	348.0
$E_{elz}(Wh)$	481.0	217.0	642.0
$E_{bat}(Wh)$	62.2	110.2	43.1
$E_{grid}(Wh)$	−418	−661.0	−268

also in a tree structure, has effect on the computational time, the biggest among evaluated SMPC techniques in this experiment. Finally, CC-MPC has the lowest cumulative cost, with the advantage that the number of control variables has not been increased in relation to the MS-MPC approach.

Anyway, even when there are important differences among the computational time of the SMPC controllers, it is important to remark that all of them solve the optimization problems faster than the sampling period.

As a summary of the behavior of the controlled microgrid, the energy supplied by the fuel cells, the provided energy to electrolyzers, and the net exchange with the main grid and batteries throughout the 20 h of the experiments are shown in Table 7.2. According to the sign criteria in this book, a positive value in energy of batteries implies that the energy discharged is greater than that charged (see SOC evolution in Figs. 7.4, 7.5, and 7.6). Concerning the main grid, a positive value entails a purchase of energy greater than the sale.

Table 7.3 Interval of the values of the states and control inputs during the experiments

Variable	MS-MPC	TB-MPC	CC-MPC
SOC (%)	[57.61, 83.73]	[57.59, 75.30]	[48.82, 84.42]
LOH (%)	[38.51, 50]	[39.07, 50]	[37.06, 50]
P_{fc}(W)	[100, 268.13]	[100, 259.69]	[100, 250.44]
P_{ez}(W)	[100, 432.9]	[100, 202.13]	[100, 584.94]
P_{grid}(W)	[−529.4, 320.3]	[−705.02, 314.0]	[−312.8, 117.5]

Another important issue in hydrogen-based systems is the startups of the fuel cell and electrolyzer. Chapter 5 of this book has considered the repercussion of the startups in the degradation and how to include it in a cost function. In this chapter, a very simple linear discrete model has been used, and the direct economical loss and reduction of the lifespan of hydrogen equipment have not been explicitly considered. Nevertheless, in this experiment, the number of startups is the same for all the controllers and is a very limited number. To be precise, the fuel cell has been connected twice and the electrolyzer only once.

Finally, Table 7.3 shows the maximum and minimum values of different variables throughout the experiments using the proposed approaches. It is important to notice that the constraints presented in (7.32) and (7.33) are always satisfied independently of the SMPC method in these tests.

As a conclusion, the results of CC-MPC show a better performance, both in the value of the cost function and, as expected because of the size of the problem, in the computation time of the optimization problem. Nevertheless, regarding the cost function results and other performance indicators, these results are far to be conclusive. Notice that the number and selected scenarios in scenario-based approaches and the obtaining of the cumulative distribution function in analytical approaches (using historical data in this experiment) are key points in the controller performance.

References

1. Baringo L, Conejo AJ (2000) Correlated wind-power production and electric load scenarios for investment decisions. Appl Energy 101:475–482
2. Bemporad A, Morari M (1999) Robustness in identification and control. Lecture Notes in Control and Information Sciences, vol 245. Springer, London, chapter Robust model predictive control: A survey. Springer
3. Bernardini D, Bemporad A (2009) Scenario-based model predictive control of stochastic constrained linear systems. In: Proceedings of the 48h IEEE conference on decision and control (CDC) held jointly with 2009 28th Chinese control conference, pp 6333–6338
4. Bichi M, Ripaccioli G, Di Cairano S, Bernardini D, Bemporad A, Kolmanovsky IV (2010) Stochastic model predictive control with driver behavior learning for improved powertrain control. In: 49th IEEE conference on decision and control (CDC), pp 6077–6082
5. Birge JR, Louveaux F (2011) Introduction to stochastic programming. Springer, New York

6. Calafiore G, Dabbene F (2006) Probabilistic and randomized methods for design under uncertainty. Springer, London, England
7. Campi MC, Garatti S, Prandini M (2009) The scenario approach for systems and control design. Ann Rev Control 33(2):149–157
8. Cominesi SR, Farina M, Giulioni L, Picasso B, Scattolini R (2018) A two-layer stochastic model predictive control scheme for microgrids. IEEE Trans Control Syst Technol 26(1):1–13
9. Conejo AJ, Nogales FJ, Prieto FJ (2002) A decomposition procedure based on approximate newton directions. Math Programm 93(3):495–515
10. Dupacov J, Consigli G, Wallace SW (2000) Scenarios for multistage stochastic programs. Ann Oper Res 100:25–53
11. Garcia-Torres F, Bordons C (2015) Optimal economical schedule of hydrogen-based microgrids with hybrid storage using model predictive control. Ind Electron IEEE Trans 62(8):5195–5207
12. Giulioni L (2015) Stochastic model predictive control with application to distributed control systems. PhD thesis, Politecnico di Milano
13. Goodwin GC, Kong H, Mirzaeva G, Seron MM (2014) Robust model predictive control: reflections and opportunities. J Control Decis 1(2):115–148
14. Grosso J, Maestre J, Ocampo-Martinez C, Puig V (2014) On the assessment of treebased and chance-constrained predictive control aproaches applied to drinking water networks. 19th IFAC world congress. Cape Town, South Africa, pp 6240–6245
15. Grosso J, Ocampo-Martinez C, Puig V, Joseph B (2014) Chance-constrained model predictive control for drinking water networks. J Process Control 24(5):504–516
16. Growe-Kuska N, Heitsh H, Romish W (2003) Scenario reduction and scenario tree construction for power management problems. In: IEEE Bologna PowerTech conference, Bologna, Italy
17. Hans CA, Sopasakis P, Bemporad A, Raisch J, Reincke-Collon C (2015) Scenario-based model predictive operation control of islanded microgrids. In: 2015 54th IEEE conference on decision and control (CDC), pp 3272–3277
18. Heitsch H, Romisch W (2003) Scenario reduction algorithms in stochastic programming. Comput Optim Appl 24:187–206
19. Hooshmand A, Asghari B, Sharma R (2013) A novel cost-aware multi-objective energy management method for microgrids. In: Proceedings of the innovative smart grid technologies (ISGT), IEEE PES, Washington, DC, USA, pp 1–6
20. Hooshmand A, Poursaeidi MH, Mohammadpour J, Malki HA, Grigoriads K (2012) Stochastic model predictive control method for microgrid management. In: Innovative smart grid technologies (ISGT), Washington, USA, pp 1–7
21. Hovgaard T, Larsen L, Jorgensen J (2011) Robust economic MPC for a power management scenario with uncertainties. In: Proceedings of the 50th IEEE conference on decision and control and european control conference (CDC-ECC), Orlando, Florida, pp 1515–1520
22. Jalali AA, Nadimi V (2006) A survey on robust model predictive control from 1999-2006. In: Proceedings of the international conference on computational intelligence for modelling control and automation, and international conference on intelligent agents, web technologies and internet commerce. Sydney. Australia
23. Li S, Zheng Y (2015) Distributed model predictive control for plant-wide systems. Wiley, Incorporated
24. Lucia S, Finkler T, Basak D, Engell S (2012) A new robust nmpc scheme and its application to a semi-batch reactor example. In: Proceedings of the international symposium on advanced control of chemical processes, Singapore, pp 69–74
25. Maestre JM, Raso L, Van Overloop PJ, De Schutter B (2012) Distributed treebased model predictive control on an open water system. In: Proceedings of the American control conference (ACC), Montreal, Canada, pp 1895–1990
26. Meibom P, Barth R, Hasche B, Brand H, Weber C, O'Malley M (2011) Stochastic optimization model to study the operational impacts of high wind penetrations in ireland. IEEE Trans Power Syst 26(3):1367–1379
27. Mesbah A (2016) Stochastic model predictive control: an overview and perspectives for future research. IEEE Control Syst Mag 36(6):30–44

28. Mesbah A, Streif S, Findeisen R, Braatz RD (2014) Stochastic nonlinear model predictive control with probabilistic constraints. In: 2014 American control conference, pp 2413–2419
29. Meshbah A, Kolmanovsky IV, Di Cairano S (2018) Stochastic model predictive control. Birkhuser, Dordrecht, pp 75–97
30. Olivares DE, Lara JD, Cañizares CA, Kazerani M (2015) Stochastic-predictive energy management system for isolated microgrids. IEEE Trans Smart Grids 6(6):2681–2693
31. Ono M, Topcu U, Yo M, Adachi S (2013) Risk-limiting power grid control with an arma-based prediction model. In: Proceedings of the 52nd IEEE annual conference on decision and control (CDC), Florence, Italy, pp 4949–4956
32. Parisio A, Rikos E, Glielmo L (2016) Stochastic model predictive control for economic/environmental operation management of microgrids: an experimental case study. J Process Control 43:24–37
33. Patrinos P, Trimboli S, Bemporad A (2011) Stochastic mpc for real-time market-based optimal power dispatch. 50th IEEE conference on decision and control and European control conference (CDC-ECC). Orlando, USA, pp 7111–7116
34. Pereira M, Limon D, Muñoz de la Peña D, Valverde L, Alamo T (2015) Periodic economic control of a nonisolated microgrid. IEEE Trans Ind Electron 62(8):5247–5255
35. Petrollese M, Valverde L, Cocco D, Cau G, Guerra J (2016) Real-time integration of optimal generation scheduling with mpc for the energy management of a renewable hydrogen-based microgrid. Appl Energy 166:96–106
36. Raso L, Schwanenberg D, van de Giesen N, van Overloop PJ (2014) Short-term optimal operation of water systems using ensemble forecasts. Adv Water Resour 71:200–208
37. Reddy SS, Sandeep V, Jung C (2017) Review of stochastic optimization methods for smart grid. Front Energy 11(2):197–209
38. Ripaccioli G, Bernardini D, Di Cairano S, Bemporad A, Kolmanovsky IV (2005) A stochastic model predictive control approach for series hybrid electric vehicle power management. 2010 American control conference. Baltimore, MD, USA, pp 5844–5849
39. Schildbach G, Fagiano L, Frei C, Morari M (2014) The scenario approach for stochastic model predictive control with bounds on closed-loop constraint violations. Automatica 50(12):3009–3018
40. Valverde L, Rosa F, Bordons C (2013) Design, planning and management of a hydrogen-based microgrid. IEEE Trans Ind Inform 9(3):1398–1404
41. van Hessem D, Bosgra O (2006) Stochatic close-loop model predictive control of continuous nonlinear chemical process. J Process Control 16:225–241
42. van Overloop PJ, Weijs S, Dijkstra S (2008) Multiple model predictive control on a drainage canal system. Control Eng Pract 16(5):531–540
43. Velarde P (2017) Stochastic model predictive control for robust operation of distribution systems. PhD thesis, University of Seville
44. Velarde P, Valverde L, Maestre JM, Ocampo-Martinez C, Bordons C (2017) On the comparison of stochastic model predictive control strategies applied to a hydrogen-based microgrid. J Power Sour 343:161–173
45. Vergara-Dietrich JD, Morato MM, Mendes PRC, Cani AA, Normey-Rico JE, Bordons C (2017) Advanced chance-constrained predictive control for the efficient energy management of renewable power systems. J Process Control
46. Wang Q, Guan Y, Wang J (2012) A chance-constrained two-stage stochastic program for unit commitment with uncertain wind power output. IEEE Trans Power Syst 27(1):206–215
47. Yo M, Ono M, Adachi S, Murayama D, Okita N (2014) Power output smoothing for hybrid wind-solar thermal plant using chance-constrained model predictive control. In: 53rd IEEE conference on decision and control, pp 928–934
48. Yu Z, McLaughlin L, Jia L, Murphy-Hoye M, Pratt A, Tong L (2012) Modeling and stochastic control for home energy management. In: Proceedings of the IEEE power and energy society general meeting, San Diego, California, USA, pp 1–9
49. Zhu D, Hug G (2014) Decomposed stochastic model predictive control for optimal dispatch of storage and generation. IEEE Trans Smart Grid 5(4):2044–2053

Chapter 8
Interconnection of Microgrids

Abstract This chapter is devoted to the energy management problem of several interconnected microgrids. EMS of a network of microgrids must determine the power flows inside each microgrid and with the main grid (as in Chap. 4), but also the energy interchange among them. This is an extension of a single microgrid EMS and MPC is an alternative to solve it. The control of these systems presents mainly two problems to be solved by a global controller: first, different microgrids typically are managed by different agents making difficult or even impossible to use a unique controller for the whole system and the second problem is the computational burden due to the dimension of the system when a high number of microgrids are considered. In this situation, Distributed Model Predictive Control (DMPC) is the technique used in this chapter to reduce the complexity. This chapter describes several methods to solve the EMS using MPC in a distributed fashion. Alternatives are tested and compared in a system with three connected microgrids.

8.1 Power Networks Based on Microgrids

Microgrids can operate in two different modes, islanded or grid-connected mode. In this chapter, the second mode will be considered, with several microgrids connected among them and also with the main grid. In this scenario, the microgrids and the different involved operators can obtain benefits purchasing or selling energy directly among them and with the grid in a flexible way.

This scenario is intrinsically distributed, that is, local consumers are supplied by local generation, mainly renewable, with the support of the external grid, and including a local energy storage system to solve the known problems of mismatch between local generation and demand, and natural intermittency and uncertainties of renewable energy sources. Also, this framework can improve the performance and quality of the electrical service.

The operation of a network of microgrids brings more flexibility to the system for both market and technical operation, that is, the network configuration can find more

C. Bordons et al., *Model Predictive Control of Microgrids*,
Advances in Industrial Control, https://doi.org/10.1007/978-3-030-24570-2_8

advantageous situations for the microgrid operation. It can improve the reliability and resilience of the electrical grid due to new reconfiguration possibilities under a global blackout of the external grid or a component failure in one of the microgrids of the network.

In this paradigm, power flow is no longer in one direction from the substation transformers to the consumers, but instead is more dynamic and flowing in two directions. This structuring of the electrical system allows a better integration of the distributed agents in the market, and can also be used to increase flexibility using new grid management approaches to distribution and transmission systems. The interconnection of microgrids can improve reliability, reduce emissions, expand energy options in the future power system, add redundancy, and increase grid security.

The normal operation of the network of microgrids should be oriented to achieve a better economic return with respect to the single operation of the microgrid. One of the most common issues is that agents of the different microgrids can be different and independent, hindering the system management. So the integration of the different agents will always be aligned to reach a better performance in the energy management problem than operating as a single microgrid. But in addition, microgrid networks should be prepared to operate independently of the utility grid in case of faults and congestion.

8.1.1 Architecture of Microgrid-Based Networks

Networks of microgrids with energy storage systems require new architectures and specific techniques to address their management and control. As mentioned above, interconnection of different microgrids in a network introduces flexibility to the system for both market and technical operation, and also requires the presence of new agents with new roles and responsibilities, which must operate in a coordinated way with traditional agents, e.g., Transmission Systems Operator (TSO), Distribution System Operator (DSO), and Market Operator (MO), which also need to change their tasks in this novel microgrid-based network.

At present, distribution networks are passive, with one-way power flow and typically with a radial configuration, where the power flows from a single source to the end user by a single path, simplifying the planning and operation of the network. Nowadays, most of the Distribution Management Systems (DMS) are designed for this kind of networks, including Supervisory Control And Data Acquisition systems (SCADA) to perform real-time control and monitoring, and other functionalities as online power flow, short-circuit analysis, state estimation, fault location, alarm processing, etc. [41]. But the complexity of the tasks to be performed by the DMS will increase significantly with the concept of networks of microgrids. On one side, distributed generation requires bidirectional power flow, but also the connection or disconnection of microgrids will have an impact on quality and reliability, introducing new challenges in DMS functionalities.

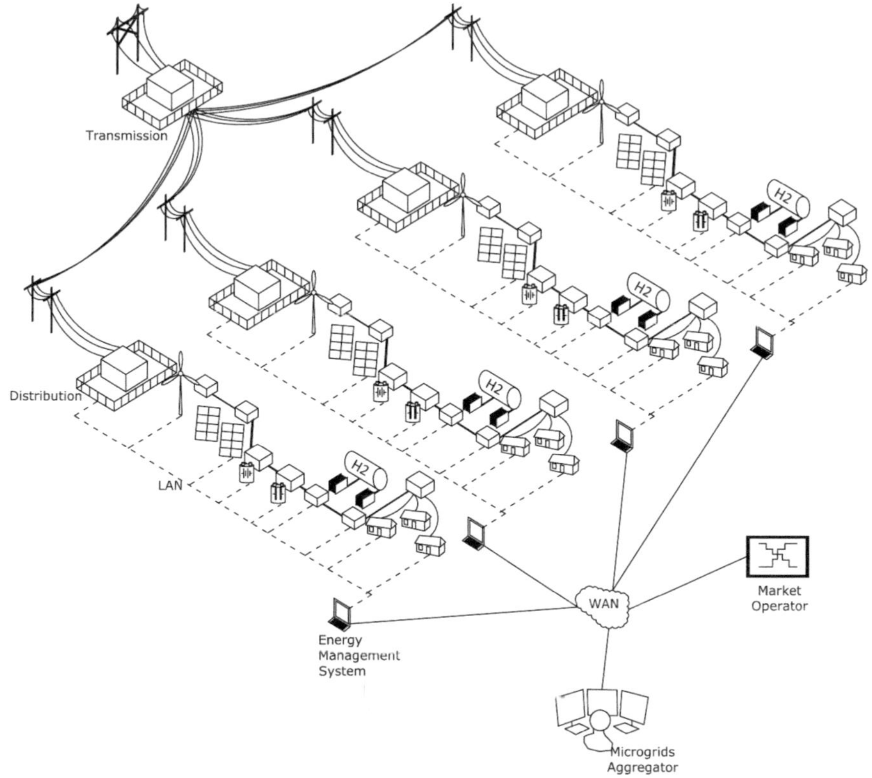

Fig. 8.1 Network of microgrids and communications system architecture object of this study

The role of the MO is also more complex in a distribution network with DERs and microgrids. To the well-known variability in the renewable generation, new sellers with a limited amount of energy to trade appear in the market. This scenario is positive for the emergence of a new agent, the Microgrid Aggregator (MA), an intermediate agent with the mission, from one side, to coordinate and participate in the management and control of a set microgrids, and on the other side, to aggregate the information of that set, acting as interlocutor with DMS or MO.

The above-indicated architecture is shown in Fig. 8.1. The control and management of the set of microgrids can be driven in several ways by the Microgrid Energy Management System (MEMS) or MA, typically in a hierarchical and/or distributed way, modifying the role of each agent. The elements of each microgrid are connected by a Local Area Network (LAN) of fieldbus. Due to the geographical distribution, the microgrids are interconnected among them and with the aggregator using a Wide Area Network (WAN). Different possibilities are as follows:

- The management and control of the network of microgrids are performed by the MA in a centralized way. The MEMS is acting as input/output data gateway, in addition to security and emergency tasks.
- The management and control of each microgrid are carried out locally by each MEMS, while the MA receives data from MEMS for monitoring, data aggregation, etc., and can also send set points for the MEMS local controllers.
- In a distributed way, the management and control are performed locally but the MEMS receives information of the decisions of other neighbor MEMSs, directly or via MA. Again, the MA can send information from a high-level controller in a hierarchical way.

These different strategies lead to centralized, decentralized, and distributed solutions for control and management problems in complex systems, which will be detailed in next section.

8.1.2 Centralized, Decentralized, and Distributed Solutions

In centralized solutions, the management and control of all the microgrids and the distribution system are managed by a single control agent. Then, it is necessary to collect measurements from different remote locations, to decide all the control actions in a single place. The advantage of this approach is that a complete knowledge of the system and a complete control over all the actuators allows a global decision-making and, theoretically, an optimization of the behavior of the system according to certain criteria. Nevertheless, as mentioned above, this solution presents important problems:

- Typically, different subsystems are operated by independent entities or organizations, making the exchange of information very complicated or even impractical.
- In large coupled networks, a centralized approach leads to a huge control problem with probably hundreds of manipulated variables, especially when the control techniques require the solution of an optimization problem, as in MPC. This issue will be faced in this chapter through distributed optimization.

Figure 8.2 shows the scheme of a centralized control system. There are two coupled microgrids, that is, with energy exchange between them, and the control system decides control actions u_1 and u_2 (i.e., storage and generation powers) with feedback measurement from both subsystems (y_1 and y_2) (typically, storage system status).

As mentioned above, due to the complexity of the management of large networks, a control system distributed among several agents seems to be a promising solution. The simpler alternative to a centralized solution is the decentralized approach, where each subsystem is managed by a different control agent (Fig. 8.3). An agent only knows and decides on his own system trying to obtain its own benefit, but the overall network behavior is not taken into account by local controllers, which could produce unpredictable results. Due to the coupling in the microgrid network system, local

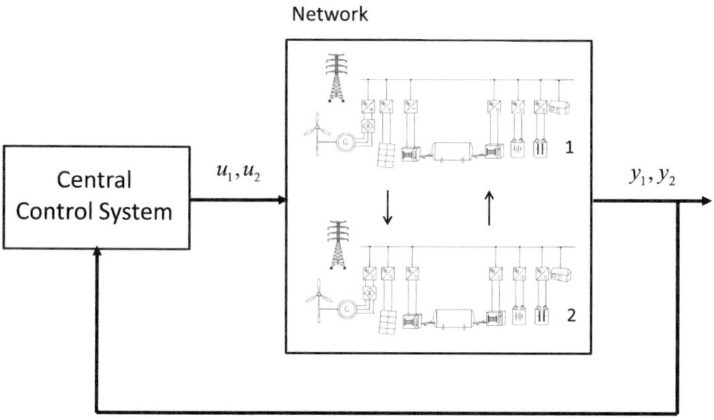

Fig. 8.2 Centralized control architecture

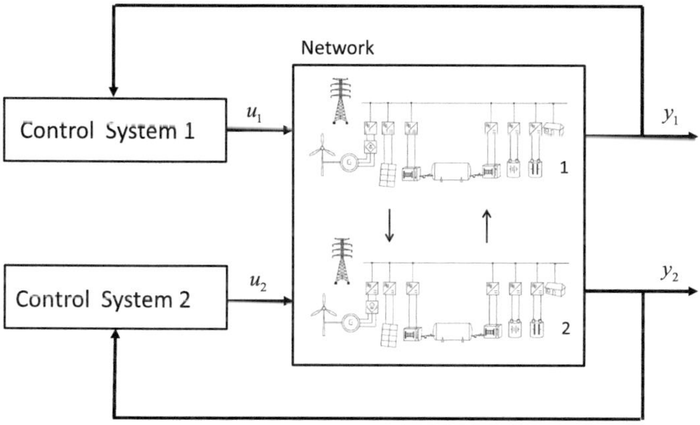

Fig. 8.3 Decentralized control architecture

actions on a microgrid have effect on the overall system. Also, with this approach, a cooperation strategy among subsystems to obtain a better overall performance is not possible (Fig. 8.3).

Between these two approaches, the distributed solution, as a decentralized approach, has different agents managing each subsystem but with a critical difference: subsystems share information, as shown in Fig. 8.4. The nature and amount of the exchanged information could be very different from one implementation to other, i.e., knowledge about the dynamic behavior of other subsystems, the decision applied on other different subsystems, or a global criterion to guide the control of the whole system. Then, cooperation and negotiation among control agents looking for an improvement of the overall system performance are possible.

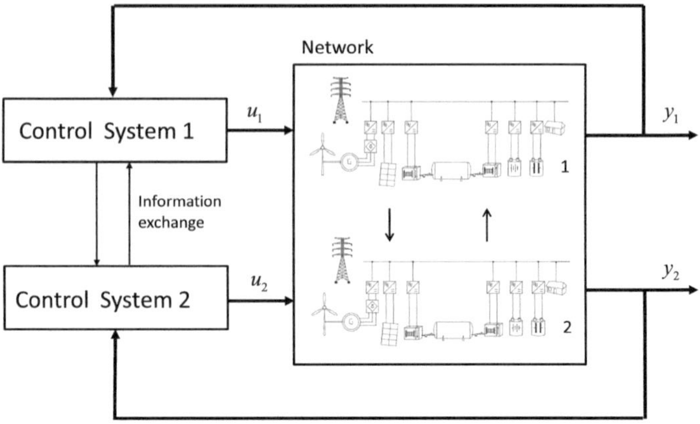

Fig. 8.4 Distributed control architecture

8.1.3 Control of Microgrid Networks

The study of interconnected microgrids is a very active research field. A centralized control model for optimal management and operation of a smart network of microgrids is presented in [32]. The works in [29, 30] address the optimal power dispatch problem considering uncertainties in load and probabilistic modeling of generated power by renewable small-scale energy resources. A game-theory coalition formulation strategy is proposed in [13, 43] to allow the microgrids autonomously cooperate and self-organize into coalitions. In [41], a bi-level stochastic solution for the coordinated operation of multiple microgrids and a distribution system is presented. A self-healing operation mode based on average consensus algorithm for optimal operation of autonomous networked microgrids is developed in [42]. In [37], a distributed EMS for the optimal operation of alternating current (AC) microgrids was designed with consideration of the underlying power distribution network and the associated constraints.

In the last years, many distributed control techniques were proposed in literature [1, 7, 22, 23]. Concerning electrical networks, a review on distributed control techniques for all hierarchy levels and a discussion of future research trends in this area were presented in [46]. In [20], a two-level architecture for distributed energy resource management for multiple microgrids is developed using multi-agent systems applying the proposed method to interconnected microgrids participating in the market with batteries as ESS. A distributed convex optimization framework is developed for energy trading between islanded microgrids in [17]. The work in [8] proposes a decentralized control architecture for microgrids for ongoing investigations into real-time, agent-based decision-making demonstrating the viability and capability of decentralized agent-based control for microgrids. A study of the problem of load demand management, with the aim of minimizing the operational cost in distributed smart grids, is presented in [12]. The article presented in [44] is focused on the decentralized optimal control algorithm for distribution management system, considering distribution network as coupled microgrids. The optimal control problem of the microgrids is designed as a decentralized partially observable Markov decision process. The work in [31] presents an advanced MPC approach for the high-level coordination of power exchanges in a network of microgrids. The MPC-based algorithm is used to determine the future scheduling of power exchanges among dispersed microgrids, as well as the charge/discharge in each local ESS for the future time horizon where the objective is to maximize the benefits at the network level.

In [19], a distributed energy management approach based on a consensus method is presented and used to coordinate local generation, flexible load, and storage devices within the microgrid. Two models (centralized and decentralized) for simulating the interaction of the MA, the entity managing a number of microgrids, and the electricity market are presented in [4]. The paper [39] studies the adoption of the multi-microgrids concept, as a potential way to facilitate large-scale integration of microgeneration. In [3], several coordination strategies for distribution grid congestion management are presented. A distributed supervisory MPC system for optimal management and operation of distributed wind and solar energy generation systems integrated into the electrical grid is carried out in [33]. The problem of the multi-microgrids management is also treated in [9, 21, 25, 34].

A DMPC technique [22] is applied to power networks by several authors. The dissertation carried out in [28] discusses how control agents have to make decisions given different constraints on the type of systems they control, the actuators they can access, the information they can sense, and the communication and cooperation they can perform. A coalitional DMPC based on Shapley value is applied in [26], and this method is also applied to a combined environmental and economical dispatch of the smart grid in [11] and also to large-scale microgrids in [10] with a method based on Lagrange multipliers. Population games are applied in [24]. Finally, a distributed management system to improve resilience is presented in [2].

8.2 Distributed Model Predictive Control

Let us consider the global system, where its dynamics is given by a linear discrete-time model. The overall system will be represented by[1]:

$$x(t+1) = Ax(t) + Bu(t) + Dd(t) \tag{8.1}$$

where $x \in \mathbb{R}^{n_x}$ denotes the vector of states variables, $u \in \mathbb{R}^{n_u}$ the vector of inputs and $d \in \mathbb{R}^{n_d}$ the disturbances. This system represents the complete network of microgrids. It can be decomposed into N coupled subsystems as follows:

[1]Matrix B_d will be renamed in this chapter as D to simplify the notation.

$$x^{(i)}(t+1) = A_{ii}x^{(i)}(t) + \sum_{i \neq j} A_{ij}x^{(j)}(t) + \sum_{j=1}^{N} B_{ij}u^{(j)}(t) + \sum_{j=1}^{N} D_{ij}d^{(j)}(t) \quad (8.2)$$

being $x^{(i)} \in R^{n_{x_i}}$, $u^{(i)} \in R^{n_{u_i}}$ and $d^{(i)} \in R^{n_{d_i}}$ with $x = [(x^{(1)})^T(x^{(2)})^T...(x^{(N)})^T]^T$ and $u = [(u^{(1)})^T(u^{(2)})^T...(u^{(N)})^T]^T$.

Notice that Eq. (8.2) considers both, state and input coupling. If there is any coupling between subsystem i and j, these two subsystems are said to be *neighbors*. Due to the nature of microgrid models, in this chapter, only input coupling will be considered in most of the cases:

$$x^{(i)}(t+1) = A_{ii}x^{(i)}(t) + \sum_{j=1}^{N} B_{ij}u^{(j)}(t) + \sum_{j=1}^{N} D_{ij}d^{(j)}(t) \quad (8.3)$$

The most direct system decomposition in network of microgrids consists of considering each one of the microgrids as a subsystem, with the same state variables, inputs, and disturbances as considered in single microgrids, but adding new input or disturbance variables: the energy exchange between neighbor subsystems.

In distributed MPC, each subsystem is controlled by a local MPC controller, but the key point is the communication and what information is transmitted among the different controllers, and also how and when the transmission is performed.

If states of neighbor subsystems are coupled, the predicted state evolution of a subsystem must be known by its neighbors. Then the predicted states must be transmitted or, if only the sequence of predicted input is transmitted, every controller must know the dynamic behavior of its neighbors, that is, the model. If the subsystems only present input coupling, the exchange of input predictions allow to compute the subsystem states prediction.

A classification based on communication issues can be found in [36]. Attending to the connection among subsystems, two cases can be considered as given below:

- Fully connection: All subsystem controllers are connected and share information. Then the information needed to make decisions is complete, but the necessary infrastructure grows exponentially with the number of subsystems. This case is only practical with a small number and tightly coupled subsystems.
- Partial connection: The information sharing of a subsystem controller is limited to a subset of the others. This is a practical solution when a large number of subsystems are involved. The communication needs are reduced but also the information is not complete.

According to the number of times that local control agents evaluate its optimization problem within the sampling time (and transmit the information to neighbor agents), two strategies can be considered as given below:

- Non-iterative solutions: The exchange of information is done only once per sampling time. This solution reduces the computation burden and the transmitted information but it does not permit any negotiation process among agents.

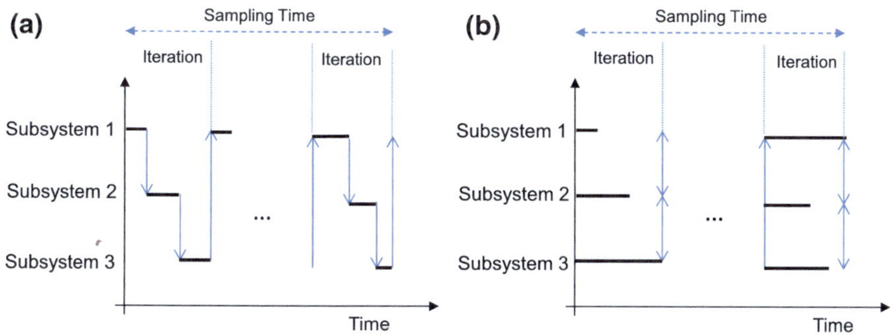

Fig. 8.5 **a** Serial communication. **b** Parallel communication

- Iterative solutions: A number of evaluation and data transmissions within each sampling time are permitted, looking for a consensus among local controllers.

Notice that in any of these solutions, a protocol to manage the communication among agents is needed. Several options can be considered. The first one is a parallel strategy, i.e., all control agents compute at the same time and send information in a synchronous fashion. A second option is a sequential strategy, where an order is established among local agents and the controllers are evaluated sequentially, i.e., agent i will compute after the i–1-th agent has computed, using information transmitted from all previous agents (See Fig. 8.5).

A Centralized MPC computes at each sampling time the sequence of the inputs of all subsystems that minimizes a global performance index. Then, as the decision is computed in a global way, all subsystem agents are collaborating to reach the optimal solution to the complete system. This is a perfect collaboration among subsystems to reach a global optimum objective and a Pareto optimal solution will be found.

But this situation changes in non-centralized approaches, where each local agent is computing a local objective function. Obviously, these approaches find optimal local objectives, but in general, a global optimum performance is not guaranteed. This idea is easy to understand in Decentralized MPC, but one of the key points in Distributed MPC is to determine how local controllers can collaborate to reach a global optimum solution, or at least an approximation to it. Setting communication issues and even establishing negotiation among local agents are usual topics in Distributed MPC approaches.

The global objective function can be defined as follows:

$$J_{global} = \sum_{i=1}^{N} J_{local}^{(i)}(\hat{\boldsymbol{x}}, \boldsymbol{u}) \tag{8.4}$$

Notice that in a decentralized approach, local objective functions only depend on local states $x^{(i)}$ and local inputs $u^{(i)}$, forming a decoupled optimization problem, but in distributed approaches, due to communication, the objective function can also depend on state variables and inputs of other subsystems, but with the usually partial information submitted by its neighbors.

In this context, an important classification can be made taking into account the objective function to be optimized in each one of the local MPC controllers:

- Noncooperative algorithms: Local controllers optimize a local objective function, sometimes with conflicting objectives. These algorithms are expected to find a Nash equilibrium.
- Cooperative algorithms: The same global objective function is optimized by each one of the local controllers. Nevertheless, usually incomplete information is used, and then a Pareto equilibrium cannot be guaranteed.

8.3 Distributed MPC Approaches

This section describes several approaches found in literature that can be implemented in control of networks of microgrids.

Considering N_p as the prediction horizon, the following notation represents input and state variables of subsystem i over that prediction horizon:

$$\begin{aligned} \mathbf{u^{(i)}}(t) &= [u^{(i)}(t|t)\ u^{(i)}(t+1|t)\ ...\ u^{(i)}(t+N_p-1|t)]^T \\ \mathbf{x^{(i)}}(t) &= [x^{(i)}(t|t)\ x^{(i)}(t+1|t)\ ...\ x^{(i)}(t+N_p-1|t)]^T \end{aligned} \tag{8.5}$$

and in the same way the global input and state variables over the prediction horizon are defined as follows:

$$\begin{aligned} \mathbf{u}(t) &= [u(t|t)\ u(t+1|t)\ ...\ u(t+N_p-1|t)]^T \\ \mathbf{x}(t) &= [x(t|t)\ x(t+1|t)\ ...\ x(t+N_p-1|t)]^T \end{aligned} \tag{8.6}$$

The same notation will be applied to other variables.

A subsystem i will be connected to a subset of other subsystems (neighbors of i), being $\mathcal{N}_i$ the set of neighbors of subsystem i, with n_i elements. Only communication between neighbors will be considered. If only input coupling is considered, Eq. (8.3) can be rewritten as follows:

$$x^{(i)}(t+1) = A_{ii}x^{(i)}(t) + B_{1,i}u^{(i)}(t) + B_{2,i}v^{(i)}(t) + D_i d^{(i)}(t) \tag{8.7}$$

where $v^{(i)}(t)$ is the set of interconnection variables with other subsystems affecting the dynamic of subsystem i. In this chapter, $v^{(i)}(t)$ will be the set of power exchanges among microgrids, that is,

$$v^{(i)}(t) = [u^{(1i)}\ u^{(2i)}\ ...\ u^{(n_i i)}] \tag{8.8}$$

where $u^{(ji)}$ is the power exchange between neighboring subsystems i and j computed by subsystem i.

An agreement on the values of $v^{(i)}(t)$, that is, on $u^{(ji)}$ and $u^{(ij)}$ among neighbors is necessary, requiring communication between pairs of controllers. The simplest strategy is to decide the power exchange on one subsystem and consider that power exchange as a disturbance on the other subsystem. Otherwise, negotiation becomes necessary.

8.3.1 Noncooperative MPC Approach

In this section, noncooperative approaches will be considered, that is, each subsystem will optimize its own cost function. Both iterative and non-iterative solutions will be considered and tested in next section.

In this approach, each one of the interconnection variables will be considered as an input variable in one of the systems and a disturbance variable in the others. Given a system i, the set of neighbors $\mathcal{N}_i$, is divided into two subsets, one including the neighbors where the interconnection variables are considered as input variables $\mathcal{N}_i^1$ and other $\mathcal{N}_i^2$, where interconnections are considered as disturbances, with $\mathcal{N}_i = \mathcal{N}_i^1 \cup \mathcal{N}_i^2$. Then, the model of the system is

$$\begin{aligned} x^{(i)}(t+1) = A_{ii}x^{(i)}(t) + B_i u^{(i)}(t) + \sum_{k \in \mathcal{N}_i^1} B_{1,ik} u^{(ki)}(t) + D_i d^{(i)}(t) + \\ + \sum_{k \in \mathcal{N}_i^2} D_{1,ik} u^{(ki)}(t) \end{aligned} \tag{8.9}$$

A non-iterative approach applied to multi-microgrids systems without storage and controlled generation can be found in [45]. This work is based on Camponogara et al. [7], where a stability constraint is included in the formulation. The objective of the iterative approach is to find a consensus on interconnection variables. This iterative process must be executed in each sampling period. At each iteration, an optimization problem using local cost functions and communication between neighbors in order to exchange the local computed values of interconnection variables is performed. Notice that the sampling time limits the number of iterations. Steps of the iterative algorithm performed by every control agent at each sampled period are described in Algorithm 8.1

8.3.2 Cooperative MPC Approach

In a cooperative approach, the global objective function J_{global} is assumed to be known by agents, in an attempt to optimize the global problem, and consequently

Algorithm 8.1 Noncooperative iterative algorithm: Agent i

1: Obtain the current state $x^{(i)}(t)$
2: Initialize iteration parameters and interconnection variables: $p = 0$ and $u_0^{(ji)}(t) = 0$. Another option for interconnection variables is a *warm start*, initializing them with the value corresponding to the last sampling time, that is $u_0^{(ji)}(t) = u^{(ji)}(t-1)$
3: **repeat**
4: $p = p + 1$
5: Solve the local optimization problem, cost function $J_{local}^{(i)}$ with constraints
6: Send $\mathbf{u}_p^{(ji)}$ to neighbor agents j in $\mathcal{N}_i^1$ and receive $u_p^{(ij)}$ computed in agent j from neighbor agents in $\mathcal{N}_i^2$
7: Evaluate end-iteration conditions

$$\begin{aligned} &p \geq p_{max} \\ &\|v_p^{(i)}(t) - v_{p-1}^{(i)}t)\| < \varepsilon \quad \forall i \end{aligned} \tag{8.10}$$

8: **until** one of the iteration conditions is TRUE
9: Apply the computed control action $u_i(t)$

find a Pareto equilibrium. The problem is the lack of information to compute this function in each local controller, needing the values of states, input, and disturbances of all other subsystems. The communication of those data among agents is imperative in this approach.

Algorithm 8.2 Cooperative iterative algorithm: Agent i

1: Obtain the current state $x_i(t)$
2: Initialize iteration parameters and interconnection variables: $p = 0$, $u_0^{(ji)}(t) = 0$, $x_0^{(j)}(t) = 0$ and $u_0^{(j)}(t) = 0$, $\forall j \in \mathcal{N}_i$. Another option to interconnection variables is a *warm start*, initializing them with the value corresponding to the last sampling time
3: **repeat**
4: $p = p + 1$
5: Solve the global optimization problem, J_{global} with constraints, subnetworks dynamics and the information $\mathbf{x}_j^{p-1}$ and $\mathbf{u}_j^{p-1}$, $\forall j \in \mathcal{N}_i, j \neq i$
6: Send $\mathbf{x}_p^{(j)}$, $\mathbf{u}_p^{(j)}$ and $\mathbf{u}_p^{(ji)}$ to neighbor agents j in $\mathcal{N}_i^1$ and receive the equivalent information computed in agent j from neighbor agents.
7: Evaluate end-iteration conditions

$$\begin{aligned} &p \geq p_{max} \\ &\|v_p^{(i)}(t) - v_{p-1}^{(i)}t)\| < \varepsilon \quad \forall i \end{aligned} \tag{8.11}$$

8: **until** one of the iteration conditions is TRUE
9: Apply the computed control action $u^{(i)}(t)$

A basic iterative, synchronous, and parallel method [40] is presented in this section, but other alternatives are also possible. Every sampling time, an iterative process is launched, where at a given iteration p, an agent i has received information

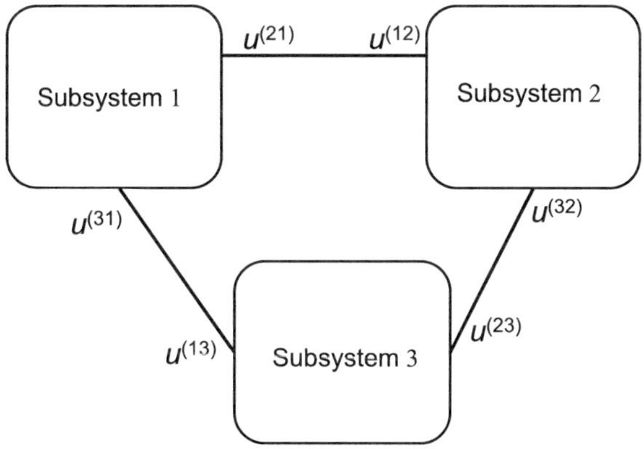

Fig. 8.6 Interconnection variables

of state and input trajectories computed at previous iteration in neighbor agents, that is, $\mathbf{x}_{p-1}^{(j)}$ and $\mathbf{u}_{p-1}^{(j)}$, $\forall j \in \mathcal{N}_i$.

Again, as in the noncooperative approach, each one of the interconnection variables will be considered as an input variable in one of the systems and a disturbance variable in the other one. The main difference is that local agents minimize the global function, using the available information at that moment, typically values of the previous iteration regarding state variables and inputs of neighboring subsystems. The cooperative approach is described in Algorithm 8.2.

8.3.3 Lagrange-Based MPC Approach

This section is based on a simplified version of the work of Negenborn et al. in [27]. In this approach, interconnecting variables are computed in both agents, i.e., $u^{(ji)}$ is computed by agent i and $u^{(ij)}$ by agent j, and consequently, a consensus process between agents is needed. If there is no connection between two subsystems, the interconnection variable value is zero. Notice that these two variables represent physically the same quantity, but must be represented as two different variables because both controllers compute it independently, although communication between them is used to find an agreement on these variables (see Fig. 8.6).

Then, a new set of constraints must be added because common interconnecting variables of two subsystems i and j must be equal, as shown in Fig. 8.6. From the point of view of agent i:

$$u^{(ji)}(t) = u^{(ij)}(t), \quad \forall j \subset \mathcal{N}_i \tag{8.12}$$

In this formulation, communication among neighbors is assumed. The presented Lagrange-based MPC is an iterative approach. A single iteration finishes when all subsystems have concluded their computations. During one sampling time, several iterations are possible. An augmented Lagrangian formulation based on the global cost function (8.4) is used, where a term related to the interconnection constraints (8.12) is added to the local cost function, and then the local optimization problem is stated as follows:

$$\min_{\mathbf{u^{(i)}},\mathbf{u^{(ji)}}} \quad J_{local}^{(i)}(\hat{\mathbf{x}}^{(i)}, \mathbf{u^{(i)}}) + \sum_{j} J_{inter}^{(i)}(\mathbf{u}^{(ji)}(t)) \tag{8.13}$$

$\forall i \quad i = 1...N$, and subject to

$$\begin{aligned} x^{(i)}(t+1) &= A_{ii}x^{(i)}(t) + B_{1,i}u^{(i)}(t) + B_{2,i}v^{(i)}(t) \\ x_{min}^{(i)} &\le x^{(i)} \le x_{max}^{(i)} \\ u_{min}^{(i)} &\le u^{(i)} \le u_{max}^{(i)} \end{aligned} \tag{8.14}$$

Finally, the control steps at time t for an agent i are stated in Algorithm 8.3.

Algorithm 8.3 Lagrange-based iterative algorithm: Agent i

1: Obtain the current state $x^{(i)}(t)$
2: Initialize Lagrange multipliers $\boldsymbol{\lambda}_1^{(ji)}(t)$, arbitrarily (or using previous knowledge) and the iteration counter $p = 1$
3: **repeat**
4: Solve the optimization problem (8.13) for iteration p
5: Send interconnection variables $\mathbf{u}_p^{(ji)}(t)$ to neighbor agents $j \in \mathcal{N}_i$ and receive $\mathbf{u}_p^{(ij)}(t)$ from those agents.
6: Update the Lagrange multipliers

$$\boldsymbol{\lambda}_{p+1}^{(ji)}(t) = \boldsymbol{\lambda}_p^{(ji)}(t) + \gamma_c(\mathbf{u}_p^{(ji)}(t) - \mathbf{u}_p^{(ij)}(t)) \tag{8.15}$$

7: $p = p + 1$
8: Evaluate end-iteration conditions

$$\begin{aligned} &p \ge p_{max} \\ &\|\boldsymbol{\lambda}_p^{(ji)}(t) - \boldsymbol{\lambda}_{p-1}^{(ji)}t)\| < \varepsilon \quad \forall i \end{aligned} \tag{8.16}$$

9: **until** one of the iteration conditions is TRUE
10: Apply the computed control action

In this parallel approach, cost function J_{inter} is defined as follows:

$$\begin{aligned} J_{inter,p}^{(i)} &= \boldsymbol{\lambda}_p^{(ji)}(t) \cdot \mathbf{u}_p^{(ji)}(t) + \frac{\gamma_c}{2}\left\|\mathbf{u}_p^{(ji)}(t) - \mathbf{u}_{p-1}^{(ij)}(t)\right\|_2^2 + \\ &+ \frac{\gamma_b - \gamma_c}{2}\left\|\mathbf{u}_p^{(ji)}(t) - \mathbf{u}_{p-1}^{(ji)}(t)\right\|_2^2 \end{aligned} \tag{8.17}$$

where γ_b and γ_c are positive scalars. This algorithm is repeated in every sampling time.

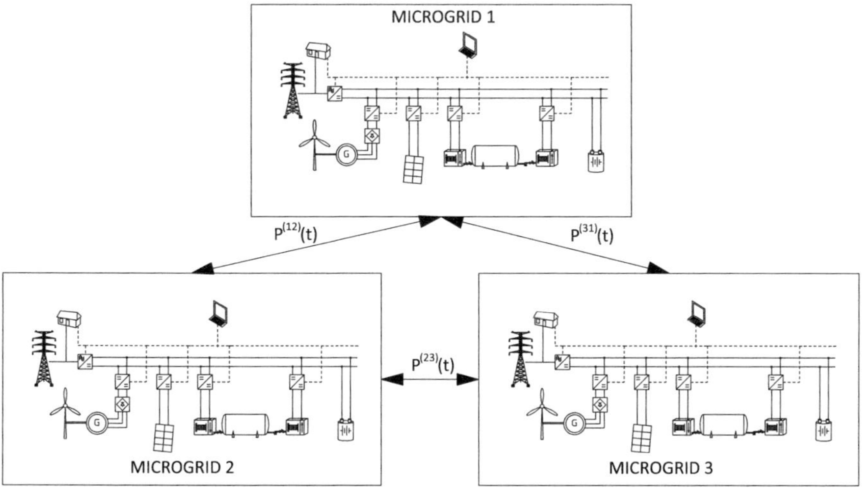

Fig. 8.7 Microgrid scheme

8.3.4 Case Study 1: Centralized and Distributed EMS Controllers

In order to evaluate different algorithms, a basic management system for three interconnected microgrids (MG_1, MG_2 and MG_3) will be considered. The system is an extension of the case study presented in Chap. 4. Each microgrid i is composed of a battery, a hydrogen system (a storage system, electrolizer, and fuel cell allowing bidirectional power flows), renewable generation (solar and wind), a local load, interconnection with other microgrids and also to the grid (see Fig. 8.7).

A control-oriented linear model will be used for each microgrid, as described in Chap. 4. $x^{(i)}(t) = [SOC^{(i)}(t)\ LOH^{(i)}(t)]^T$ is the state vector of microgrid i, where $SOC^{(i)}$ is the state of charge of the battery and $LOH^{(i)}$ is the level of hydrogen in the hydride tanks of microgrid i. A fixed value of the efficiency of the battery is used.

$$SOC^{(i)}(t+1) = SOC^{(i)}(t) - \frac{\eta_{bat}^{(i)} T_s}{C_{\max}^{(i)}} P_{bat}^{(i)}(t) \tag{8.18}$$

$$LOH^{(i)}(t+1) = LOH^{(i)}(t) + \frac{\eta_{elz}^{(i)} T_s}{V_{\max}^{(i)}} P_{elz}^{(i)}(t) - \frac{T_s}{\eta_{fc}^{(i)} V_{\max}^{(i)}} P_{fc}^{(i)}(t) \tag{8.19}$$

where $P_{bat}^{(i)}$ is the power supplied by the battery in microgrid i and $V_{\max}^{(i)}$ is the maximum volume of H_2 of the hydrogen tank in the same microgrid. As described in Chap. 4, P_{fc} and P_{elz} are complementary variables. For the sake of simplicity, these two variables will be replaced by $P_{H_2}^{(i)} = P_{fc}^{(i)} - P_{elz}^{(i)}$. Notice that if other storage units are used, they can be included by adding other equation of the type Eq. (8.18).

The following balance equation must be considered for each microgrid, where $P_{net}^{(i)}(t) = P_{gen}^{(i)}(t) - P_{load}^{(i)}(t)$:

$$P_{net}^{(i)}(t) + P_{bat}^{(i)}(t) + P_{H_2}^{(i)}(t) + P_{grid}^{(i)}(t) + \sum_{j \neq i} P^{(ij)}(t) = 0 \tag{8.20}$$

In this case study, a day-ahead control problem will be considered. The control problem can be managed using centralized or distributed techniques, but the main features of the problem are as follows:

- **Control Objectives**: The control objective in microgrids is typically a minimization of the power costs, considering the economic costs of the storage systems and the power interchanges among other microgrids or the main grid. Also, the regulation of the storage system level can be an objective. Notice that the control objective can be considered from a local microgrid point of view or considering the global network performance. In this case study, the following local and global objective functions are considered:

$$J_{global} = \sum_{i=1}^{N} J_{local}^{(i)} \tag{8.21}$$

The local objective function can be defined also in a multi-objective way considering the regulation of the storage systems levels,

$$J_{local}^{(i)}(t) = \alpha_{bat}^{(i)} P_{bat}^{(i)}(t)^2 + \alpha_{H2}^{(i)} P_{H_2}^{(i)}(t)^2 + \alpha_{grid}^{(i)} P_{grid}^{(i)}(t)^2 + \sum_{j \in \mathcal{N}_i} \alpha_{ij} P^{(ij)}(t)^2 +$$
$$\gamma_{bat}^{(i)}(SOC^{(i)}(t) - SOC_{ref}^{(i)})^2 + \gamma_{H_2}^{(i)}(LOH^{(i)}(t) - LOH_{ref}^{(i)})^2 \tag{8.22}$$

- **Manipulated Variables**: The input variables in this case study of microgrid i are $P_{H2}^{(i)}$, $P_{grid}^{(i)}$, and $P^{(ji)}$. Notice that $P_{bat}^{(i)}$ is obtained through the balance equation.

- **Exogeneous Inputs**: The power generated in the photovoltaic field and in the wind turbine, $P_{gen}^{(i)}$, the local load $P_{load}^{(i)}$, and the exchange power with other microgrid k but computed in that microgrid , $P^{(ik)}$.

- **Physical Constraints**: Physical constraints are given by upper and lower limits of energy storage systems and maximum and minimum power of the equipment and lines, that is,

$$\begin{aligned} SOC_{min}^{(i)} &\leq SOC^{(i)}(t) \leq SOC_{max}^{(i)} \\ LOH_{min}^{(i)} &\leq LOH^{(i)}(t) \leq LOH_{max}^{(i)} \\ P_{r,min}^{(i)} &\leq P_r^{(i)}(t) \leq P_{r,max}^{(i)}|_{r=bat,grid,H_2} \\ P_{min}^{(ij)} &\leq P^{(ij)}(t) \leq P_{max}^{(ij)}, \forall i,j \quad i \neq j \end{aligned} \tag{8.23}$$

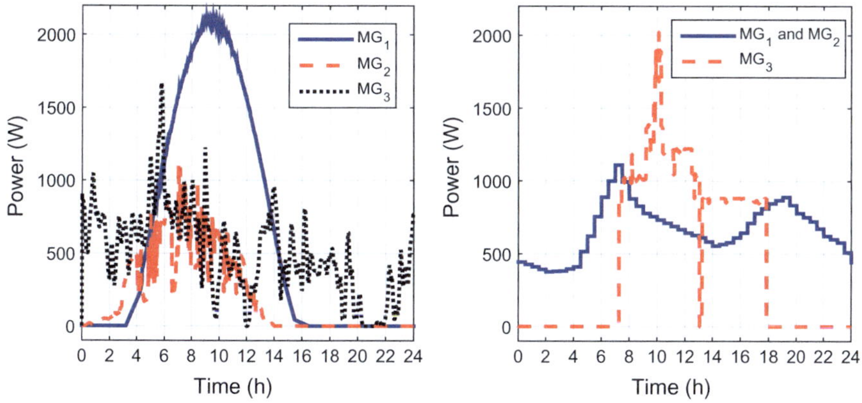

Fig. 8.8 Renewable Generation and Demand of the three microgrids

The renewable generation of MG_1 corresponds to a photovoltaic panel on a sunny day. The generation of MG_2 is also photovoltaic but on a cloudy day and only the energy of a wind turbine is considered in MG_3 with moderate wind. The load profiles of microgrids are typical of home consumption. Microgrids MG_1 and MG_2 have the same demand profile. The renewable generation and demand for the three microgrids are shown in Fig. 8.8.

The weights of cost function (8.22) are adjusted to promote self-consumption, mainly battery and then hydrogen. On the other hand, the use of energy exchanged between microgrids is preferable to the use of the grid.

The results have been obtained in simulation using software Simμgrid. Notice the different conditions of the three microgrids. MG_1 has an excess of energy during sunlight hours but a deficit in night hours (see P_{net} in Fig. 8.9). The scenario of MG_2 is completely different, with an energy deficit during practically the whole day (Fig. 8.10). This lack of energy must be provided by local storages, by other microgrid or the main grid. Finally, MG_3 generation is provided by a wind turbine, and consequently with a high variability throughout the day. The result is an excess of energy during the first hours of the day, a deficit in the central hours and again a slight excess in the final hours (P_{net} in Fig. 8.11).

Centralized Approach

First, a simulation with a centralized MPC has been performed. Figures 8.9, 8.10, and 8.11 show the main variables of the behavior of the three microgrids during 24 h. As a general conclusion and according to control objectives, the exchange of energy with the main grid is practically negligible and the energy exchange between microgrids is limited. Battery plays a predominant role both for the storage of the energy excess and to supply the deficit. Notice that this behavior can be changed by the use of other weights in the cost function.

In MG_1, the use of the battery is absolutely predominant with a limited participation of the hydrogen system when the SOC of the batteries is near the upper limit. At

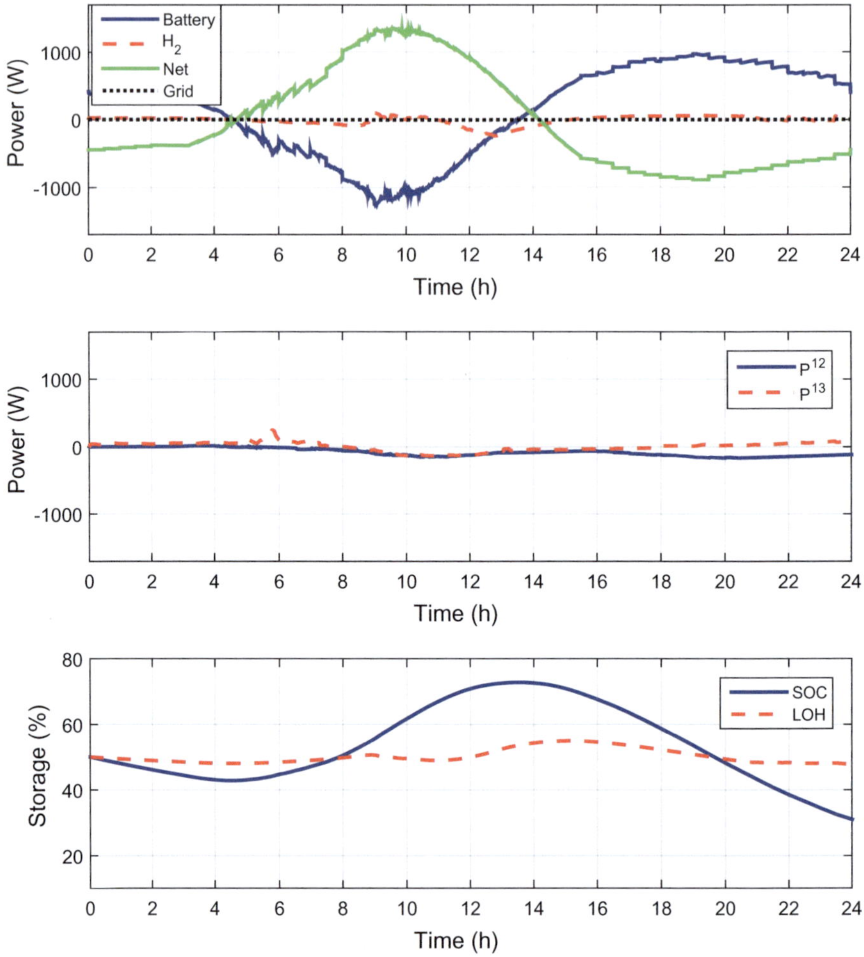

Fig. 8.9 Centralized MPC results of Microgrid 1. Power of the microgrid elements (upper figure), power exchange with other microgrids (middle), and SOC and LOH evolution (lower figure)

the end of the day, some energy is supplied to MG_2 due to the complex situation of this microgrid at these final hours. The deficit of energy in MG_2 during all the day creates problems in the management of this microgrid. As can be seen in Fig. 8.10, the energy is supplied mainly by the battery during the first 18 h, but when the SOC reaches the lower limit, the fuel cell and the other two microgrids collaborate in the energy supply. Again, the use of the main grid is practically avoided. The energy exchange between pairs of microgrids can be observed more clearly in Fig. 8.12.

In MG_3, the variability in the generation is managed mainly by the battery with a limited use of the hydrogen systems when the SOC value is high. Notice that a significant part of the excess of energy at the end of the day is sent to MG_2 to reduce its lack of energy.

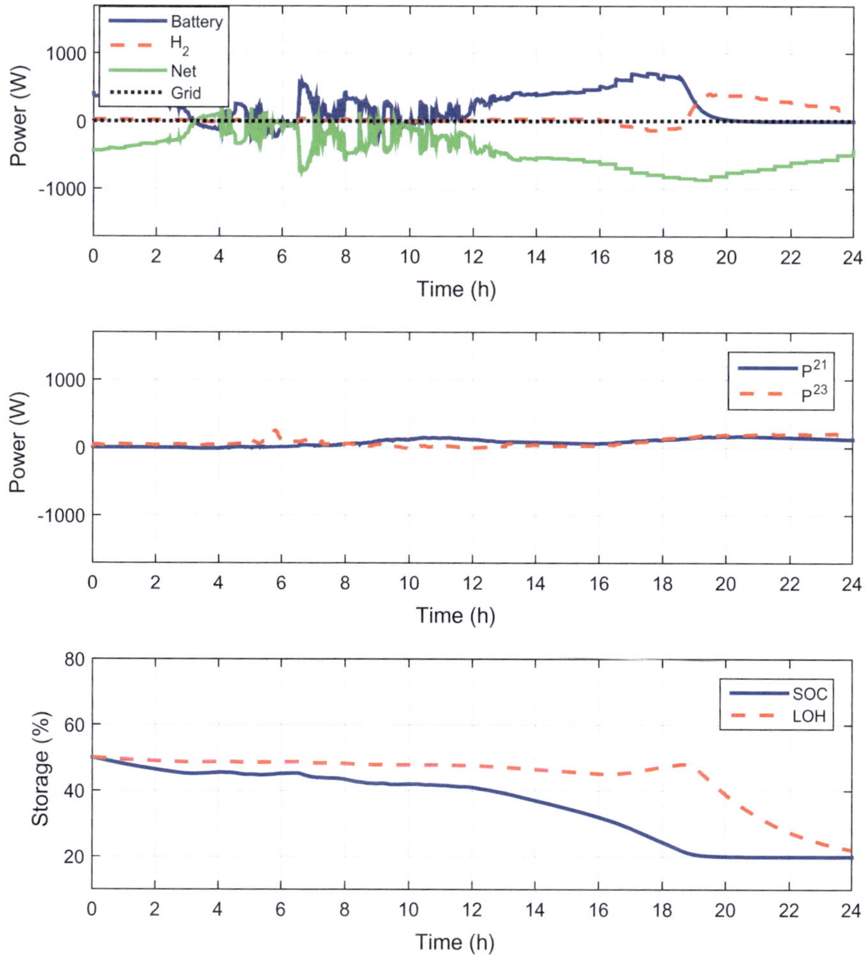

Fig. 8.10 Centralized MPC results of Microgrid 2. Power of the microgrid elements (upper figure), power exchange with other microgrids (middle), and SOC and LOH evolution (lower figure)

Distributed Approaches

Three distributed MPC approaches have been tested with the same three micro-grids system using equivalent cost functions to those in the centralized approach. Noncooperative MPC, cooperative MPC, and Lagrange-based MPC controllers are executed locally in each microgrid, but communication between them is established. The communicated data at each sampling time and the cost functions are different in the three approaches, as described in previous section.

Concerning the obtained results, all the distributed approaches and the centralized MPC present qualitatively similar behavior in all the microgrids. That is, the four methods use mainly the battery to store the excess of energy and to compensate the

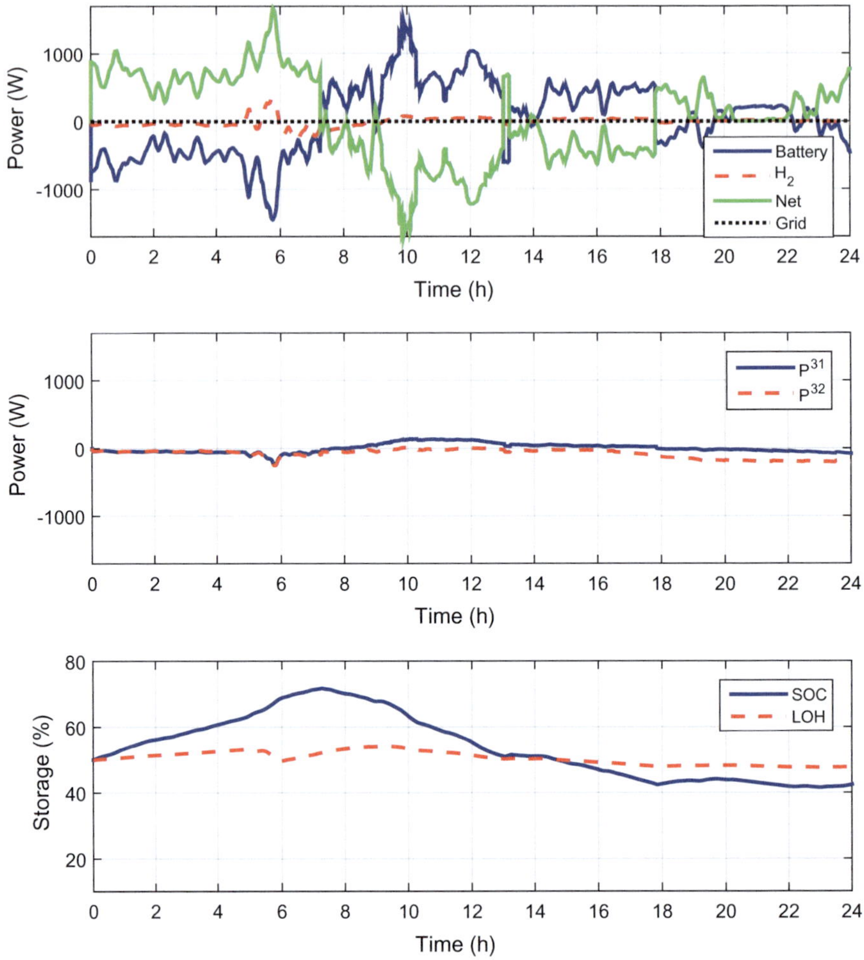

Fig. 8.11 Centralized MPC results of Microgrid 3. Power of the microgrid elements (upper figure), power exchange with other microgrids (middle), and SOC and LOH evolution (lower figure)

deficit. The hydrogen system and the power exchange among microgrids are used in a limited way, and the energy exchanged with the main grid is negligible, except in the noncooperative MPC.

To avoid repetitive figures, only the behavior of microgrid MG_2, the one with a more complicated management, is shown for each DMPC approach in Figs. 8.13, 8.14, and 8.15.

Noncooperative MPC uses a local cost function and the communication needs are more reduced than in the other distributed approaches. The result is a less effective management of the energy. Notice that the lower level of the SOC of the battery is reached 2 h earlier than in the centralized solution. The effect of this situation is an

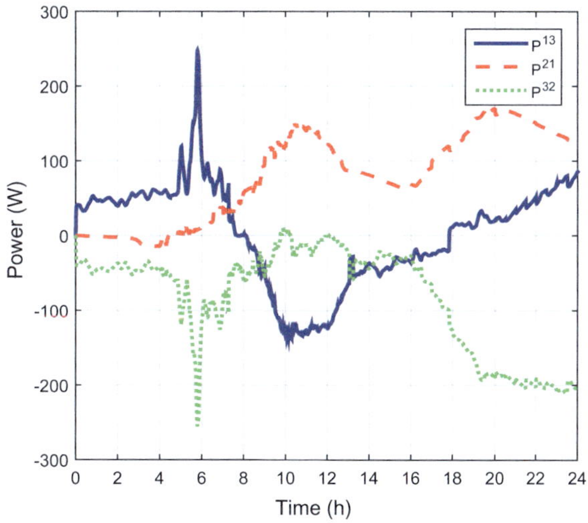

Fig. 8.12 Power exchange among microgrids in centralized MPC

Table 8.1 Cumulative cost function values for the experiments using centralized and distributed controllers

	Centralized	Noncoop.	Cooperative	Lagrange
Cost function	4.2269	67.7239	4.7315	5.1171

increase in the deficit of energy in the later hours, and important energy supply from the main grid, the most expensive type of energy in the cost function used.

Cooperative MPC uses a global cost function and consequently each local agent has to know additional information from their neighbors such as the values of states and interconnection variables at each sampling time. As can be seen in Fig. 8.14, results are very close to centralized MPC. Very similar qualitative results are also obtained in Lagrange-based approach, as can be seen in Fig. 8.15.

Table 8.1 shows the quantitative results of the cumulative cost function. Taking into account that each algorithm uses a different cost function (global or local) in its optimization problem, the centralized cost function has been computed for results of all the approaches in order to have a common comparison framework. As expected, the best value corresponds to the centralized approach, the worst to the noncooperative algorithm, while the other two methods present cumulative cost function values close to the centralized approach.

The energy supplied by the fuel cells and the provided energy to electrolyzers throughout 24 h are presented in Table 8.2. Notice that only the use of the fuel cell in MG_2 is significant, the microgrid with an important energy deficit. Nevertheless, the centralized approach makes a more reduced use of hydrogen system in all situations, and on the contrary, noncooperative distributed method presents the more extended use of hydrogen equipment.

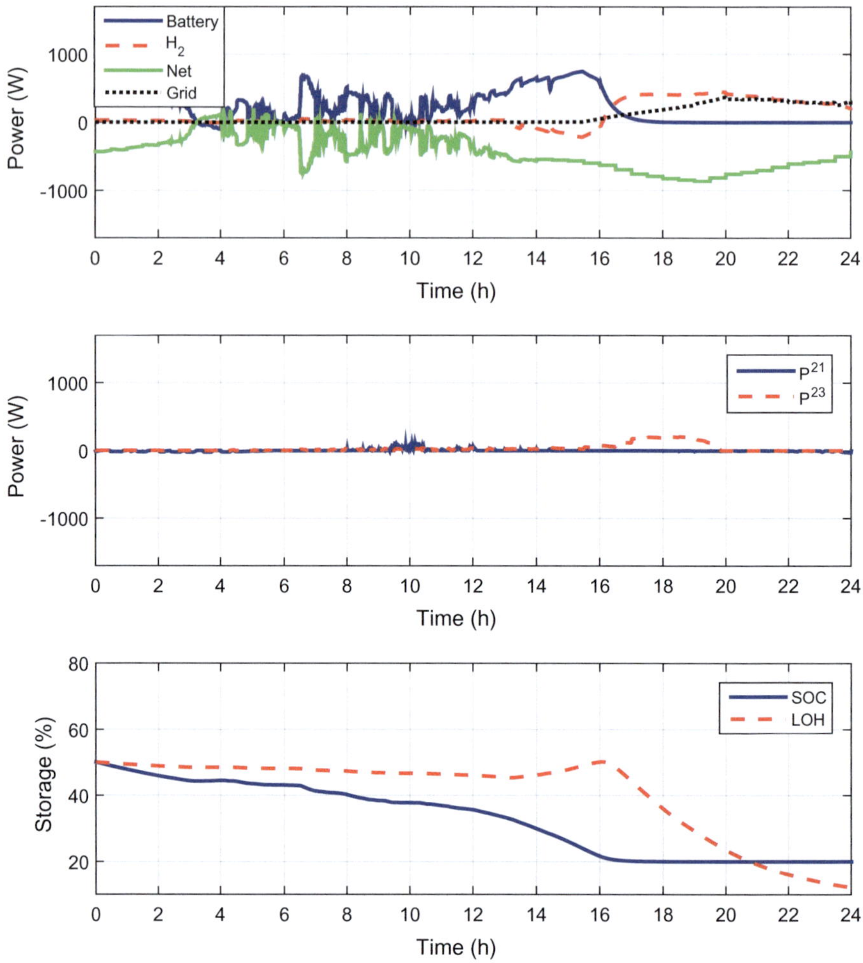

Fig. 8.13 Noncooperative MPC results of Microgrid 3. Power of the microgrid elements (upper figure), power exchange with other microgrids (middle), and SOC and LOH evolution (lower figure)

Finally, the energy exchange with the main grid throughout the simulation time can be observed in Table 8.3. The main problem with the noncooperative approach is the excessive use of energy from the main grid instead of alternative and more economical sources. Notice the negligible use of the main grid in other approaches, especially in the centralized one. On the other hand, Table 8.3 also shows that the exchange of energy between microgrids in the noncooperative approach is poor compared to

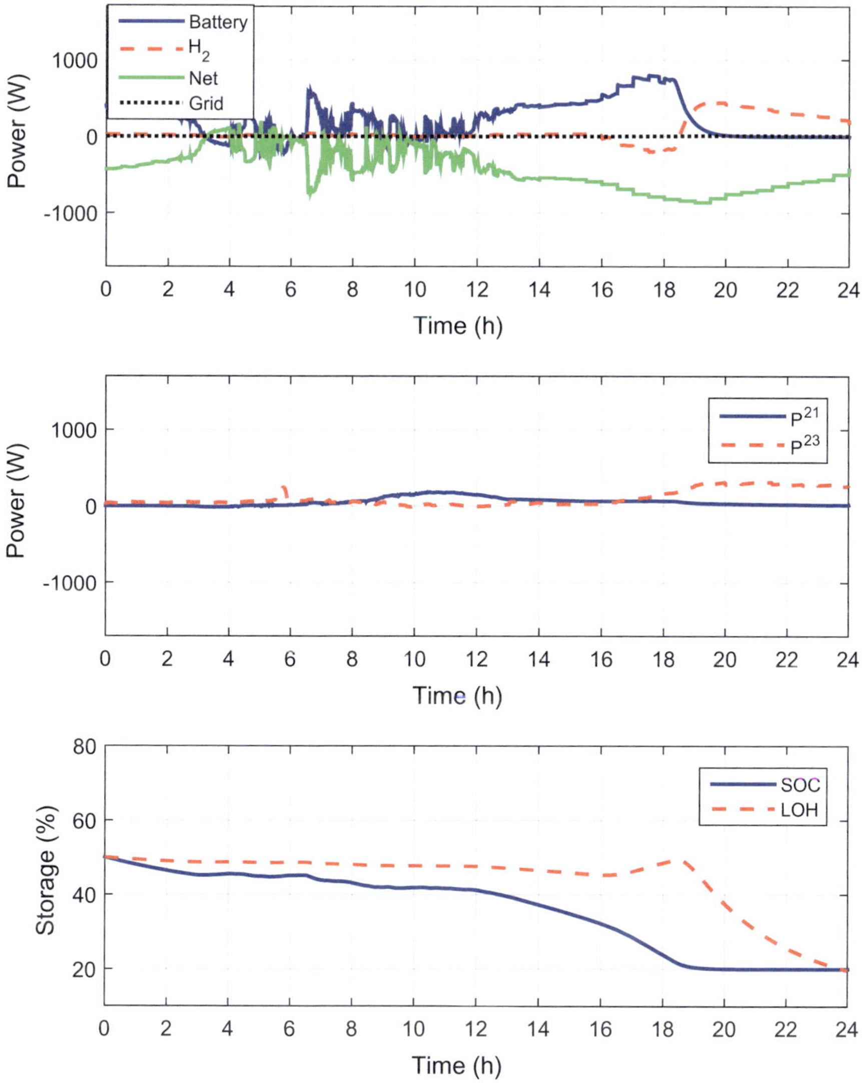

Fig. 8.14 Cooperative MPC results of Microgrid 3. Power of the microgrid elements (upper figure), power exchange with other microgrids (middle), and SOC and LOH evolution (lower figure)

the same magnitudes in centralized MPC, due to the lack of communication and the use of a local cost function. In the other two distributed approaches, those exchanges increase to values close to the centralized technique.

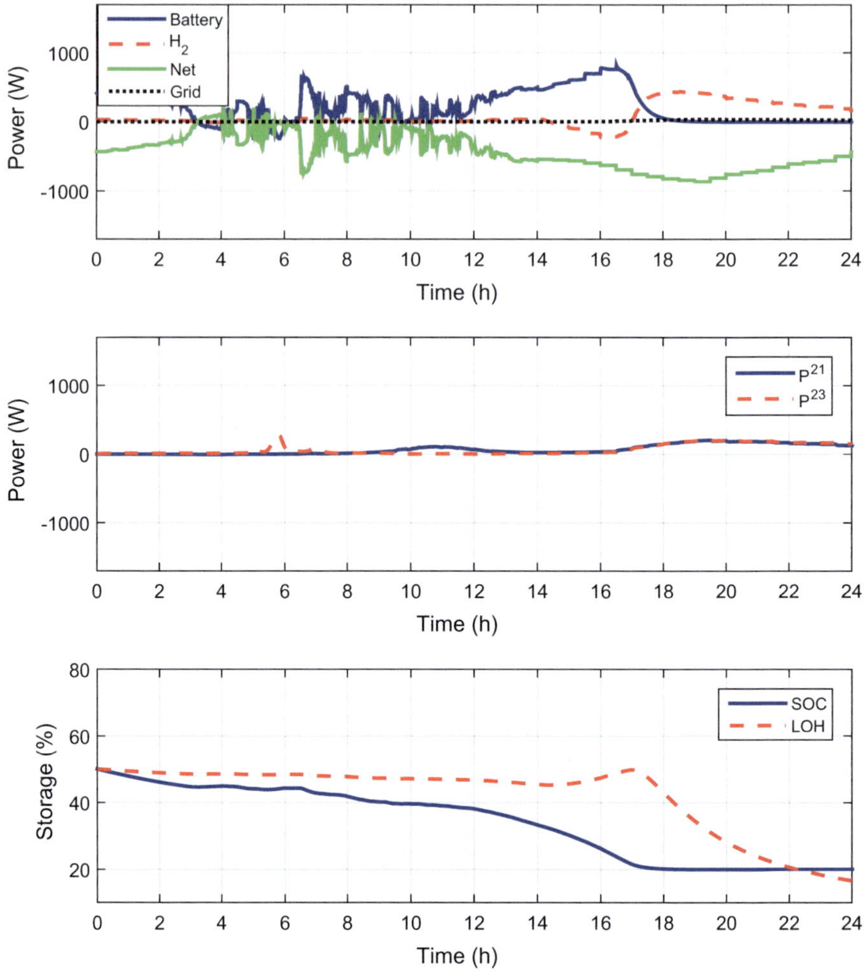

Fig. 8.15 Lagrange-based MPC results of Microgrid 3. Power of the microgrid elements (upper figure), power exchange with other microgrids (middle), and SOC and LOH evolution (lower figure)

Table 8.2 Energy exchanges (kWh) with hydrogen devices during the experiments using centralized and distributed controllers

	Centralized	Noncoop.	Cooperative	Lagrange
Fuel Cell MG_1	0.4584	0.6173	0.6080	0.6017
Fuel Cell MG_2	1.7736	2.9982	2.0461	2.3986
Fuel Cell MG_3	0.4407	0.5271	0.4892	0.5087
Electrolyzer MG_1	0.6973	0.9500	0.8242	0.9861
Electrolyzer MG_2	0.2401	0.3677	0.3406	0.3589
Electrolyzer MG_3	0.6840	0.8990	0.7126	0.8512

Table 8.3 Energy (kWh) exchanged with the main grid and among microgrids during the experiments using centralized and distributed controllers

	Centralized	Noncoop.	Cooperative	Lagrange
Grid—MG_1	0.0367	0.5346	0.0361	0.0715
Grid—MG_2	0.0624	2.0840	0.0839	0.2402
Grid—MG_3	0.0247	0.09091	0.0281	0.0484
P^{12}	1.9603	0.1627	1.3051	1.6354
P^{13}	1.3460	0.5017	1.1068	0.6339
P^{23}	1.9938	0.7798	2.4700	1.6019

8.4 Centralized MPC with Distributed Optimization

The architecture of the network is a critical issue in the decision of what kind of algorithm, centralized or distributed, can be implemented. As mentioned above, the main drawback of a centralized implementation is the computational burden with a high number of agents or when a complex objective function has to be solved in the MPC problem.

When the agent in charge of the management is a Microgrid Aggregator (MA), the most natural option is the centralized solution. A different approach with respect to the centralized algorithm proposed previously in this chapter is the decomposition of the optimization problem in smaller and more tractable ones. Several solutions can be found in literature about distributed optimization with application in the context of MPC, as Jacobi and Gauss–Seidel algorithms [5, 6], dual decomposition methods [16], and multiple shooting methods [35], among others.

In this section, a centralized control problem, solved using a distributed optimization algorithm is presented, specifically, an economic schedule of a network of interconnected microgrids with hybrid ESS. The laboratory microgrid of Chap. 4 will be used again.

Concerning the system architecture presented previously, the MA is in charge of this control problem solved in a centralized based way. In this approach, the decision of each microgrid about the exchange of energy with other microgrids must be taken. A microgrid will only establish connection with other microgrids if it acquires an equal or better economic benefit than acting as a single microgrid.

This optimization problem is solved using a distributed optimization technique, where the cost function, as in Chap. 5, maximizes the economic benefit of the microgrids, including the minimization of degradation of the storage systems, while fulfilling the different system constraints. The method is developed with the aim to be applied to microgrids with complex cost functions formulated as MIQP or MINLP.

8.4.1 Day-Ahead Controller Description

This section presents a method for a 24 h ahead scheduling for an EMS problem of a network of microgrids using distributed optimization. The method is composed of three main parts: (i) the optimization of each microgrid as a single system, (ii) the improvement of the economic benefit of each microgrid using a peer-to-peer (P2P) procedure, (iii) evaluation of all the possible combinations of pairs of microgrids, searching for the most economical exchange of energy through the network of microgrids.

Step 1: Single Microgrid Optimization

The day-ahead market optimization local problem for a single microgrid i can be defined as follows (see Chap. 5, Eq. (5.71) and [14]):

$$J_{local}^{(i)} = J_{grid,local}^{(i)} + J_{bat,local}^{(i)} + J_{H_2,local}^{(i)} \tag{8.24}$$

where the grid, batteries, and hydrogen cost functions are defined in Chap. 5, Eqs. (5.72), (5.73) and (5.74), respectively. The solver will provide the optimal set of input variables, as defined in Chap. 5, when energy exchange among microgrids is not considered.

Step 2: Peer-to-Peer Optimization

In a second step, the energy interconnections between pairs of microgrids are considered. The single microgrid cost function (8.24) is augmented to a peer-to-peer cost function for two coupled microgrids as follows:

$$J_{p2p} = \sum_{i=A,B} J_{local}^{(i)} \tag{8.25}$$

An additional constraint must be included in the optimization problem to establish that the power exchange from A to B computed in agent A or B has the same physical magnitude.

$$P^{(AB)}(t) + P^{(BA)}(t) = 0 \tag{8.26}$$

where $P^{(ij)}(t)$ is the power flow between microgrid (i) and microgrid (j) computed in agent j. For the sake of simplicity, transport losses will not been considered in this book. The interested reader can find a development of the method including those losses in [15].

When both microgrids are working in coupling mode, a power balance constraint of the whole system must be included as given below:

$$\begin{aligned} &\sum_{i=A,B} P_{pv}^{(i)}(t) + P_{wt}^{(i)}(t) = \\ &\sum_{i=A,B} \left(P_{grid}^{(i)}(t) + P_{bat}^{(i)}(t) + P_{load}^{(i)}(t) + P_{elz}^{(i)}(t) - P_{fc}^{(i)}(t) \right) \end{aligned} \tag{8.27}$$

$$P_j^{min,(i)} \leq P_j^{(i)}(t) \leq P_j^{max,(i)}|_{j=grid,elz,fc,bat}^{i=A,B} \tag{8.28}$$

$$SOC_{bat}^{min,(i)} \leq SOC_{bat}^{(i)}(t) \leq SOC_{bat}^{max,(i)}|^{i=A,B} \tag{8.29}$$

$$LOH^{min,(i)} \leq LOH^{(i)}(t) \leq LOH^{max,(i)}|^{i=A,B} \tag{8.30}$$

$$0 \leq \delta_j^{(i)}(t) \leq 1|_{j=elz,fc}^{i=A,B} \tag{8.31}$$

Also, the energy balance equation for each microgrid can be formulated as follows (formulated for microgrid (A), similar balance equation for microgrid (B)):

$$\begin{aligned} &P_{pv}^{(A)}(t) + P_{wt}^{(A)}(t) - P_{load}^{(A)}(t) = P_{grid}^{(A)}(t) \\ &+ P_{elz}^{(A)}(t) - P_{fc}^{(A)}(t) + P_{bat}^{(A)}(t) + P^{(BA)}(t) \end{aligned} \tag{8.32}$$

The solution of the control problem for a network of two microgrids (microgrid (A) and microgrid (B)) provides the set of optimal control variables $u_{p2p}^{opt} = [u_{p2p}^{opt,(A)}, u_{p2p}^{opt,(B)}, P^{(AB)}]$. Both microgrids agree to collaborate if the solution given by the peer-to-peer optimization problem provides a more advantageous situation with respect to operate individually, for both microgrids, that is,

$$J_{local}^{(A)}(u_{p2p}^{opt,(A)}) <= J_{local}^{(A)}(u_{local}^{opt,(A)}) \tag{8.33}$$

$$J_{local}^{(B)}(u_{p2p}^{opt,(B)}) <= J_{local}^{(B)}(u_{local}^{opt,(B)}) \tag{8.34}$$

To accomplish the constraint given by Eqs. (8.33) and (8.34), the problem should be formulated as a MIQQ (Mixed-Integer Quadratic programming problem with Quadratic constraints). However, this kind of formulation takes higher complexity and solving time for the control problem. In [15], a method is presented to simplify the complexity of the problem by reformulating the cost function without quadratic terms.

Step 3: Network Optimization

In this last step, the complete network of microgrids is considered. The proposed algorithm provides a suboptimal but easy to obtain solution. It is based on the computation of the power exchange among microgrids selecting sequentially the most promising couple of microgrids as follows:

First, all possible couples of microgrids are evaluated using the algorithm described in Sect. 8.4.1. The number of couple evaluations is $(N-1)!$ where N is the number of microgrids in the network. Only couples of microgrids where the value of the peer-to-peer function is lower than the sum of its local cost functions are considered (feasible couples). Then, among these feasible couples, the one with the highest decrement is selected. If microgrids i and j form the selected couple, this step fixed the power exchange between microgrid i and microgrid j, $P^{(ij)}$.

In the next step, the evaluations of the couples is performed again eliminating the couple (i, j), that is, $(N-1)!-1$ combinations are considered. A new couple (k, l) is selected and $P^{(kl)}$ is fixed. So the energy balance constraint for each microgrid mentioned in Eq. (8.32) has to be replaced as follows:

$$\begin{aligned} P_{pv}^{(i)}(t) + P_{wt}^{(i)}(t) - P_{load}^{(i)}(t) - \sum_{M=1,\ldots,N_{iter}-1} \sum_{\alpha=1,\ldots,N} P^{(i\alpha)}(t)|_{iter-M} = \\ P_{grid}^{(i)}(t) + P_{elz}^{(i)}(t) - P_{fc}^{(i)}(t) + P_{bat}^{(i)}(t) + P_{uc}^{(i)}(t) + P^{(ij)}(t) \end{aligned} \quad (8.35)$$

The term $P^{(i\alpha)}(t)|_{iter-M}$ refers to the exchange of energy between the microgrids of a selected couple in the previous iteration. This algorithm continues until a new feasible couple cannot be found in a given step or all the links have been selected.

$$P^{(ij)}(t) = 0 \ \forall i, j \in N \quad (8.36)$$

8.4.2 *Results and Discussion*

Four connected microgrids are going to be considered in the simulation validation tests. The main parameters of the devices of the microgrids are shown in Table 8.4. The values considered for the parameters of the cost functions are shown in Table 8.5, with data based on U.S. Department of Energy (DOE) [18, 38].

Peer-to-Peer Optimization

First, the P2P optimization algorithm is tested, with two identical systems (microgrid (1) in table 8.4). These microgrids will be referenced as (1-A) and (1-B). Figure 8.16-left presents the prices in the day-ahead market used in this test. First, the microgrids have to be scheduled independently, as single microgrids, obtaining the behavior shown in Fig. 8.16-right. A comparative study of this scheduling problem using different optimization techniques can be found in [15].

Table 8.4 Main parameters of the components of the microgrids. ©[2019] IEEE. Reprinted, with permission, from IEEE Transactions on Industrial Electronics [15]

Microgrid	1	2	3	4
PV panels	3 kWp	1 kWp	2 kWp	2 kWp
Wind turbine	1 kW	3 kW	3 kW	2 kW
Electrolyzer	900 W	900 W	1000 W	2 kW
H2-tank	7 Nm^3	7 Nm^3	7 Nm^3	14 Nm^3
Fuel cell	600W	600 W	1000 W	2 kW
Batteries	2.7 kW	2.7 kW	6 kW	2.7 kW
	11 kWh	11 kWh	24 kWh	11 kWh
Grid connection	2.5 kW	2.5 kW	6 kW	2.5 kW
	−6 kW	−6 kW	−6 kW	−6 kW

Table 8.5 Values of the controller

Electrolyzer: $Cost_{degr,elz} = 0.0577$ €/W, Lifetime $= 10000$ h $\varsigma = 0.23$ Nm^3/kWh, CC = 8.22 €/kW, $Cost_{o\&m,elz} = 0.002$ €/h $Cost_{startup,elz} = 0.123$ €, $Cost_{shutdown,elz} = 0.0062$ €
Fuel cell: $Cost_{degr,fc} = 0.0018$ €/W, Lifetime $= 10000$ h $\varsigma = 1.320$ kWh/Nm^3, CC = 30 €/kW, $Cost_{o\&m,fc} = 0.001$ €/h $Cost_{startup,fc} = 0.01$ €, $Cost_{shutdown,fc} = 0.005$ €,
Batteries: $\eta_{ch} = 0.90$, $\eta_{dis} = 0.95$, CC = 125 €/kWh, Life cycles = 3000, $Cost_{degr,dis} = 10^{-9}$ €/W^2h, $Cost_{degr,ch} = 10^{-9}$ €/W^2h

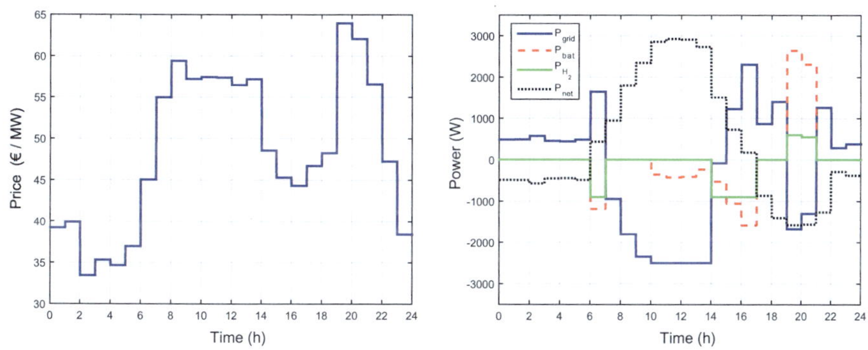

Fig. 8.16 Price evolution per hour (left) and day-ahead optimization results obtained for Microgrid (1) as single system (right)

Next, in step 2, the P2P procedure is applied to the pair of microgrids (1-A) and (1-B) to compare the cost functions operating as single microgrids or in connected mode, that is, with energy exchange between them. The behavior in connected mode is shown in Fig. 8.17 for both microgrids. For the interested readers, an exhaustive

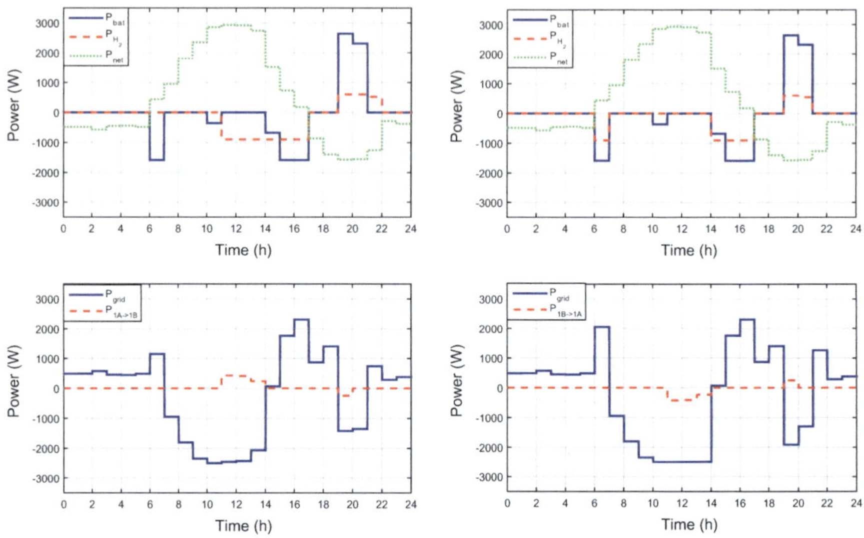

Fig. 8.17 Day-ahead optimization results obtained for microgrid (1-A) (right) and (1-B) (left) after P2P optimization

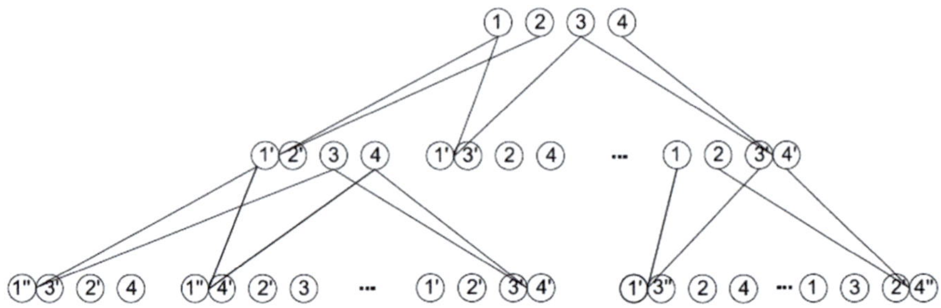

Fig. 8.18 Tree with microgrid's combination in the network optimization algorithm

comparison of the terms of the cost function (degradation cost of batteries, hydrogen equipments, etc.) is presented in [15].

Network Optimization

The results of the global optimization of the network of microgrids are presented in Figs. 8.19 and 8.20. Values obtained at different iterations of the network optimization algorithm can be observed in Table 8.6. In the first iteration (ITER: 0), the cost function of each microgrid acting individually is shown. In the second iteration (ITER: 1), the different couples of microgrids are evaluated showing the local cost function for each one of the two microgrids, as well as the increment of benefit of working as a couple of microgrids.

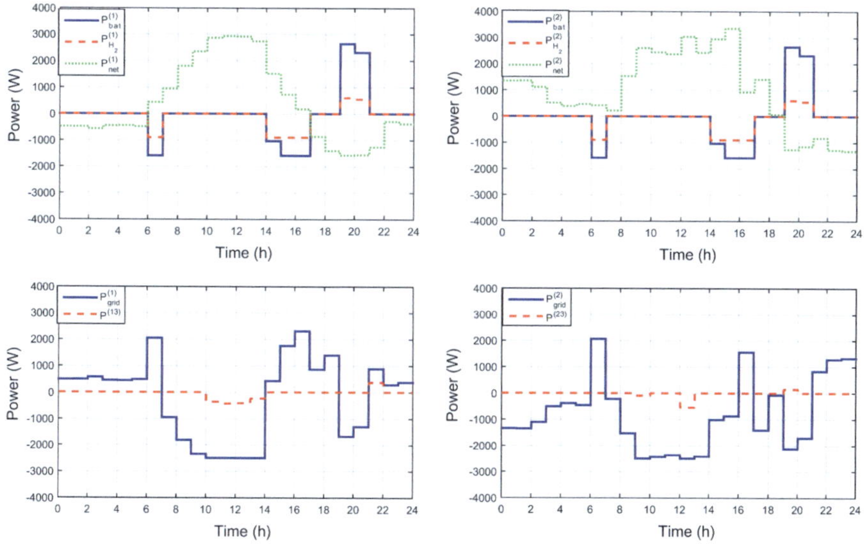

Fig. 8.19 Day-ahead optimization results obtained for microgrid (1) (left) and microgrid (2) (right) after network optimization

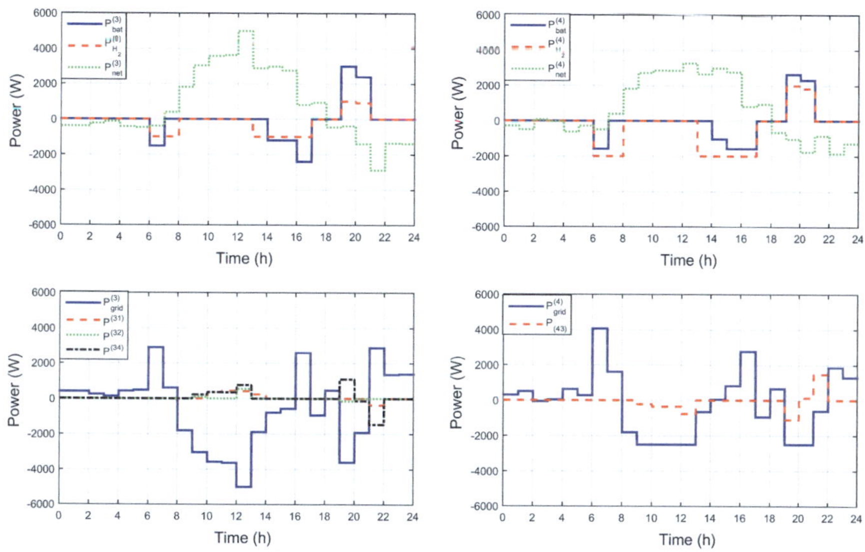

Fig. 8.20 Day-ahead optimization results obtained for microgrid (3) (left) and microgrid (4) (right) after network optimization

Table 8.6 Cost function values at the different iteration steps of the DMPC algorithm applied to the network (Data expressed in €as unit). ©[2019] IEEE. Reprinted, with permission, from IEEE Transactions on Industrial Electronics [15]

ITER: 0	(1)	(2)	(3)	(4)		
J_{single}	0.4207	−1.3100	−0.2122	−0.0904		
ITER: 1	(1) (2)	(1) (3)	(1) (4)	(2) (3)	(2) (4)	(3) (4)
$J_{local,(A)}$	0.4172	0.4207	0.4012	−1.3100	−1.3243	−0.2376
$J_{local,(B)}$	−1.3100	−0.2202	−0.904	−0.2154	−0.0904	−0.0905
ΔJ_{global}	−0.0035	−0.0080	−0.0195	−0.0033	−0.0143	−0.0253
ITER: 2[3,4]	(1) (2)	(1) (3′)	(1) (4′)	(2) (3′)	(2) (4′)	(3′) (4′)
$J_{local,(A)}$	0.4172	0.4207	0.4207	−1.3100	−1.3100	−0.2376
$J_{local,(B)}$	−1.3100	−0.2457	−0.0919	−0.2408	−0.0907	−0.0905
ΔJ_{global}	−0.0035	−0.0082	−0.0014	−0.0033	−0.0002	0
ITER: 3[1,3′]	(1′) (2)	(1′) (3″)	(1′) (4′)	(2) (3″)	(2) (4′)	(3″) (4′)
$J_{local,(A)}$	0.4201	0.4207	0.4207	−1.3100	−1.3100	−0.2457
$J_{local,(B)}$	−1.3100	−0.2457	−0.0919	−0.2490	−0.0907	−0.0905
ΔJ_{global}	−0.006	0	0	−0.0033	−0.0002	0
ITER: 4[2, 3″]	(1′) (2′)	(1′) (3‴)	(1′) (4′)	(2′) (3‴)	(2′) (4′)	(3‴) (4′)
$J_{local,(A)}$	0.4207	0.4207	0.4207	−1.3100	−1.3100	−0.2490
$J_{local,(B)}$	−1.3100	−0.2490	−0.0919	−0.2490	−0.0905	−0.0905
ΔJ_{global}	0	0	0	0	0	0

In the third iteration shown in section ITER:2 of the table, the path given by the node of the couple [3,4] is followed by the optimization algorithm as best path. But there exist different branches of the trees given by all the combination possibilities given in ITER:1 (see Fig. 8.18). In ITER:3, the followed node is the one given by the couple [1,3′]. Notice that microgrid (3′) is the mutation of microgrid (3) after the commitment to energy exchange with microgrid (4) at the previous step. Similar procedure is followed with the mutation of microgrid (4) toward (4′). The terminal node given in the fifth iteration (ITER:4[2,3′]) does not find any new couple because $\Delta J_{global} = 0$ for all the cases. Although only the optimal path of the tree is shown, all the branches and nodes should be calculated in order to find the optimal solution. Notice that not all the microgrids have to exchange energy and there does not exist any energy exchange between microgrids (2)–(4) or their mutations along the combinatorial tree in the optimal path given by the algorithm. Only microgrid (3) exchanges energy with all the microgrids of the network having three mutations.

References

1. Alvarado I, Limon D, Muñoz de la Peña D, Maestre JM, Ridao MA, Scheu H, Marquardt W, Negenborn RR, Schutter BD, Valencia F, Espinosa J (2011) A comparative analysis of distributed mpc techniques applied to the hd-mpc four-tank benchmark. J. Process Control 21(5):800–815. Special Issue on Hierarchical and Distributed Model Predictive Control
2. Ananduta W, Maestre JM, Ocampo-Martinez C, Ishii H (2018) Resilient distributed energy management for systems of interconnected microgrids. In: 2018 IEEE conference on decision and control (CDC), pp 3159–3164
3. Andersen PB, Hu J, Heussen K (2012) Coordination strategies for distribution grid congestion management in a multi-actor, multi-objective setting. In: Innovative smart grid technologies (ISGT Europe), 2012 3rd IEEE PES international conference and exhibition on, pp 1–8. IEEE
4. Asimakopoulou GE, Dimeas AL, Hatziargyriou ND (2013) Leader-follower strategies for energy management of multi-microgrids. IEEE Trans Smart Grid 4(4):1909–1916
5. Bertsekas DP, Tsitsiklis JN (1989) Parallel and distributed computation: numerical methods. Prentice-Hall
6. Camponogara E (2014) Distributed optimization for MPC of linear dynamic networks. Springer, Netherlands, Dordrecht, pp 193–208
7. Camponogara E, Jia D, Krogh BH, Talukdar S (2002) Distributed model predictive control. IEEE Control Syst 22(1):44–52
8. Colson CM, Nehrir MH (2013) Comprehensive real-time microgrid power management and control with distributed agents. IEEE Trans Smart Grid 4(1):617–627
9. Dall'Anese E, Zhu H, Giannakis GB (2013) Distributed optimal power flow for smart microgrids. IEEE Trans Smart Grid 4(3):1464–1475
10. Del Real AJ, Arce A, Bordons C (2014) An integrated framework for distributed model predictive control of large-scale power networks. IEEE Trans Ind Inform 10(1):197–209
11. Del Real AJ, Arce A, Bordons C (2014) Combined environmental and economic dispatch of smart grids using distributed model predictive control. Int J Electr Power Energy Syst 54:65–76
12. Fathi M, Bevrani H (2013) Statistical cooperative power dispatching in interconnected microgrids. IEEE Trans Sustain Energy 4(3):586–593
13. Fele F, Maestre JM, Camacho EF (2017) Coalitional control: cooperative game theory and control. IEEE Control Syst Mag 37(1):53–69
14. Garcia-Torres F, Bordons C (2015) Optimal economical schedule of hydrogen-based microgrids with hybrid storage using model predictive control. Ind Electron IEEE Trans 62(8):5195–5207
15. Garcia-Torres F, Bordons C, Ridao MA (2019) Optimal economic schedule for a network of microgrids with hybrid energy storage system using distributed model predictive control. IEEE Trans Ind Electron 66(3):1919–1929
16. Giselsson P, Dang Doan M, Keviczky T, Schutter BD, Rantzer A (2013) Accelerated gradient methods and dual decomposition in distributed model predictive control. Automatica 49(3):829–833
17. Gregoratti D, Matamoros J (2015) Distributed energy trading: the multiple-microgrid case. IEEE Trans Ind Electron 62(4):2551–2559
18. Howel D (2012) Battery status and cost reduction prospects. in ev everywhere grand challenge battery workshop. Department of Energy (DOE)
19. Hug G, Kar S, Wu C (2015) Consensus + innovations approach for distributed multiagent coordination in a microgrid. IEEE Trans Smart Grid 6(4):1893–1903
20. Kumar Nunna HSVS, Doolla S (2013) Multiagent-based distributed-energy-resource management for intelligent microgrids. Ind Electron IEEE Trans 60(4):1678–1687
21. Logenthiran T, Srinivasan D, Khambadkone AM (2011) Multi-agent system for energy resource scheduling of integrated microgrids in a distributed system. Electr Power Syst Res 81(1):138–148
22. Maestre JM, Negenborn RR (2014) Distributed model predictive control made easy, vol 69. Springer Science & Business Media

23. Maestre JM, Ridao MA, Kozma A, Savorgnan C, Diehl M, Doan MD, Sadowska A, Keviczky T, De Schutter B, Scheu H, Marquardt W, Valencia F, Espinosa J (2015) A comparison of distributed mpc schemes on a hydro-power plant benchmark. Optimal Control Appl Methods 36(3):306–332
24. Mojica-Nava E, Barreto C, Quijano N (2015) Population games methods for distributed control of microgrids. IEEE Trans Smart Grid 6(6):2586–2595
25. Morstyn T, Hredzak B, Agelidis VG (2015) Distributed cooperative control of microgrid storage. IEEE Trans Power Systems 30(5):2780–2789
26. Muros FJ, Maestre JM, Algaba E, Alamo T, Camacho EF (2014) An iterative design method for coalitional control networks with constraints on the shapley value. IFAC Proc Volumes 47(3):1188–1193
27. Negenborn RR, Houwing M, Schutter BD, Hellendoorn J (2009) Model predictive control for residential energy resources using a mixed-logical dynamic model. In: International conference on networking, sensing and control, pp 702–707
28. Negenborn RR (2007) Multi-agent model predictive control with applications to power networks. PhD thesis, University of Delft
29. Nikmehr N, Ravadanegh SN (2015) Optimal power dispatch of multi-microgrids at future smart distribution grids. IEEE Trans Smart Grid 6(4):1648–1657
30. Nikmehr N, Ravadanegh SN (2016) Reliability evaluation of multi-microgrids considering optimal operation of small scale energy zones under load-generation uncertainties. Int J Electr Power Energy Syst 78:80–87
31. Ouammi A, Dagdougui H, Dessaint L, Sacile R (2015) Coordinated model predictive-based power flows control in a cooperative network of smart microgrids. IEEE Trans Smart grid 6(5):2233–2244
32. Ouammi A, Dagdougui H, Sacile R (2015) Optimal control of power flows and energy local storages in a network of microgrids modeled as a system of systems. IEEE Trans Control Syst Technol 23(1):128–138
33. Qi W, Liu J, Christofides PD (2013) Distributed supervisory predictive control of distributed wind and solar energy systems. IEEE Trans Control Syst Technol 21(2):504–512
34. Salinas S, Li M, Li P (2013) Multi-objective optimal energy consumption scheduling in smart grids. IEEE Trans Smart Grid 4(1):341–348
35. Savorgnan C, Romani C, Kozma A, Diehl M (2011) Multiple shooting for distributed systems with applications in hydro electricity production. J Process Control 21(5):738–745. Special Issue on Hierarchical and Distributed Model Predictive Control
36. Scattolini R (2009) Architectures for distributed and hierarchical model predictive control a review. J Process Control 19(5):723–731
37. Shi W, Xie X, Chu CC, Gadh R (2015) Distributed optimal energy management in microgrids. IEEE Trans Smart Grid 6(3):1137–1146
38. Spendelow J, Marcinkoski J, Satyapal S (2014) Doe hydrogen and fuel cells program record 14012. Department of Energy (DOE)
39. Vasiljevska J, Peças Lopes JA, Matos MA (2012) Evaluating the impacts of the multi-microgrid concept using multicriteria decision aid. Electric Power Syst Res 91:44–51
40. Venkat AN, Rawlings JB, Wright SJ (2007) Distributed model predictive control of large-scale systems. Springer, Berlin Heidelberg, Berlin, Heidelberg, pp 591–605
41. Wang Z, Chen B, Wang J, Begovic MM, Chen C (2015) Coordinated energy management of networked microgrids in distribution systems. IEEE Trans Smart Grid 6(1):45–53
42. Wang Z, Chen B, Wang J, Chen C (2016) Networked microgrids for self-healing power systems. IEEE Trans Smart Grid 7(1):310–319
43. Wei C, Fadlullah ZM, Kato N, Takeuchi A (2014) Gt-cfs: a game theoretic coalition formulation strategy for reducing power loss in micro grids. IEEE Trans Parallel Distrib Syst 25(9):2307–2317
44. Wu J, Guan X (2013) Coordinated multi-microgrids optimal control algorithm for smart distribution management system. IEEE Trans Smart Grid 4(4):2174–2181

45. Xu J, Zou Y, Niu Y (2013) Distributed predictive control for energy management of multi-microgrids systems. 13th IFAC symposium on large scale complex systems: theory ANS applications. Shanghai, China, pp 551–556
46. Yazdanian M, Mehrizi-Sani A (2014) Distributed control techniques in microgrids. IEEE Trans Smart Grid 5(6):2901–2909

Chapter 9
Microgrids Power Quality Enhancement

Abstract Power quality is one of the major issues in electrical grids due to the extended use of power electronics, renewable energy generation, and nonlinear electronic loads. Usually, the elements of the microgrid are interfaced with power converters which are finally responsible for the power quality levels in the microgrid. For this reason, control of power converters becomes a relevant topic in microgrids. This chapter introduces the main aspects of power quality in microgrids and the basic principles of operation of MPC applied to power converters. The two main MPC methods for power converters, Continuous Control Set MPC (CCS-MPC), and Finite Control Set MPC (FCS-MPC) are described and their application to a Voltage Source Inverter (VSI) is shown in order to demonstrate their capabilities. Finally, an MPC-based algorithm to enhance the power quality in microgrids in presence of nonlinear and unbalanced loads is introduced. It works in both modes, islanded and grid-connected, providing the capability of fast transition between modes when required.

9.1 Control of Power Quality in Microgrids

Power quality is of paramount importance for the operation of microgrids. The generation should meet the demand cleanly, reliably, sustainably, and at low cost [8]. In electric power systems, any deviation with respect to the theoretical sinusoidal waveform (produced in the generation centers) is considered to be a disturbance in the power quality of the electrical grid. The deviation can alter any of the parameters of the wave: frequency, amplitude, waveform, and symmetry among phases. Adequate quality supply provides the necessary compatibility between all the devices connected to the same grid.

While reliability indexes are not yet standardized, voltage characteristics of European public distribution system concerning the supplier's side are regulated by Standard EN 50160, which defines the main features of voltage supplied by public distribution grids under normal exploitation conditions. Although there are some

C. Bordons et al., *Model Predictive Control of Microgrids*,
Advances in Industrial Control, https://doi.org/10.1007/978-3-030-24570-2_9

standards as IEC 61000 3-4 or 3-12, there is still shortage in the regulatory framework of the reliability and commercial dimensions of quality, but this will acquire further importance in the smart grid [21].

Power electronics for interfacing DERs must minimize the effects of intermittency of renewable energy generation and compensate the presence of nonlinear and unbalanced loads. Fast transition between grid-connected and islanded mode should be included in order to mitigate the effects of faults in the main grid. Model predictive control applied to microgrids and power electronics emerges as a promising alternative to enhance the power quality aspects. MPC can handle converters with multiple switches and operation modes such as islanded, grid-connected, or transition between both modes. MPC has the potential to integrate multi-objective purposes using a unique cost function.

9.1.1 Control Layers

As described in Chap. 1, microgrids are based on three control levels [6, 18]. Tertiary control level is dedicated to the schedule and economic management of the microgrid. The secondary one receives this schedule set points being in charge of real-time power sharing in the microgrid. Primary control level is related to the power quality issues. The primary control tracks the set points for current and voltage of the distributed generators (DG) connected to the microgrid. In order to avoid frequency and voltage deviations, the secondary control acts restoring the microgrid voltage and frequency and compensates for the deviations caused by the primary control [18]. In Fig. 9.1, a global scheme of the microgrid with its control structure is shown. The secondary controllers send their control set points to the primary control inside the power converters of each component, which are connected through a Local Area Network

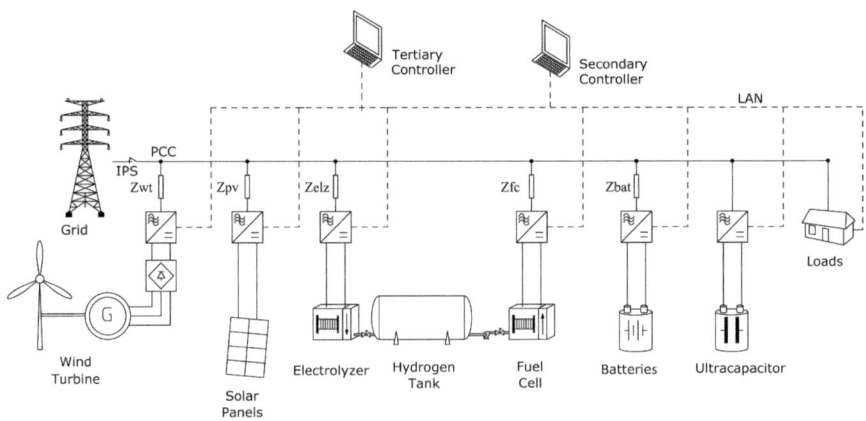

Fig. 9.1 Microgrid Control Scheme

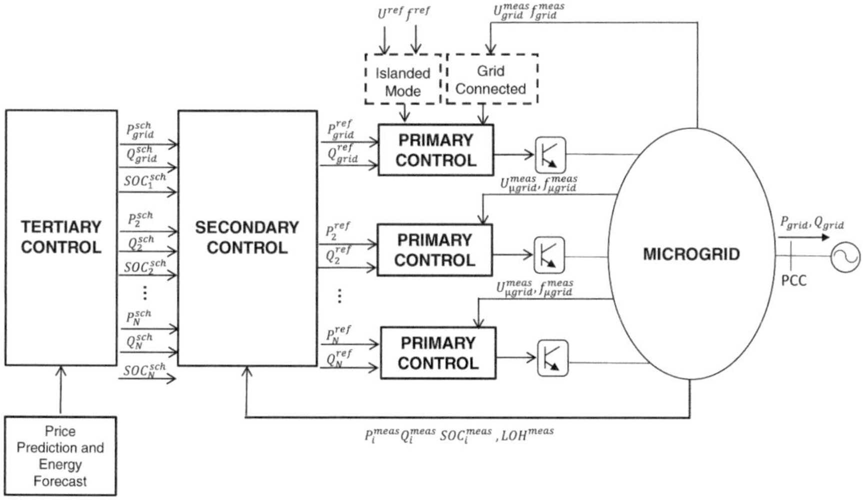

Fig. 9.2 Global overview of the three hierarchical control levels of the microgrid

(LAN). All these power converters are connected to the Point of Common Coupling (PCC). Sometimes these components are located at a considerable distance from the PCC so the line impedances (represented by Z_j in Fig. 9.1) have to be considered in the controllers. As can be seen in the figure, the microgrid and the grid are connected through an Intelligent Power Switch (IPS). A global overview of the block diagram for the three hierarchical control levels of the microgrid is shown in Fig. 9.2, where P and Q are active and reactive power, respectively, U is voltage, f is frequency, superscript sch stands for the scheduled values calculated by tertiary control, ref for the set points to the primary controllers, and $meas$ for the measured values.

9.1.2 Operation Modes

The power quality in the microgrid is a result of the interaction between the power flow of the main grid and those of the different devices connected to the microgrid. As is well known, microgrids have essentially two operation modes: grid-connected and islanded mode. The transition between both modes is an important issue to be considered by the controller. The main difference in operation between both modes is related to voltage and frequency control. In the case of the grid-connected mode, the voltage and frequency references are imposed by the main grid, while in the islanded mode they are set up by the microgrid itself (see Fig. 9.2). In the block diagram of this figure, it is considered that one of the inverters acts as master power converter of the microgrid. This inverter will be responsible for maintaining the voltage and frequency of the microgrid in case of islanded mode operation. Notice that in this case, the other inverters operate in grid-connected mode or microgrid-connected mode, being synchronized with the voltage waveform imposed by the master of the microgrid in case the microgrid works disconnected from the main grid. This master inverter is responsible for the accurate tracking of the reference given for P_{grid}^{ref} and Q_{grid}^{ref} by the secondary controller.

In grid-connected mode, the microgrid must be capable of exporting/importing energy at the voltage amplitude and frequency imposed by the main grid. The microgrid power exchange with the main grid is scheduled in the tertiary control layer of the microgrid. One of the main problems in this mode is the slow response of the control signals when a change in the reference of the exchanged power occurs. The absence of synchronous machines connected to the low-voltage power grid requires power balancing during transients that could be provided by the ESSs of the microgrid.

As mentioned before, the microgrid can work in islanded mode due to different situations, such as the nonexistence of the main grid, a fault in the main grid, or a period of maintenance. The IPS is open during this working mode and closed when the microgrid is connected to the main grid. In this mode, the microgrid itself is responsible for the balance among all its components and must supervise that the power flows within the microgrid fulfill the power quality levels that its components require. During restoration after a power supply shutdown, reactive power balance, commutation of the transient voltages, balancing power generation, starting sequence, and coordination of generation units have to be considered [40].

The IPS is continuously supervising the status of the main grid and the microgrid and, if a fault in the main grid is detected, it must disconnect the microgrid. In such a case, this switch can readjust the power reference at nominal values. The IPS is kept in closed state in grid-connected mode if voltage and frequency are inside admissible ranges (typically, 2% for frequency and 5% for amplitude), changing to open state otherwise. When switching from islanded to grid-connected mode, the IPS must achieve the synchronization with voltage, amplitude, phase, and frequency [40].

9.1.3 Methods for Quality Control

The primary control level of the microgrid manages the voltage and frequency delivered by the inner loop of each power inverter of the microgrid. Droop control is the most commonly used method at the primary control level. This method assigns to each inverter of the microgrid a droop characteristic based on its generation capabilities [18]. An extensive review of the control methods applied to primary and secondary control of the microgrid can be found in [19, 31]. Advanced droop methods are presented in several papers, such as [19, 38]. These methods have the drawback of active and reactive power coupling, which has been addressed by several authors making an approach based on the virtual output impedance method. As a result, the expected voltage can be modified [17]. New approaches are being proposed with algorithms based on the concept of construction and compensation-based method [20]. As previously mentioned, a secondary control level is usually required to correct the frequency and voltage. Additionally, it can be used for reactive power compensation [28] and to reduce the harmonics content of the voltage waveform [36]. Secondary control can be achieved with a centralized or a distributed strategy [7, 37]. Most of the existing literature for primary control in microgrids are based on classical Proportional–Integral control with Pulse Width Modulation (PI-PWM). These kind of controllers may produce poor results in the transient response.

The use of advanced control techniques such as MPC can help to overcome some of these problems. Although MPC was born in the framework of industrial process control, in the last years its application to electronic power converters and electrical drives is significantly increasing [41, 42], since it can help to fulfill the growing demands in performance, efficiency, and safety demanded by the industrial electronics applications.

Recent studies have developed MPC techniques for primary and secondary control levels of microgrids. In [2], an MPC-based controller and a Smith-predictor-based controller are applied to the secondary level of the microgrid. In the mentioned paper, secondary control of the microgrid voltage is not developed. An MPC-based microgrid control with supplementary fault current limitation and smooth transition mechanisms is presented in [3]. Finite Control Set (FCS)-MPC is used as the primary controller to regulate the output power of each DG (in grid-connected mode) or the voltage of the point of DG coupling (in islanded) mode. A frequency predictive model for each generator of the microgrid is developed in [14]. A distributed MPC-based secondary voltage control scheme for autonomous droop-controlled microgrids is developed in [24] using a secondary distributed cooperative control approach.

9.2 Control of Power Converters

Energy storage systems and most renewable generators are usually conversion sources producing DC voltages/currents. Their low-voltage output typically requires a booster power converter followed by an inverter in order to achieve the required voltage levels. Therefore, power quality control in microgrids is closely related to control of power converters. Power converters integrate binary signals that command the transitions of the switches as well as continuous variables, such as voltages and currents, therefore exhibiting a hybrid nature. This makes modeling a difficult task but, in spite of their complexity, models are available and they are accurate.

The classical way of controlling power converters has been based on the combination of PID control and PWM. The interest of using a PWM module is twofold: it allows fixed frequency operation and it also guarantees a decoupling between switching and sampling times. Since the model of the converter is usually available, several methods that use a model of the plant to compute the control action are employed, like sliding-mode control or deadbeat control. Examples of application of improved deadbeat control using disturbance observers in several power electronics applications such as rectifiers, inverters, active filter, and power supplies can be found in [26].

9.2.1 Power Converters in Microgrids

The main control classification regarding power electronics is based on the variable to be controlled: current or voltage; therefore, power converters may be classified as a voltage or a current source. Regarding inverters, the voltage source inverters' design is imposed in industry due to their efficiency, fast dynamic response, and high reliability.

The current loop is responsible for the injected-current power quality and overcurrent protection. The current controller must guarantee accurate current tracking with a fast transient and force the VSI to equivalently act as a current source amplifier within the current loop bandwidth. Grid voltage harmonics, unbalance, and transients directly affect the current control performance and might impair the power quality and even the stability of the converter. Total Harmonic Distortion (THD) in the injected current can be affected by distortions in the grid voltage, and eventually inverter instability can occur due to possible interaction between the grid and the inverter power circuit filter. Consequently, grid disturbance rejection is one of the important features that must be considered in current controllers in the design of inverter-based distributed generation systems [29]. The major techniques to regulate the output of a current-controlled VSI include either a variable switching frequency, such as the hysteresis control scheme or fixed-switching schemes, such as the ramp comparison stationary and synchronous frame PI, Stationary Reference Frame Proportional Resonant (SRF-PR) and deadbeat predictive current control schemes [29].

Voltage controllers in VSI-based DG units can operate in different manners depending on the microgrid operation mode. In grid-connected mode, voltage control can be used to regulate the grid voltage at the PCC. There are certain requirements in this mode of operation: first, a low THD of the output voltage must be achieved under varying load conditions. This is an important goal since the harmonics produced in the current by nonlinear loads can distort the output voltage of the inverter, modifying the power quality delivered to other loads. Second, in the case of load transients and grid disturbances, low-voltage sag and fast recovery must be achieved by the controller. These voltage regulation requirements become more challenging when the unbalanced nature of grid or load voltages appear, as in the case of fluctuating output power of wind and photovoltaic generation, which can cause severe and random voltage disturbances [29].

9.2.2 MPC and Power Converters

MPC presents several attractive features that make it suitable for the control of power converters. The controller can take into account the available complex dynamics while considering several design criteria and constraints. The cost function can include multiple criteria, allowing the optimization of important parameters like number of commutations, switching losses, harmonics reduction, reactive power control, or ripple minimization. Some examples of several cost functions for power converters can be found in [34, 35, 43]. The choice of the prediction horizon can influence the behavior of the converter. As mentioned in Chap. 2, a long prediction horizon can improve stability and performance. For motor drives, it has been shown that long prediction horizons lead to a significant performance improvement at steady-state operating conditions, lowering the current distortions and/or the switching frequency [16]. However, the price to pay for the use of long horizons is an increase in the computational complexity.

The computational cost is of crucial importance in the case of power converters, due to the small sampling time and the reduced computational capabilities of the existing hardware. Since MPC solves an open-loop optimal problem at each sampling instant, the computational cost is high compared to classical controllers. Therefore, different varieties of MPC have been proposed in literature to face this problem [12]. Additionally, increasing use of powerful microprocessors is enabling application of MPC in this field with significant success.

Anyway, many methods have been proposed to reduce the computational burden of QP problems. One important research line is to use multiparametric quadratic programming [4] or linear programming [5], so that the MPC is transformed into a simple piecewise affine (PWA) controller. Then, the solution can be precomputed offline for the space of all possible states and the implementation is reduced to a search within a set of solutions that depend on the current state. This is known as explicit MPC. An example of application for this approach to a PWM-based inverter with an LCL filter is presented in [25]. Based on a model that accounts for the switched behavior of the converter, an explicit MPC is derived in order to provide a fast response, making it very suitable for applications where a large bandwidth is required.

However, the obtained explicit solution may be excessively complex if the dimension of the problem, associated to the number of decision variables (which depends on the horizon), is too big. In those circumstances, fast online optimization solvers, as the fast gradient method, can be used. An implementation of this method for the control of an AC–DC power converter is presented in [33]. Computation times in the order of few tens of μs are achieved, making the approach ideal for power electronics applications. The hybrid MPC methodology used in the book can also be applied to boost converters, as shown in [30]. In case of nonlinear models, the optimization can be even more costly, giving rise to a nonlinear programming (NLP) problem. An application of nonlinear MPC to a DC–DC converter can be found in [9].

9.2.3 MPC Methods for Power Converters

The main aspects and features of MPC are exposed in Chap. 2. As done for the rest of applications, MPC needs a model of the system to predict the future evolution of the output along the prediction horizon. Power electronics devices have the particularity

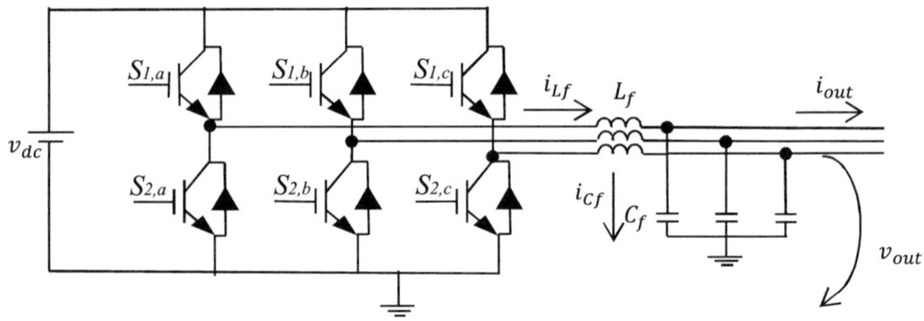

Fig. 9.3 Voltage source inverter with output LC filter

of their switching nature coupled with continuous state variables, given by the current of the inductors and the voltage of the capacitors.

Predictive control methods for power converters can be divided into several categories, according to the operating principle and other characteristics, see [23, 35] for detailed classifications. A general classification of methods can be established depending on the way of managing the control actions: those that calculate the gate signal, that is, that control the inverter directly, and those methods that determine a continuous control signal, which is synthesized by a modulator. In the first case, the problem combines continuous variables (voltages and currents) with binary variables (switches), leading to a Mixed-Integer Program (MIP), which can be computationally expensive, especially in case that long horizons are used. This classification gives rise to two main branches that will be analyzed below: Direct or Finite Control Set MPC (FCS-MPC) and Continuous Control Set MPC (CCS-MPC).

In the following sections, the development of a FCS-MPC and a CCS-MPC controller will be explained, illustrated through their application to control a VSI of the type shown in Fig. 9.3. The theory of these controllers is explained for both cases: islanded and grid-connected modes. Although in the experimental application, it is assumed that the VSI operated in islanded mode being voltage controlled. These experiments were presented in [10, 43].

Finite Control Set MPC

Finite control set (or direct) MPC, which is introduced in [34] for the current control of VSI, is a conceptually simple way of solving the optimization problem. It takes advantage of the finite number of switching states, which drives to an optimization problem that can be simplified to the prediction and evaluation of the cost function only for those possible switching states. Therefore, the problem can be solved using an exhaustive search, where the set of switching sequences is enumerated, the output is predicted, and the cost function is evaluated for each sequence. The switching sequence that yields the minimal cost is chosen as the optimal one. This is a straightforward way of computing the control action which is not computationally expensive in case short horizons are used. Notice that a modulator is not needed. This method has been successfully applied to a wide range of power converters and

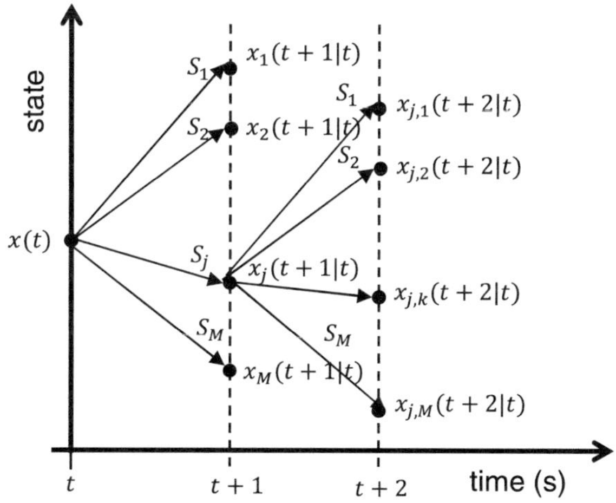

Fig. 9.4 FCS-MPC operating principle

drive applications [39], as voltage control of inverters with an output LC filter [13, 15].

The methodology is simple and intuitive and is depicted in Fig. 9.4, where each of the M admissible switching actions S_i gives rise to a system state in the next sampling instant. The state is given by the current of the inductors and voltage of the capacitors and can be predicted using a dynamical model of the power converter.

As can be seen in Fig. 9.5, a cost function J is minimized in order to track a desired reference $w(t+1)$ for the next sample instant. Figure 9.3 represents one of the most common topologies for power inverters, a two-level voltage source inverter (2L-VSI). It is composed of three legs, with two power switches each, whose states are function of the associated gate signals. One switch is connected to the positive terminal of the DC voltage source (or DC-link capacitor) and the other connected to the negative terminal. The gate signals applied to the switches situated in the same leg

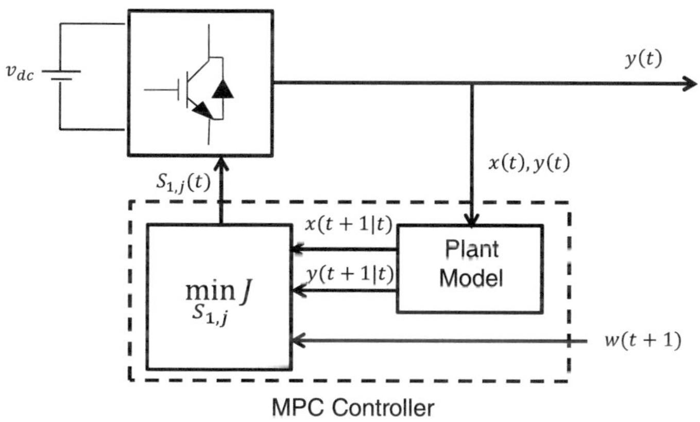

Fig. 9.5 Block diagram of the FCS-MPC

of the inverter must take opposite values. Each gate signal just can adopt either one of two values: "0" if the power switch is at OFF-state and "1" when the power switch is at ON-state. Since there are three legs, the number of possible states is $2^3 = 8$. The output of each leg of the inverter is connected to an LC filter. Particularizing for the topology given in Fig. 9.3, the following model based on its state variables can be obtained, discretized using a sampling period T_s [34]:

$$v_{C,j}(t+1) = S_{1,j}(t)v_{dc} - L_f \frac{i_{L_f,j}(t+1) - i_{L_f,j}(t)}{T_s}\Bigg|_{j=a,b,c} \tag{9.1}$$

$$i_{C_f,j}(t+1) = C_f \frac{v_{C_f,j}(t+1) - v_{C_f,j}(t)}{T_s}\Bigg|_{j=a,b,c} \tag{9.2}$$

$$v_{C,j}(t) = v_{C_f,j}(t)\big|_{j=a,b,c} \tag{9.3}$$

$$i_{out,j}(t) = i_{L_f,j}(t) - i_{C_f,j}(t)\big|_{j=a,b,c} \tag{9.4}$$

where the state variables are the inductors currents per phase $i_{L_f,j}$ and capacitors voltages, the input signals are the switching actions given to the gates $S_{1,j} \in \{0, 1\}$, and the output variables are represented by the output currents per phase $i_{out,j}$ or the output voltages per phase $v_{out,j}$ depending if it is a current-controlled or voltage-controlled VSI. In case of a current-controlled VSI, $v_{out,j}$ is introduced as a disturbance to the controller and has to be estimated. Similar procedure has to be considered for the case of a voltage-controlled VSI with the $i_{out,j}$.

The model can be used to compute the state predictions along the horizon. For each value j of the switching action, a prediction of the state at instant $t + 1$ is obtained. For a prediction horizon bigger than one, the procedure is replicated for each value of $x_i(t + 1|t)$, that is, from this state, each of the M switching actions will lead to a new value of the state in $t + 2$ and so on as can be seen in Fig. 9.4. For a generic state $x_j(t + 1|t)$ resulting from the switching action $S_j(t + 1)$, M new switching actions with their corresponding states $x_{j,n}(t + 2|t)|_{n=1,\ldots,M}$ will be found. This strategy is very intuitive and easy to implement provided the horizon is small, but if the horizon is large it can lead to combinatorial explosion.

A cost function (usually quadratic) is minimized as can be seen in Fig. 9.5 which selects the optimal switches' combination for the gate signals of the inverter $S_{1,j}$ in order to follow a given reference at the output of the inverter.[1]

$$\min_{S_{1,a}, S_{1,b}, S_{1,c}} \quad J(\mathbf{x}(t), \mathbf{u}(t), \mathbf{y}(t), \mathbf{w}(t)) \tag{9.5}$$

As done in classical PI-PWM-based controllers applied for inverters, Clarke's or Park's transformation is used with the objective of simplifying the analysis of three-phase circuits. In electrical engineering, Park's transformation or *dqo* transformation

[1] Boldface letters indicate vectors composed of elements along the horizon.

is a mathematical transformation that projects the three separate sinusoidal phase quantities onto two axes rotating with the same angular velocity. These axes are known as the direct or d-axis and the quadrature or q-axis, that is, with the q-axis being at an angle of 90 degrees from the direct axis. Usually, the d-axis is aligned with the voltage reference. The mathematical expression for this transformation can be expressed by

$$\begin{bmatrix} v_d(t) \\ v_q(t) \\ v_o(t) \end{bmatrix} = \frac{1}{\sqrt{3}} \begin{bmatrix} \cos(\omega t) & \cos(\omega t - \frac{2\pi}{3})) & \cos(\omega t + \frac{2\pi}{3})) \\ \sin(\omega t) & \sin(\omega t - \frac{2\pi}{3})) & \sin(\omega t + \frac{2\pi}{3})) \\ 1 & 1 & 1 \end{bmatrix} \begin{bmatrix} v_a(t) \\ v_b(t) \\ v_c(t) \end{bmatrix} \tag{9.6}$$

A voltage controller can be designed with a simple cost function like this:

$$J = \sum_{k=1}^{k=N_p} \sum_{j=d,q,o} \left(v_{out,j}(t+k|t) - v_j^{ref} \right)^2 \tag{9.7}$$

where the references v_d^{ref}, v_q^{ref}, and v_o^{ref} for the next sample instant, due to the inherited property of Park's transformation of being a rotational reference frame of three-phase systems, can be fixed to a constant value for all the instants. In case of balanced systems, the third term (corresponding with the o-axis) of the cost function given in (9.7) can be eliminated. Notice that if the control horizon is increased, the number of possible solutions increases exponentially, which may not be affordable since low-computational time is required. For this reason, the control horizon is usually limited to one, which might lead to poor performance. In the case of a current-controlled inverter, an active power reference and a reactive power reference are imposed to the controller.

Using the dqo framework, the expressions for the instantaneous active and reactive power can be expressed as follows:

$$P_{inst}(t) = \frac{3}{2}(v_d(t)i_d(t) + v_q(t)i_q(t) + v_o(t)i_o(t)) \tag{9.8}$$

$$Q_{inst}(t) = \frac{3}{2}(v_q(t)i_d(t) - v_d(t)i_q(t)) \tag{9.9}$$

Notice that if $\vec{v}$ is aligned with the d-axis, the components $v_q(t)$ and $v_o(t)$ can be neglected. Considering that the grid voltage magnitude and frequency are constant for all the sample instants, the component v_d will be also a constant. It is easy to obtain from (9.8) and (9.9) a reference for the current vector expressed in dqo-axis i_d^{ref} and i_q^{ref} from the given values for P^{ref} and Q^{ref} being $i_o^{ref} = 0$ if it is considered a three-phase harmonics-free and balanced system. The cost function in this case for the grid-connected mode can be expressed as follows:

$$J = \sum_{k=1}^{k=N_p} \sum_{j=d,q,o} \left(i_{out,j}(t+k|t) - i_j^{ref} \right)^2 \tag{9.10}$$

Notice that the prediction model of the power inverter given by Eqs. (9.1)–(9.4) is composed of four equations per each phase of the inverter while there exist six unknowns $i_{L_f,j}$, $i_{C_f,j}$, $i_{out,j}$, $v_{L_f,j}$, $v_{C_f,j}$, and $v_{out,j}$. The fifth equation is obtained depending on the configuration. In case of a voltage controller, the currents per phase for the next sample instant $i_{out,j}(t+1|t)$ can be estimated as detailed in [13, 15]. In case of a current controller, it can be assumed that the values in *dqo*-coordinates for the output phase voltages are constant between two sample instants because they are imposed by the main grid where the power inverter is connected. The basic steps needed to implement the FCS-MPC are depicted in Algorithm 9.1.

Algorithm 9.1 FCS-MPC

Input: $u_{C,j}(t)$, $i_{L,j}(t)$, $i_{out,j}(t)$ and $v_{out,j}(t)$
Output: S_i

```
for t = t_i to t_f do
    for i = 0 to 7 do
        Evaluate cost function J using the predictive model
        Choose the S_i that gives the smallest value of J
        Set S* = S_i
    end for
    Apply S*
end for
```

The main advantage of FCS-MPC is its low-computational cost, which allows the consideration of complex dynamics (such as nonlinearities) of power converters in an easy way. On the other hand, steady-state errors and widespread spectra are often found [32].

MPC with Continuous Control Set

In this type of controller, the control actions are continuous-time signals which are sent to a modulator. The optimization problem can be solved analytically by setting the derivative of the cost function equal to zero in the unconstrained case or solving a QP in the constrained case. Long horizons can be employed without reaching combinatorial explosion. The most extended method of this type is Generalized Predictive Control (GPC) [11], so this branch is also known as GPC-type control, although other methods can be used. A basic application for this kind of controllers can be found in [43]. The block diagram of the CCS-MPC is shown in Fig. 9.6. As can be seen, instead of using the switching actions given by $S_{1,j}$ directly, a duty cycle $d(t)$ for a PWM signal is obtained as result of the controller.

As mentioned in Chap. 2, GPC uses a transfer function model of the system with an integrated white noise (the so-called CARIMA model) and a quadratic cost function. The solution of the optimization problem provides the set of changes in the

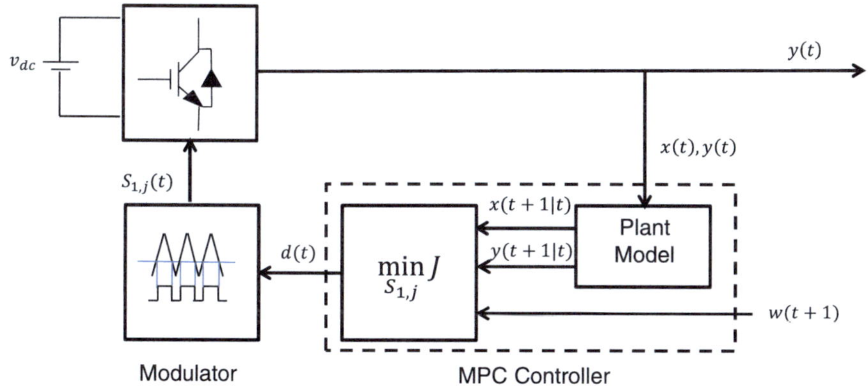

Fig. 9.6 Block diagram of the CCS-MPC

control actions along the horizon $\mathbf{u} = [\Delta u(t), \Delta u(t+1), \ldots, \Delta u(t+N_c-1)]$. In this case, the model is continuous, based on the following equations, where, for the sake of simplicity, the associated resistance of the capacitor R_{C_f} is neglected, being: $v_{out,j}(t) = v_{C_f,j}(t)$ and $j = a, b, c$:

$$i_{L,j}(t) = C_f \frac{dv_{out,j}}{dt} + i_{out,j}(t) \tag{9.11}$$

$$v_{dc} d_j(t) = L_f \frac{di_{L_f,j}}{dt} + v_{out,j}(t) \tag{9.12}$$

As it occurs for the FCS-MPC, the output voltages depend on the power semiconductors switching functions and the state variables $i_{L_f,j}$ and $v_{C_f,j}$. Therefore, the output voltages of the three-phase power converter have been chosen as the control signal (input), $u(t) = v_{dc}[S_{1,a}(t), S_{1,b}(t), S_{1,c}(t)]^T$, and the final switching sequence will be generated through a PWM modulation technique. The model can be expressed as a function of the control variable $u_j(t) = v_{dc} d_j(t)$, which represents the pulse sent to the three-phase power converter.

Particularizing for the case of voltage control, the output current $i_{out,j}(t)$ should be expressed as a function of $v_{out,j}(t)$ using an estimator. This can be done following the methods explained in [13, 15]. In order to simplify, in this case, it will be supposed that the VSI has a resistor connected whose value R_{load} is known. The output $v_{out,j}(t)$ of the inverter can be expressed as function of the control action $u_j(t)$ in the time domain as follows:

$$u_j(t) = L_f \cdot C_f \frac{d^2 v_{out,j}(t)}{dt^2} + \frac{L_f}{R_{load}} \frac{dv_{out,j}(t)}{dt} + v_{out,j}(t) \tag{9.13}$$

Using Laplace's transform, the transfer function is given by

$$H(s) = \frac{Y(s)}{U(s)} = \frac{1}{L_f \cdot C_f s^2 + \frac{L_f}{R_{load}} s + 1} \tag{9.14}$$

where the system output $Y(s)$ is the output voltage per phase $v_{out,j}(t)$. Since GPC is a discrete-time controller, the transfer function in the z domain, $H(z^{-1}) = \dfrac{B(z^{-1})}{A(z^{-1})}$, is used. GPC uses this model in the form of a CARIMA model [11]:

$$A(z^{-1})y(t) = B(z^{-1})u(t-1) + \frac{C(z^{-1})}{\Delta}v(t) \tag{9.15}$$

where $u(t)$ and $y(t)$ are the control and outputs sequences, respectively, $v(t)$ is an additive white Gaussian noise, $\Delta = (1 - z^{-1})$ and A, B, and C are polynomials in the backward shift operator z^{-1}.

The cost function defined over a prediction horizon usually has the following format:

$$J = \sum_{j=1}^{N_p} [y(t+j \mid t) - w(t+j)]^2 + \sum_{j=1}^{N_c} \lambda_j [\Delta u(t+j-1)]^2 \tag{9.16}$$

In GPC, the Diophantine equation is used with the purpose of separating the predicted outputs in terms that depend only on past values (known at the current instant t) and others which depend on the future control signals. Following the steps described in [11], the vector of predicted outputs along the horizon can be expressed as a function of a certain matrix $\mathbf{G}$ (which is obtained from the Diophantine equation and whose dimensions are $N_p \times (N_p - N_c + 1)$) and of the free response of the system, $\mathbf{f}$:

$$\mathbf{y} = \mathbf{G}\mathbf{u} + \mathbf{f} \tag{9.17}$$

The cost function given by (9.16) can be rewritten using (9.17) as follows:

$$J = (\mathbf{G}\,\mathbf{u} + \mathbf{f} - \mathbf{w})^T(\mathbf{G}\,\mathbf{u} + \mathbf{f} - \mathbf{w}) + \lambda \mathbf{u}^T\mathbf{u} \tag{9.18}$$

In order to reduce the computational cost of the power electronics control platform, the problem is considered as non-constrained, so the optimal set of control actions is found when the gradient of J is zero, $\frac{\partial J}{\partial u} = 0$, leading to:

$$\mathbf{u} = (\mathbf{G}^T\mathbf{G} + \lambda \mathbf{I})^{-1}\mathbf{G}^T(\mathbf{w} - \mathbf{f}) \tag{9.19}$$

which represents the increment of the control signal which is applied at each sampling instant. Notice that the control signal that is actually sent to the process is the first element of vector $\mathbf{u}$, which given by

$$\triangle u(t) = K(\mathbf{w} - \mathbf{f}) \tag{9.20}$$

where K is the first row of matrix $(\mathbf{G}^T\mathbf{G} + \lambda\mathbf{I})^{-1}\mathbf{G}^T$, which can be computed beforehand from the transfer function. The steps needed to implement the GPC are described in Algorithm 9.2.

Algorithm 9.2 GPC

Input: $u_{C,j}(t)$, $i_{L,j}(t)$, $i_{out,j}(t)$ and $v_{out,j}(t)$
Output: $u(t)$
1: Compute the value of gain K
2: **for** $t = t_i$ to t_f **do**
3: Compute the control action $u(t)$ from (9.20)
4: Apply $u(t)$ as the duty cycle to the PWM
5: **end for**

The CCS-MPC methodology has as main advantages that long horizons can be used and its tracking error is zero average and it also exhibits a concentrated spectra. Nevertheless, the tractable convex formulations are limited to linear models.

Experiments and Assessment

This section illustrates the implementation of both MPC algorithms to an experimental VSI (similar to the one displayed in Fig. 9.3) connected to an LC filter feeding an RL load. The objective of VSI control is to generate a three-phase sinusoidal output voltage tracking a desired reference of 120 V RMS at 50 Hz. Although the model used by the controllers is given by the parameters of Table 9.1, the experiments are performed with an actual load of $R_{load} = 15\,\Omega$, $L_{load} = 10\,\text{mH}$ in order to test them under model mismatch. The sampling frequency is set to 12 kHz in the GPC case and is changed between 20 and 40 kHz in the FCS-MPC case. The THD of the output voltages and Root Mean Square (RMS) value of the voltage error have been selected as quality performance indices.

The gate signals in the GPC are generated by a PWM modulator, so the switching frequency matches the sampling frequency. However, for the FCS-MPC the switches are directly manipulated and the switching frequency is variable depending on the operating point and the sampling frequency. The effective switching frequency is

Table 9.1 Parameters of the power inverter

Parameter	Value
Filter inductance L_f	2 mH
Filter capacitor C_f	50 mF
Load resistance R_{load}	60 Ω
DC-Link voltage v_{dc}	400 V
RMS reference voltage V_{RMS}	120 V

smaller than the sampling frequency, so it is necessary to work with a higher sampling frequency in order to get a similar performance to GPC.

The GPC is designed with $N_c = 1$, $N_p = 6$ and a control-weighting factor of $\lambda = 0.1$. The system is represented using its CARIMA model with a sampling time of 83.33 μs (since the sampling frequency is 12 kHz, similar switching frequency is used). The transfer function given in (9.14) is discretized for the values of the inverter parameters given in Table 9.1, obtaining the polynomials $A(z^{-1})$ and $B(z^{-1})$, considering the noise polynomial $C(z^{-1})$ equal to 1.

$$A(z^{-1}) = 1 - 1.854z^{-1} + 0.921z^{-2} \tag{9.21}$$

$$B(z^{-1}) = 0.033 + 0.032z^{-1} \tag{9.22}$$

Following the standard steps in GPC, vector K is given by (see a detailed description of all the GPC matrices and parameters in [43])

$$K = \begin{bmatrix} 0.021 \ 0.081 \ 0.173 \ 0.288 \ 0.417 \ 0.550 \end{bmatrix} \tag{9.23}$$

During execution, the only computation that has to be done is the evaluation of the increment in the control action:

$$\begin{aligned} \Delta u(t) = {} & 0.021w(t+1) + 0.081w(t+2) + 0.173w(t+3) + \\ & + \ 0.288w(t+4) + 0.417w(t+5) + 0.550w(t+6) - \\ & - \ 0.534\Delta u(t-1) - 21.102y(t) + 34.738y(t-1) - 15.167y(t-2) \end{aligned} \tag{9.24}$$

where $w(t+j)$ is the reference trajectory, which is supposed to be known in advance.

The FCS-MPC is implemented initially with a sampling frequency $f_s = 20$ kHz and an effective switching frequency (f_{esw}) of approximately 3 kHz. Performance can be improved increasing these values, but with an upper limit imposed by the microcontroller instruction cycle of 6.6 ns. For executing this control algorithm properly, the microcontroller is able to reach a maximum sampling frequency of 40 kHz, which results in an effective switching frequency of about 4.5 kHz. Notice that the usual range of reported FCS-MPC implementations is between 15 kHz (see [22] for an asymmetric flying capacitor converter) and 40 kHz (see [35] for a three-phase inverter). In the GPC case, the frequency is fixed only at 12 kHz since, as will be shown later, this sampling frequency is enough to obtain good performance.

Several experiments were conducted for both controllers and the values of output voltage and harmonic distortion are shown in the following figures in order to compare their performance. Figure 9.7 shows experimental waveforms, captured with a three-phase power analyzer, and the corresponding THD of the output voltages of the three phases for the FCS-MPC controller using a prediction horizon equal to 1 and 20 kHz sampling frequency. The performance is rather poor and it can be improved by increasing the sampling frequency until the limit of 40 kHz that results in an approximately 4.5 kHz effective switching frequency. Figure 9.8 shows how

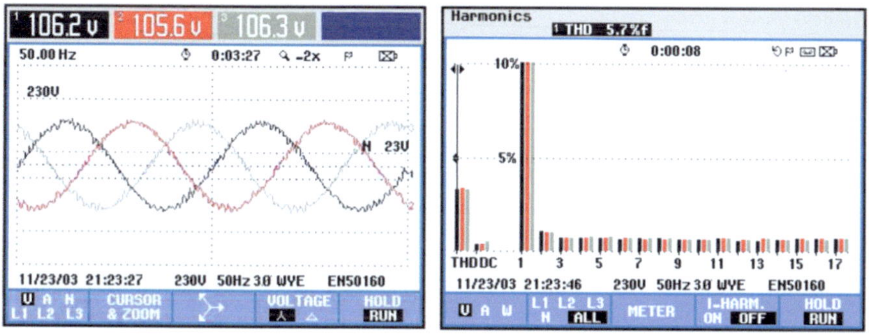

Fig. 9.7 Experimental results for FCSMPC controller with $f_s = 20\,\text{kHz}$ (effective switching frequency of 3 kHz). Left: Output voltages, right: Harmonics and THD values. ©[2015] IEEE. Reprinted, with permission, from IEEE Industrial Electronics Magazine [10]

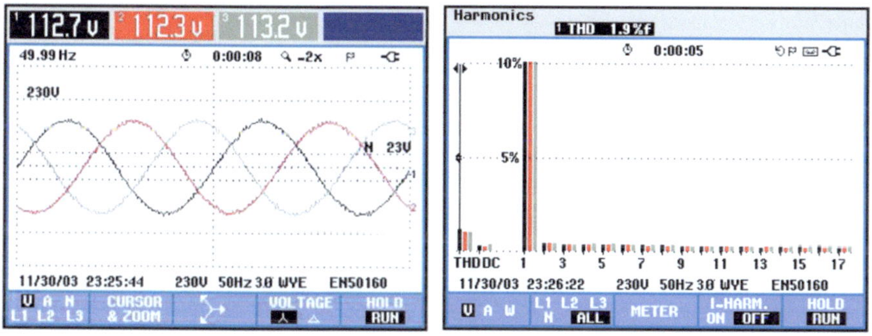

Fig. 9.8 Experimental results for FCSMPC controller with $f_s = 40\,\text{kHz}$ (effective switching frequency of 4.5 kHz). Left: Output voltages, right: Harmonics and THD values. ©[2015] IEEE. Reprinted, with permission, from IEEE Industrial Electronics Magazine [10]

the performance is improved. The output tracks the reference voltage with a good behavior in spite of the model mismatch. A frequency of 40 kHz can be considered a recommended value for the sampling frequency in FCS-MPC implementations; this is the one chosen in [35] for a similar VSI since a smaller value of 15 kHz shown bad results even in simulations.

The FCS-MPC has been designed with a control horizon of 1 since the computational problem for longer values is untractable in the available time imposed by the microcontroller. However, in GPC, this value can be easily increased without a noticeable increase in the computational load. The tuning value of $N_p = 6$ provides the results shown in Fig. 9.9, with a sampling (and switching) frequency of 12 kHz.

The performance of both methods is compared in Table 9.2 for different values of the prediction horizon, the control-weighting factor, and the switching frequency (actual and effective). A detailed assessment of the effect on the proper tuning of the prediction horizon and the control-weighting factor on system performance can be

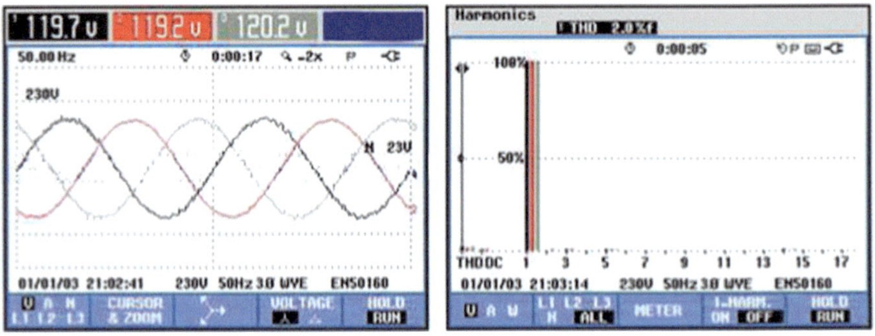

Fig. 9.9 Experimental results for GPC controller with $N_p = 6$ and $\lambda = 1.05$. Left: Output voltages, right: Harmonics and THD values. ©[2015] IEEE. Reprinted, with permission, from IEEE Industrial Electronics Magazine [10]

Table 9.2 Performance comparison

Controller	N_p	f_s (kHz)	f_{esw} (kHz)	λ	THD (%)	Error$_{RMS}$ (%)
$FCS-MPC$	1	20	3	–	5.7	15.4
$FCS-MPC$	1	40	4.5	–	1.9	7.8
GPC	5	12	12	1.05	2.2	4.2
GPC	6	12	12	1.05	2.0	2.6

found in [43]. Notice that it is not easy to compare FCS-MPC and GPC methods in the same conditions since in one case (GPC) the switching frequency is fixed (and equal to the sampling frequency) while in the FCS-MPC this value may change. So, the sampling frequency in FCS-MPC must be considerably higher than GPC (in this case 40 kHz versus 12 kHz) in order to get an appropriate switching frequency.

The best results for GPC running at $f_s = 12$ kHz are obtained for a set with values of $N_p = 6$ and $\lambda = 1.05$, although a GPC with $N_p = 1$ was also tested to show that worse performance is obtained for short horizons, independently of the method. The choice of the tuning parameters affects the THD and RMS value of the error between the measured and reference voltages. The proposed GPC exhibits low THD in the output voltages and, although the FCS-MPC presents similar THD results for $f_s = 40$ kHz (effective switching frequency of 4.5 kHz), the RMS value of the error is worse than in the GPC. Notice that this poor result stems from the steady-state tracking error, which can be significant in the FCS-MPC, especially when operating with low switching frequencies or small current reference amplitudes, as analyzed in [1].

Table 9.3 Controller comparison

Controller	Modulator	Horizon	Derivation	Computation	Theory
$FS - MPC$	No	Short	Simple	Low (N = 1)	Scarce
GPC	Yes	Any	Complex	Low (uncons.)	Well established

The main outstanding features of both methods are represented in Table 9.3. Both methods are able to handle the mismatch between the model and the actual load. The experiments reveal that prediction horizon length and control-weighting factor value directly affect the performance of the system. As mentioned above, the complexity (and therefore computing time) of GPC is almost independent of the prediction horizon. Notice that most of the computation of this unconstrained case can be done beforehand and the computational burden is small. It is seen that FCS-MPC requires a high sampling frequency (which means that the microcontroller is working near its operation limit) to behave like GPC (which can easily work with longer horizons and smaller sampling frequencies).

These experiments have illustrated the application of both methods to power converters, comparing some performance indices for the case of an experimental VSI. Other comparisons among controllers can be found in published papers, such as a comparison among several MPC schemes and techniques based on PWM, Space Vector Modulation (SVM), and optimized pulse patterns for the control of a medium-voltage drive presented in [16]. A comparative assessment of FCS-MPC with a linear current controller in two-level voltage source inverters is also presented in [44].

9.3 Power Quality Management in Microgrids Using MPC

One of the final goals of microgrids is to mitigate the effect of the disturbances in power quality at the low-voltage level of the smart grid. They should minimize the effects that the high penetration of distributed generation, the presence of nonlinear and unbalanced loads cause over the smart grids. MPC controllers can enhance the behavior that droop controls or PI-based controllers present for transient response. This section presents a methodology based on FCS-MPC to address quality control in microgrids, and it is demonstrated on a generic microgrid as shown in Fig. 9.10, which is composed of the power inverter shown in Fig. 9.11 and several DERs. The microgrid is connected with the main grid by an IPS to manage the transition between grid-connected and islanded mode at the PCC. As can be seen in Fig. 9.10, the microgrid integrates nonlinear loads/sources and unbalanced loads which hinders the voltage and current control problem in the microgrid.

This converter acts as final responsible for tracking the reference that is established in the secondary control level for the energy exchange with the main grid. The controller can be designed to equilibrate whichever kind of power consumption

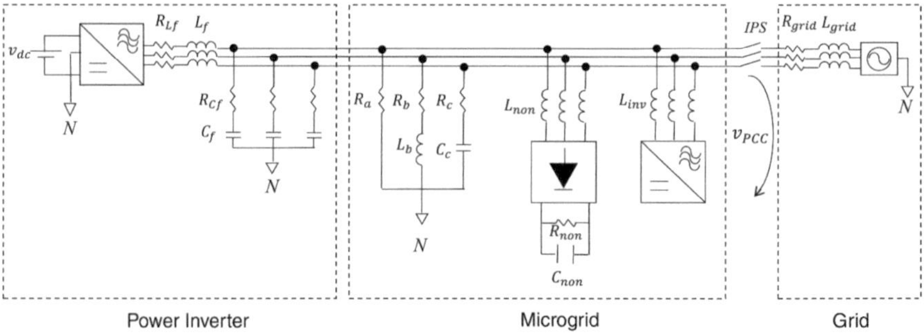

Fig. 9.10 Microgrid scheme

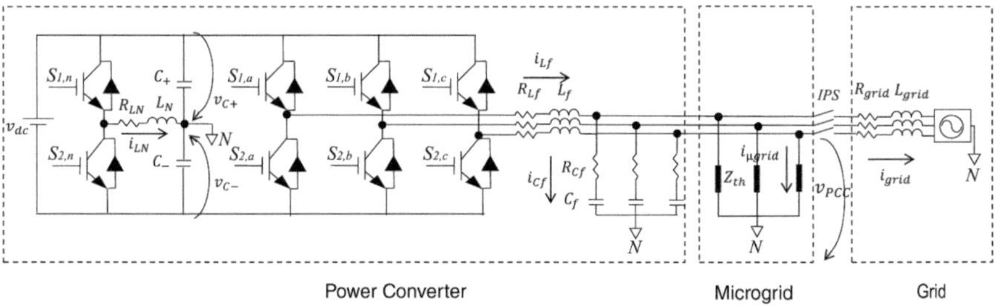

Fig. 9.11 Inverter connected to both the main grid and a microgrid with an equivalent Thevenin's impedance

or generation in the microgrid (unbalanced or nonlinear). The VSI will be also responsible for generating a voltage waveform according to the standard EN-50165 in islanded mode despite the presence of harmonics and unbalanced loads. Due to the presence of this type loads, a four-wire three-phase VSI with active control of the neutral point is considered. A controller based on the FCS-MPC strategy presented in [15] will be described in this section, showing an immediate response for the transition between grid-connected and islanded mode and vice versa. Different experiments to demonstrate the enhanced power quality operation of the microgrid such as harmonic mitigation, unbalanced and nonlinear loads in both modes, thanks to MPC, are shown.

The objectives to be considered in both operation modes and that will be included in the cost function are as follows:

- Islanded mode: In this mode, the inverter is in charge of voltage and frequency waveforms. Therefore, the cost function must include optimal tracking for the voltage reference at each sample instant. The voltage harmonics must also be minimized, providing a balanced neutral point, and the equilibrium of the voltage magnitude between phases must be imposed.

- Grid-connected mode: The scheduled active and reactive power (P^{sch}, Q^{sch}) must be achieved, independently of the power consumption or generation in the microgrid. Besides, the controller must balance the active and reactive powers per phase.

The first step of the controller is to perform the Fourier analysis of the current and voltage output at the current sample instant t. With these measurements, the Thevenin-equivalent impedance is calculated at the output of the power inverter. This equivalent impedance is used to calculate the output current and voltage predictions, which are included in the cost function to be minimized in islanded mode. In grid-connected mode, the values for active and reactive powers of the voltage–current pairs at fundamental frequency are calculated. Finally, the optimal gate signal combination is calculated by the FCS-MPC.

9.3.1 Fourier Analysis

As is well known, a signal $y(t)$ can be expressed by a Fourier series of the following form:

$$y(t) = \frac{a_0}{2} + \sum_{n=0}^{\infty}(a_n^y \cos(n\omega t) + b_n^y \sin(n\omega t)) \tag{9.25}$$

where n represents the rank of the harmonics ($n = 1$ corresponds to the fundamental component). The magnitude and phase of the selected harmonic component can be calculated by the following equations:

$$|\mathscr{Y}_n| = \sqrt{(a_n^y)^2 + (b_n^y)^2};\ \underline{/\mathscr{Y}_n} = \arctan\left(\frac{b_n^y}{a_n^y}\right) \tag{9.26}$$

where $\mathscr{Y}_n$ is the Fourier's expression of the signal $y(t)$, which can be expressed in Cartesian coordinates with the following expressions:

$$\mathrm{Im}(\mathscr{Y}_n(t)) = a_n^y = \frac{2}{T}\int_{t-T}^{t} y(t)\cos(n\omega t)dt \tag{9.27}$$

$$\mathrm{Re}(\mathscr{Y}_n(t)) = b_n^y = \frac{2}{T}\int_{t-T}^{t} y(t)\sin(n\omega t)dt \tag{9.28}$$

being $T = 1/f$ the corresponding period to the fundamental frequency. The upper index y is related to the signal $y(t)$ in which the Fourier analysis is developed. Using expression (9.26), for the voltage and current signals, the value of this signal expressed in the Fourier's domain $\mathscr{U}(t)$ and $\mathscr{I}(t)$ can be obtained. The equivalent Thevenin's impedance calculated for the fundamental frequency can be estimated in polar coordinates:

$$\left|Z_1^{th}(t)\right| = \frac{\sqrt{(a_1^v(t))^2 + (b_1^v(t))^2}}{\sqrt{(a_1^i(t))^2 + (b_1^i(t))^2}} \tag{9.29}$$

$$\varphi_1^{th}(t) = \angle Z_1^{th}(t) = \arctan\left(\frac{b_1^v(t)}{a_1^v(t)}\right) - \arctan\left(\frac{b_1^i(t)}{a_1^i(t)}\right) \tag{9.30}$$

The superscripts v and i are related to the output voltage and output current of the inverter. Expressing $Z_{th,1}(t)$ in Cartesian coordinates, the equivalent resistance and the equivalent impedance can be obtained as

$$Z_1^{th}(t) = R_1^{th}(t) + jX_1^{th}(t) \tag{9.31}$$

Depending on the value of $X_1^{th}(t)$, an equivalent inductance (L) or capacitance (C) is obtained. Equation (9.32) is used when $\text{sign}(X_1^{th}(t)) = \text{sign}(R_1^{th}(t)))$ and Eq. (9.33) when $\text{sign}(X_1^{th}(t)) \neq \text{sign}(R_1^{th}(t))$:

$$L^{th}(t) = \frac{X_1^{th}(t)}{2\pi f};\ C^{th}(t) = 0 \tag{9.32}$$

$$L^{th}(t) = 0;\ C^{th}(t) = -X_1^{th}(t) \cdot 2\pi f \tag{9.33}$$

The expression for active and reactive powers of the voltage–current pairs calculated at fundamental frequency can be computed (being $\varphi_{j,1}^{th}(t+1|t)$ the phase of Thevenin's impedance evaluated at $n = 1$) as follows:

$$[P(t), Q(t)] = \frac{|\mathscr{U}(t)||\mathscr{I}(t)|}{2}[\cos(\varphi_1^{th}(t)), \sin(\varphi_1^{th}(t))] \tag{9.34}$$

9.3.2 Model of the System

The model of the system (see Fig. 9.11) can be obtained as a function of its decision variables (gate signals of each leg S_{1a}, S_{1b}, S_{1c}, and S_{1n}), the set of state variables composed of the inductor currents and capacitor voltages, and the output currents and voltages of the inverter per phase $[v_{C+}, v_{C-}, i_{LN}, v_{Cf}, i_{Lf}]$ using forward Euler method for discretization:

$$v_{dc}(t+1) = v_{C+}(t+1) - v_{C-}(t+1) \tag{9.35}$$

$$i_{Cj}(t+1) = C_j \frac{v_{Cj,N}(t+1) - v_{Cj,N}(t)}{T_s}\Big|_{j=+,-} \tag{9.36}$$

$$v_{C+}(t+1)\cdot S_{1n}(t+1)+v_{C-}(t+1)\cdot(1-S_{1n}(t+1)) \\ = R_{L_N}\cdot i_{L_N}(t+1)+L_N\frac{\Delta i_{L_N}(t+1)}{T_s} \tag{9.37}$$

$$\sum^{\alpha=C+,C-,LN} i_\alpha(t)+\sum_{j=a,b,c}^{\beta=C_f,grid,\mu grid} i_{\beta,j}(t)=0 \tag{9.38}$$

The values of the inductor currents of the LC filter ($i_{L_{fj}}(t+1)$) can be predicted with the following equations:

$$v_{out,jN}(t+1)=v_{C+}(t+1)\cdot S_{1j}(t+1) \\ +v_{C-}(t+1)\cdot(1-S_{1j}(t+1))-L_f\frac{\Delta i_{L_{fj}}(t+1)}{T_s} \\ -R_{L_f}\cdot i_{L_{fj}}(t+1)|_{j=a,b,c} \tag{9.39}$$

$$i_{C_{fj}}(t+1)=C_f\frac{\Delta v_{out,j}(t+1)-R_{C_f}\Delta i_{C_{fj}}(t+1)}{T_s} \tag{9.40}$$

$$i_{grid,j}(t+1)+i_{\mu grid,j}(t+1)=i_{L_f,j}(t+1)-i_{C_f,j}(t+1) \tag{9.41}$$

Under the assumption that $Z_n^{th}(t+1)=Z_n^{th}(t)$ and approaching the equivalent Thevenin's impedance by the fundamental frequency, the relationship (9.42) is obtained in case that $X_j^{th}(t)\geq 0$ and (9.43) when $X_j^{th}(t)<0$:

$$v_{PCC,j}(t+1)=R_j^{th,\mu grid}(t)\cdot i_{\mu grid,j}(t+1) \\ +L_j^{th,\mu grid}(t)\frac{i_{\mu grid,j}(t+1)-i_{\mu grid,j}(t)}{T_s}\Bigg|_{j=a,b,c} \tag{9.42}$$

$$i_{\mu grid,j}(t+1)\Big|_{j=a,b,c}= \\ C_j^{th,\mu grid}(t)\frac{\Delta[v_{PCC,j}(t+1)-R_j^{th,\mu grid}(t)\cdot i_{\mu grid,j}(t+1)]}{T_s} \tag{9.43}$$

In those cases, when the inverter works tied to the main grid:

$$v_{grid,j}(t+1)-v_{PCC,j}(t+1)=R^{grid}\cdot i_{grid,j}(t+1) \\ +L_j^{grid}\frac{i_{grid,j}(t+1)-i_{grid,j}(t)}{T_s}\Bigg|_{j=a,b,c} \tag{9.44}$$

9.3.3 Islanded Mode MPC-Based Controller

When the microgrid works in islanded mode, the inverter manages voltage and frequency waveforms in order to achieve standard EN 50160. For this purpose, the MPC calculates the equivalent Thevenin's impendance $Z_j^{th}(t)$ per phase. Under the assumption that it will be the same at the next sample instant, the controller minimizes the following cost function:

$$
\begin{aligned}
J = & \sum_{j=a,b,c} w_{ref}^{inst} \left(v_{PCC,j}(t+1|t) - v_{PCC}^{ref}(t+1|t) \right)^2 + \\
& + \sum_{j=a,b,c} w_{ref,j}^{cycle} \left(\mathscr{U}_{PCC,j}(t+1|t) - \mathscr{U}_{PCC,j}^{ref} \right)^2 + \\
& + w_{balance} \left(v_{C+}(t+1|t) - v_{C-}(t+1|t) \right)^2 + \\
& + \sum_{j=a,b,c} w_{v,j} \left(\Delta v_{PCC,j}(t+1|t) \right)^2 + \sum_{j=a,b,c} w_{i,j} \left(\Delta i_{\mu grid,j}(t+1|t) \right)^2 + \\
& + w_{ab}(|\mathscr{U}_{PCC,a}(t+1|t))| - |\mathscr{U}_{PCC,b}(t+1|t)|)^2 + \\
& + w_{ac}(|\mathscr{U}_{PCC,a}(t+1|t))| - |\mathscr{U}_{PCC,c}(t+1|t)|)^2 + \\
& + w_{bc}(|\mathscr{U}_{PCC,c}(t+1|t))| - |\mathscr{U}_{PCC,b}(t+1|t)|)^2
\end{aligned}
\tag{9.45}
$$

The first term in (9.45) is added to obtain an optimal tracking for the voltage reference at each sample instant. The second term is used to compensate errors in the tracking of the voltage reference calculated at the fundamental frequency using the Fourier's transform. The terms from third to fifth are used to minimize the voltage harmonics at the PCC, providing a balanced neutral point. While the terms fourth and fifth minimize the variation of current and voltage acting as virtual LC filter. Finally, the last terms impose the equilibrium for the voltage magnitude between phases.

Each load or generator connected to the microgrid imposes different requirements regarding harmonics. For this reason, the terms $w_{ref,j}^{cycle}$, $w_{i,j}$, and $w_{v,j}$ are dynamically adjusted based on a heuristic strategy depending on $Z_j^{th}(t)$.

9.3.4 Grid-Connected MPC-Based Controller

When the microgrid works in grid-connected mode, it is governed by the power inverter in order to achieve the scheduled active and reactive powers (P^{sch}, Q^{sch}) at the tertiary control level of the microgrid. Independently of the power consumption or power generation in the microgrid, this schedule has to be satisfied. The first step of the control algorithm is to calculate the value $\mathscr{I}_{PCC,j}^{ref}(t+1|t)|_{j=a,b,c}$ for the first harmonic using (9.34). Once this value is obtained, the current reference for the next sample instant $i_j^{ref}(t+1|t)$ can be easily evaluated following the next expression:

$$i_j^{ref}(t+1) = |\mathscr{I}_{PCC,j}^{ref}(t+1)| \sin(\omega(t+1+D_j) + \varphi_j^i(t+1)) \tag{9.46}$$

where $\varphi_j^i(t) = \underline{/\mathscr{I}(t)}$. A digital delay D_j in the response of the inverter is found, which depends on the equivalent impedance seen by the power inverter. The following cost function is used:

$$\begin{aligned}
J = {} & w_{ref}^{inst}\left[\sum_{j=a,b,c}\left(i_{PCC,j}(t+1|t) - i_j^{ref}(t+1|t)\right)^2\right] + \\
& + w_{ref}^{cycle}\left[\sum_{j=a,b,c}\left(\mathscr{I}_{PCC,j}(t+1|t)(t+1|t) - \mathscr{I}_{PCC,j}^{ref}(t+1|t)\right)^2\right] + \\
& + w_{balance}\left(v_{C+}(t+1|t) - v_{C-}(t+1|t)\right)^2 + \\
& + \sum_{j=a,b,c} w_{v,j}\left(\Delta v_{PCC,j}(t+1|t)\right)^2 + \sum_{j=a,b,c} w_{i,j}\left(\Delta i_{grid,j}(t+1|t)\right)^2 + \\
& + w_{ab}\left[(P_a - P_b)^2 + (Q_a - Q_b)^2)\right] + w_{ac}\left[(P_a - P_c)^2 + (Q_a - Q_c)^2)\right] + \\
& + w_{bc}\left[(P_b - P_c)^2 + (Q_b - Q_c)^2)\right]
\end{aligned} \tag{9.47}$$

As done in the previous case, in order to minimize the steady-state error, two references are used: the instantaneous current and the Fourier's value at the fundamental frequency (first and second terms). The terms from third to fifth have a similar functionality than in the previous mode, as well as the sixth but equilibrating the active and reactive power obtained per phase.

9.3.5 Simulation Results

In order to validate the proposed controller, several simulations are done using Simpower© toolbox of MATLAB. The sampling period for the simulation is 1 μs using a sampling period for the controller of $T_s = 20\,\mu$s. The values of the system components can be found in Table 9.4.

Islanded Mode

For the validation of the correct behavior in islanded mode, the controller has been exposed to different operating conditions. The slave inverter is connected to the microgrid at instants $t \in [0\,\text{s}, 0.7\,\text{s}]$ having the values of $[P_{ref}, Q_{ref}] =$ [10000 W, 5000 Var] for $t < 0.15$ s and $[P_{ref}, Q_{ref}] = [-10000\text{ W}, -5000\text{ Var}]$ otherwise. The nonlinear load is connected at $t = 0.3$ s and the unbalanced load is connected at $t = 0.5$ s.

The evolution of the currents per phase in islanded mode can be found in Fig. 9.12. Although the microgrid has imbalance and harmonics in the current, the voltage waveforms comply with the standard EN-50160. The spectral analysis at $t = 0.7$ s

Table 9.4 Parameters of the power inverter, the microgrid, and the connection with the main grid

Parameter	Value
Filter inductance L_f	1 mH
Filter inductance resistance R_{L_f}	0.1 Ω
Filter capacitor C_f	0.5 mF
Filter capacitor resistance R_{C_f}	0.1 Ω
DC-link voltage U_{dc}	950 V
Neutral inductance L_N	2.5 μF
Neutral inductance resistance R_{L_N}	0.1 Ω
Neutral balancing capacitors C_+, C_-	6600 μF
Grid connection line inductance L_{grid}	0.1 mH
Grid connection line resistance R_{grid}	0.1 Ω
Slave inverter line inductance L_{inv}	0.1 mH
Slave inverter line resistance R_{inv}	0.1 Ω
Nonlinear load line inductance L_{non}	0.1 mH
Nonlinear load line resistance $R_{L_{non}}$	0.1 Ω
Nonlinear load dc resistance R_{non}	60 Ω
Nonlinear load dc capacitor C_{non}	6.6 mF
Unbalanced load phase a resistance R_a	1 MΩ
Unbalanced load phase b resistance R_b	10 Ω
Unbalanced load phase c resistance R_c	10 Ω
Unbalanced load phase b inductance L_b	1 mH
Unbalanced load phase c capacitor C_c	0.1 mF

for one of the phases is exposed in Fig. 9.13 (the other phases are similar, but are committed for the sake of clarity). The maximum harmonics content is lower than 0.45% with a $THD < 1\%$ for the three phases. As can be seen in Fig. 9.14, the values of the magnitude and phase obtained for the voltage of the three phases track the desired reference, showing low harmonics content during all the experiments.

Grid-Connected Mode and Transition Between Modes

The simulation results for grid-connected operation and the transition between modes are shown in the following. For this simulation, the nonlinear and the unbalanced loads are connected to the microgrid in all the sample instants. The schedule is given for the active and reactive power exchange between the microgrid and the main grid $[P^{sch,j}, Q^{sch,j}] = [-15000\text{ W}, -9000\text{ Var}]$ at $t < 0.5$ s and $[P^{sch,j}, Q^{sch,j}] = [15000\text{ W}, 9000\text{ Var}]$ at $t \geq 0.5$ s. Between $t \in [1\text{ s}, 1.5\text{ s}]$ a fault in the main grid occurs, so the transition to islanded mode is required, restoring the connection of the microgrid with the main grid for $t > 1.5$ s.

The waveforms for the current in the different phases obtained for the microgrid and the main grid are exposed in 9.15. As can be seen, despite of the unbalance and harmonics found in the phase currents of the microgrid the exchange currents with

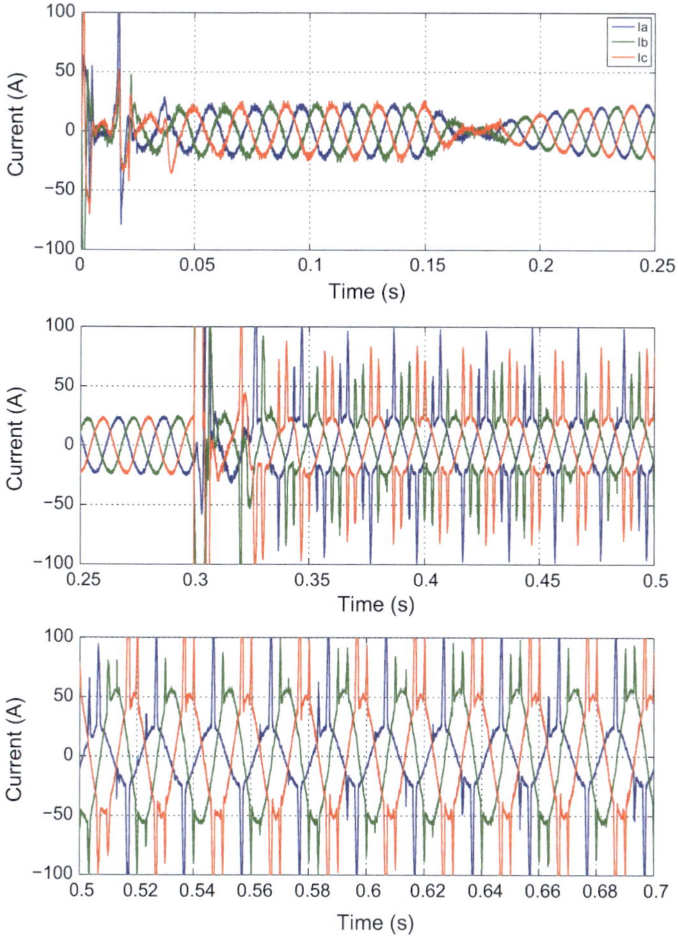

Fig. 9.12 Currents per phase in islanded mode

the main grid are balanced with low THD content as can be observed in Fig. 9.16, where low harmonics content is also obtained for the voltage waveforms at the PCC. It is seen how MPC can adapt to changes in active and reactive power requested to the microgrid and can manage the transition to islanded mode (and back) when a fault in the main grid occurs. During these situations, the quality of the involved signal is maintained at satisfactory values.

The values of the magnitude and phase, as well as the waveforms of the voltage per phase can be found in Figs. 9.17 and 9.18. The waveforms for the voltage in the transition between modes are shown in Fig. 9.18.

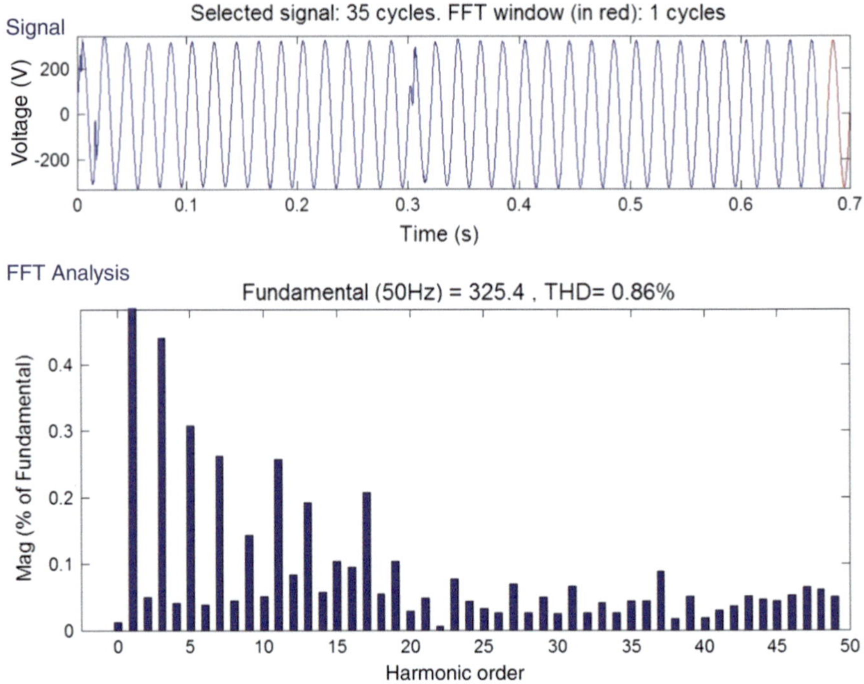

Fig. 9.13 Spectral analysis for the voltage values

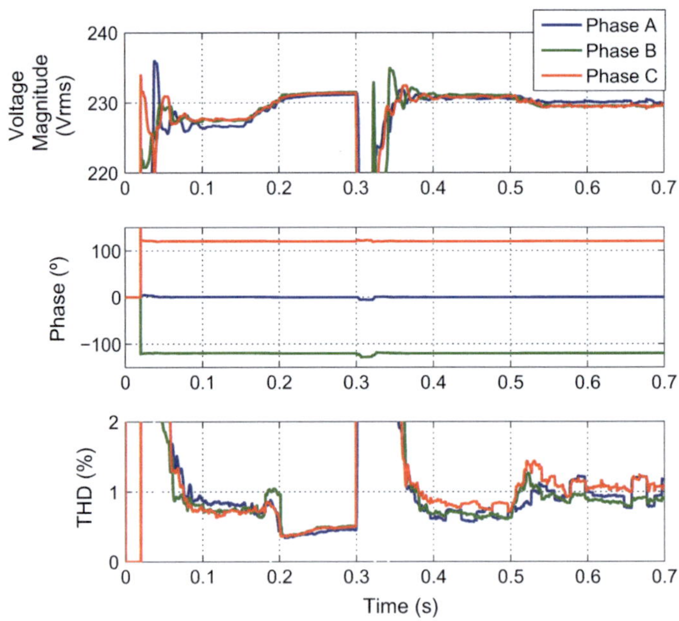

Fig. 9.14 Magnitude, phase, and THD of the voltages per phase

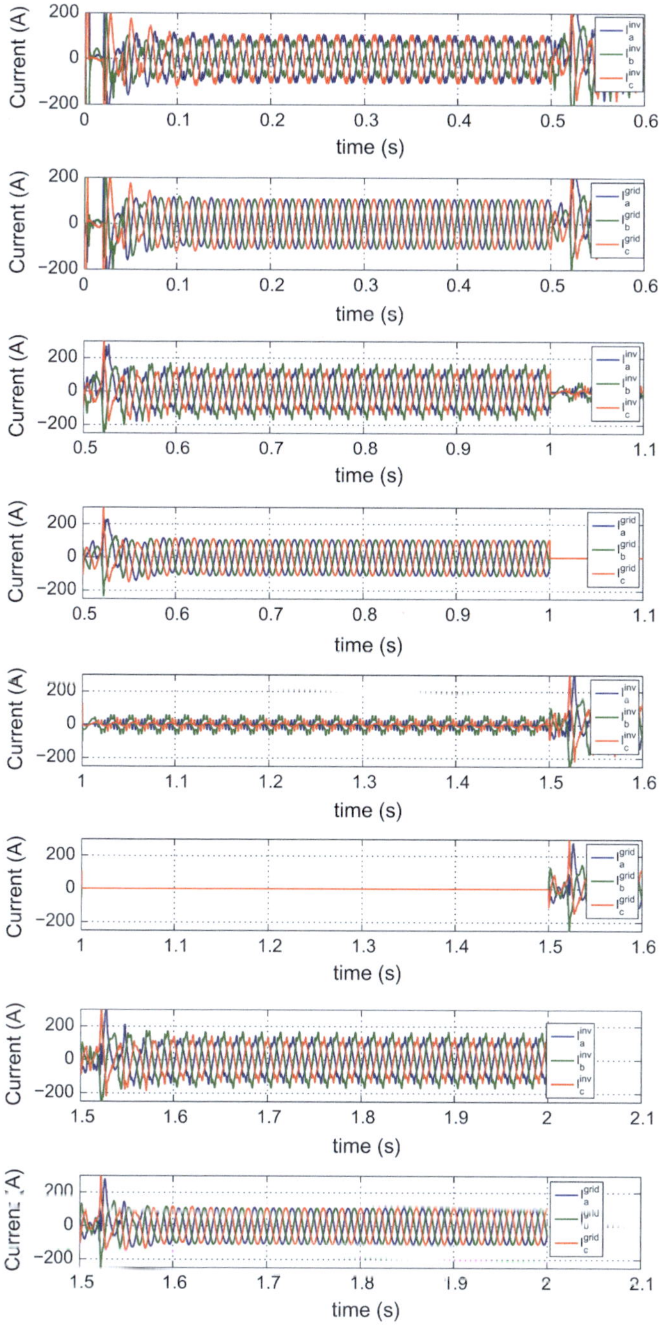

Fig. 9.15 Current supplied by the inverter and current injected to the grid

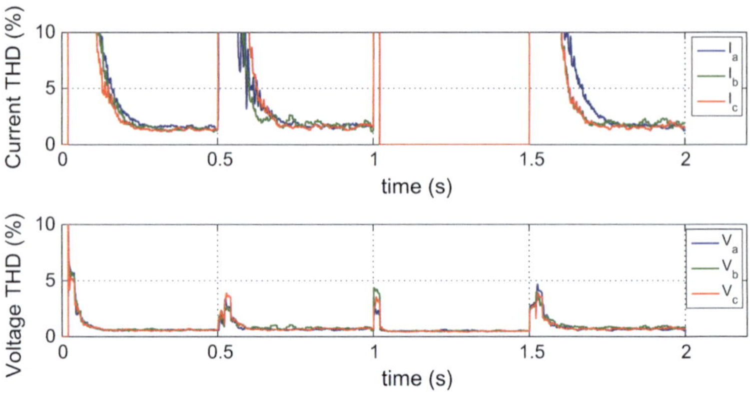

Fig. 9.16 Total harmonics distortion for the current exchange with the main grid and voltage of the microgrid

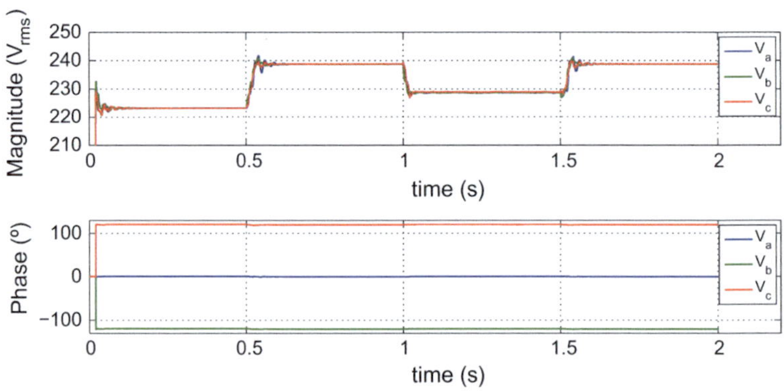

Fig. 9.17 Magnitude and phase values for the voltages at the PCC

The results show an excellent behavior for the output variables of the inverter, with a low THD in voltage in the case of islanded and in the current exchanged with the grid in case it works as a grid-tied inverter.

Despite the switching between grid-connected to islanded mode and vice versa, the voltage at the PCC is maintained without affection. Finally, the tracking of the reference given for active and reactive power can be found in Fig. 9.19, where the reference is reached just in two cycles of the fundamental frequency. In Fig. 9.20, the spectral analysis for the current exchanged with the grid is displayed for the cases when current is consumed from the grid (at $t = 0.5$ s) and when current is injected to the grid (at $t = 1$ s) for the phase C (similar results are obtained for the rest of phases). As can be seen, the content of all the harmonics in both cases are below 0.35%, complying with standard IEC 61000-3-2 for class A and IEC 61000-3-4, despite the presence of nonlinear and unbalanced loads in the microgrid.

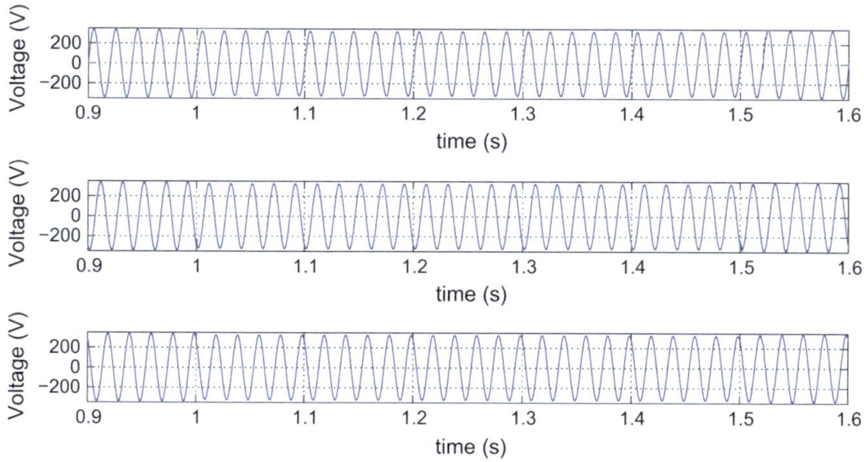

Fig. 9.18 Voltage waveform at the PCC at the transition instants between modes

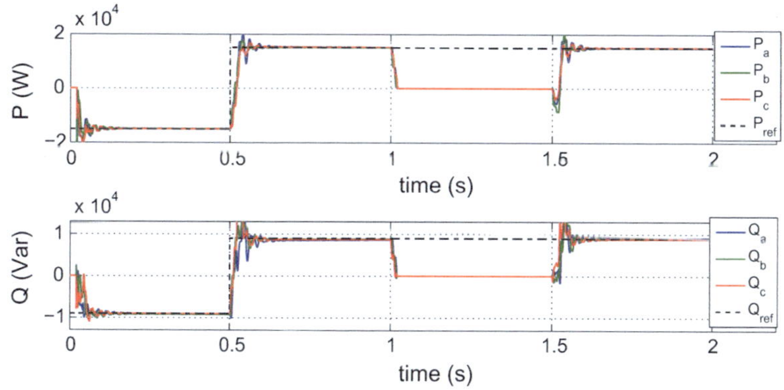

Fig. 9.19 Active and reactive power exchange with the grid

These simulations have demonstrated that MPC can enhance power quality in microgrids in both operation modes and during the transition. Besides, the proposed controller can deal with different situations, such as nonlinear and unbalanced loads, and failures in the main grid. The controller, based on FCS-MPC, has low-computational requirements since the horizon has been set to 1, which allows its real-time implementation.

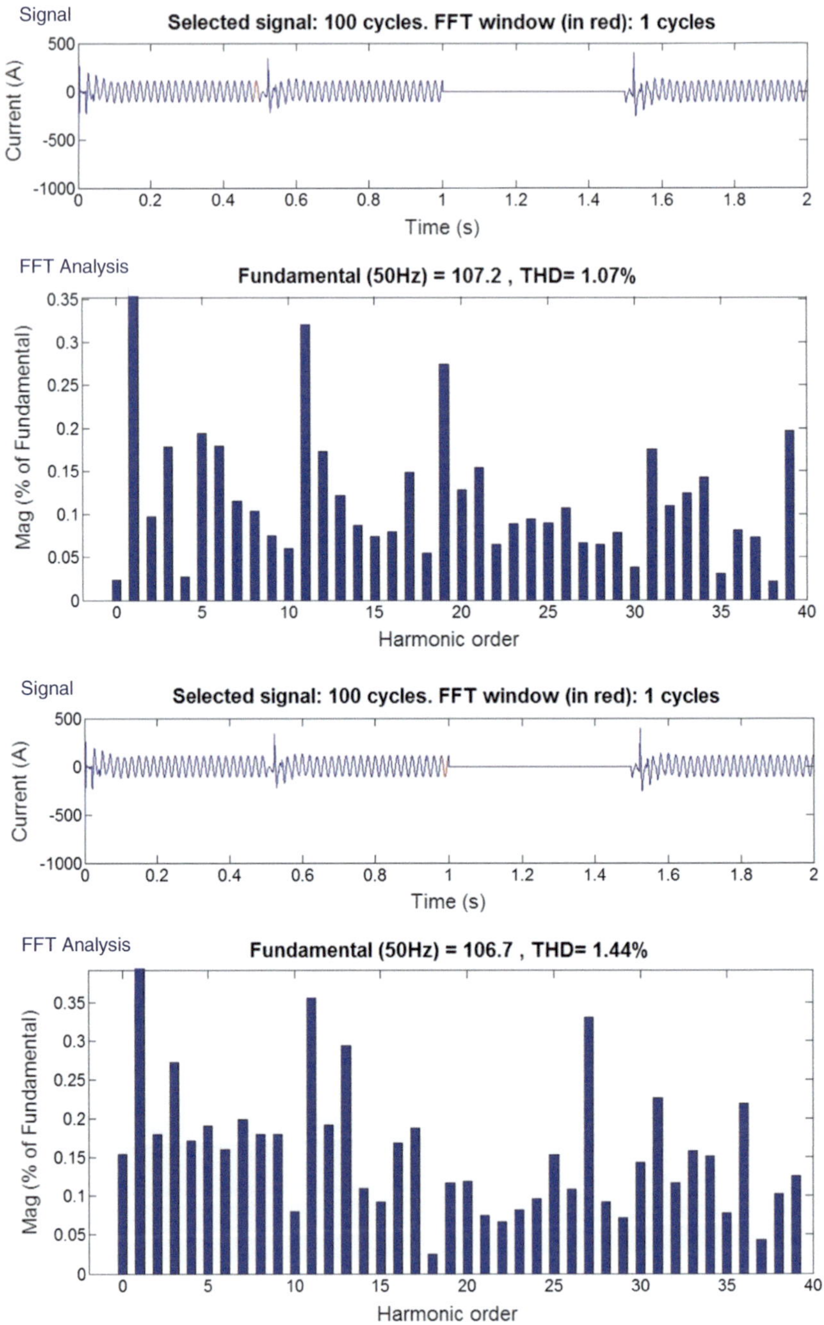

Fig. 9.20 Spectral analysis for the current waveforms at $t = 0.5$ s and $t = 1$ s

References

1. Aguilera RP, Lezana P, Quevedo DE (2013) Finite-control-set model predictive control with improved steady-state performance. IEEE Trans Ind Inf 9(2):658–667
2. Ahumada C, Cárdenas R, Sáez D, Guerrero JM (2016) Secondary control strategies for frequency restoration in islanded microgrids with consideration of communication delays. IEEE Trans Smart Grid 7(3):1430–1441
3. Babqi AJ, Etemadi AH (2017) MPC-based microgrid control with supplementary fault current limitation and smooth transition mechanisms. IET Gener Trans Distrib 11(9):2164–2172
4. Bemporad A, Morari M, Dua V, Pistikopoulous E (2002) The explicit linear quadratic regulator for constrained systems. Automatica 38(1):3–20
5. Bemporad A, Borrelli F, Morari M (2002) Model predictive control based on linear programming. The explicit solution. IEEE Trans Autom Control 47(12):1974–1985
6. Bidram A, Davoudi A (2012) Hierarchical structure of microgrids control systems. IEEE Trans Smart Grid 3:19631976
7. Bidram A, Davoudi A, Lewis FL, Guerrero JM (2013) Distributed cooperative secondary control of microgrids using feedback linearization. IEEE Trans Power Syst 28(3):3462–3470
8. Bollen MHJ, Zhong J, Zavoda F, Meyer J, McEachern A, Lopez F (2010) Power quality aspects of smart grids. In: Procedings of international conference on renewable energies and power quality
9. Bonilla J, De Keyser R (2007) Nonlinear predictive control of a DC-DC converter: a NEPSAC approach. In: European control conference, Kos, Greece, July 2007, pp 1753–1758
10. Bordons C, Montero C (2015) Basic principles of MPC for power converters: bridging the gap between theory and practice. IEEE Ind Electron Mag 9(3):31–43
11. Clarke DW, Mohtadi C, Tuffs PS (1987) Generalized predictive control part i. The basic algorithm. Automatica 23(2):137–148
12. Cortés P, Kazmierkowski MP, Kennel RM, Quevedo DE, Rodriguez J (2008) Predictive control in power electronics and drives. IEEE Trans Ind Electron 55(12):4312–4324
13. Cortés P, Ortiz G, Yuz JI, Rodríguez J, Vazquez S, Franquelo LG (2009) Model predictive control of an inverter with output filter for UPS applications. IEEE Trans Ind Electron 56(6):1875–1883
14. Delfino F, Rossi M, Ferro G, Minciardi R, Robba M (2015) MPC-based tertiary and secondary optimal control in islanded microgrids. In: 2015 IEEE international symposium on systems engineering (ISSE), Sept 2015, pp 23–28
15. Garcia-Torres F, Bordons C, Vazquez S (2015) Voltage predictive control for microgrids in islanded mode based on fourier transform. In: ICIT. IEEE, pp 2358–2363
16. Geyer T (2011) A comparison of control and modulation schemes for medium-voltage drives: emerging predictive control concepts versus PWM-based schemes. IEEE Trans Ind Appl 47(3):1380–1389
17. Guerrero JM, Garcia de Vicuña L, Matas J, Castilla M, Miret J (2005) Output impedance design of parallel-connected UPS inverters with wireless load-sharing control. IEEE Trans Ind Electron 52(4):1126–1135
18. Guerrero JM, Vasquez JC, Matas J, García de Vicuña L, Castilla M (2011) Hierarchical control of droop-controlled ac and dc microgrids–a general approach toward standardization. IEEE Trans Ind Electron 58(1):158–172
19. Han H, Hou X, Yang J, Wu J, Su M, Guerrero JM (2016) Review of power sharing control strategies for islanding operation of AC microgrids. IEEE Trans Smart Grid 7(1):200–215
20. He J, Li YW (2012) An enhanced microgrid load demand sharing strategy. IEEE Trans Power Electron 27(9):3984–3995
21. Janjic A, Stajic Z, Radovic I (2011) Power quality requirements for the smart grid design. Int J Circ Syst Sig Process
22. Lezana P, Aguilera RP, Quevedo DE (2009) Model predictive control of an asymmetric flying capacitor converter. IEEE Trans Ind Electron 56(6):1839–1846

23. Linder A, Kanchan R, Kennel R, Stolze P (2012) Model-based predictive control of electric drives. Cuvillier Verlag, Gottingen
24. Lou G, Gu W, Xu Y, Cheng M, Liu W (2017) Distributed MPC-based secondary voltage control scheme for autonomous droop-controlled microgrids. IEEE Trans Sustain Energy 8(2):792–804
25. Mariethoz S, Morari M (2009) Explicit model-predictive control of a PWM inverter with an LCL filter. IEEE Trans Ind Electron 56(2):389–399
26. Mattavelli P (2005) An improved deadbeat control for UPS using disturbance observers. IEEE Trans Ind Electron 52(1):206–212
27. Mayne DQ, Rawlings JB, Rao CV, Scokaert POM (2000) Constrained model predictive control: stability and optimality. Automatica 36:789–814
28. Micallef A, Apap M, Spiteri-Staines C, Guerrero JM, Vasquez JC (2014) Reactive power sharing and voltage harmonic distortion compensation of droop controlled single phase islanded microgrids. IEEE Trans Smart Grid 5(3):1149–1158
29. Mohamed Y (2008) New control algorithms for the distributed generation interface in grid-connected and micro-grid systems
30. Neely J, Pekarek S, DeCarlo R, Vaks N Real-time hybrid model predictive control of a boost converter with constant power load. In: 2010 IEEE twenty-fifth annual applied power electronics conference and exposition (APEC), February 2010, pp 480 –490
31. Olivares DE, Mehrizi-Sani A, Etemadi AH, Cañizares CA, Iravani R, Kazerani M, Hajimiragha AH, Gomis-Bellmunt O, Saeedifard A, Palma-Behnke R, Jimenez-Estevez GA, Hatziargyriou ND (2014) Trends in microgrid control. IEEE Trans Smart Grid 5(4):1905–1919
32. Quevedo DE, Aguilera RP, Geyer T (2014) Advanced and intelligent control in power electronics and drives. Predictive control in power electronics and drives: basic concepts, theory, and methods. Springer International Publishing, Cham, pp 181–226
33. Richter S, Mariethoz S, Morari M (2010) High-speed online MPC based on a fast gradient method applied to power converter control. In: 2010 American control conference (ACC), pp 4737–4743
34. Rodriguez J, Pontt J, Silva CA, Correa P, Lezana P, Cortés P, Ammann U (2007) Predictive current control of a voltage source inverter. IEEE Trans Ind Electron 54(1):495–503
35. Rodriguez J, Cortés P (2012) Predictive control of power converters and electrical drives. Wiley
36. Savaghebi M, Vasquez JC, Jalilian A, Guerrero JM (2012) Secondary control for compensation of voltage harmonics and unbalance in microgrids. In: 2012 3rd IEEE international symposium on power electronics for distributed generation systems (PEDG). IEEE, pp 46–53
37. Shafiee Q, Vasquez JC, Guerrero JM Distributed secondary control for islanded microgrids—a networked control systems approach. In: 38th Annual conference on IEEE industrial electronics society, IECON 2012, October 2012, pp 5637–5642
38. Vandoorn TL, De Kooning JDM, Meersman B, Guerrero JM, Vandevelde L (2012) Automatic power-sharing modification of p/v droop controllers in low-voltage resistive microgrids. IEEE Trans Power Deliv 27(4):2318–2325
39. Vargas R, Cortés P, Ammann U, Rodriguez J, Pontt J (2007) Predictive control of a three-phase neutral-point-clamped inverter. IEEE Trans Ind Electron 54(5):2697–2705
40. Vasquez JC (2009) Decentralized control techniques applied to electric power distributed generation in microgrids. PhD thesis, Polytechnic University of Catalonia
41. Vazquez S, Leon J, Franquelo LG, Rodriguez J, Young H, Marquez A, Zanchetta P (2014) Model predictive control: a review of its applications in power electronics. IEEE Ind Electron Mag 8(1):16–31
42. Vazquez S, Rodriguez J, Rivera M, Franquelo LG, Norambuena M (2017) Model predictive control for power converters and drives: advances and trends. IEEE Trans Ind Electron 64(2):935–947

43. Vazquez S, Montero C, Bordons C, Franquelo LG (2013) Design and experimental validation of a model predictive control strategy for a VSI with long prediction horizon. IECON annual conference of the IEEE industrial electronics society. Austria, Vienna, pp 5786–5791
44. Young HA, Perez MA, Rodriguez J, Abu-Rub H (2014) Assessing finite-control-set model predictive control: a comparison with a linear current controller in two-level voltage source inverters. IEEE Ind Electron Mag 8(1):44–52

Index

C. Bordons et al., *Model Predictive Control of Microgrids*,
Advances in Industrial Control, https://doi.org/10.1007/978-3-030-24570-2

MIX
Papier aus verantwortungsvollen Quellen
Paper from responsible sources
FSC® C105338
FSC
www.fsc.org

If you have any concerns about our products,
you can contact us on
ProductSafety@springernature.com

In case Publisher is established outside the EU,
the EU authorized representative is:
Springer Nature Customer Service Center GmbH
Europaplatz 3, 69115 Heidelberg, Germany

Printed by Libri Plureos GmbH
in Hamburg, Germany